Lecture Notes in Mathematics

A collection of informal reports and seminars
Edited by A. Dold, Heidelberg and B. Eckmann, Zürich

192

Proceedings of Liverpool Singularities – Symposium I

Edited by C. T. C. Wall, University of Liverpool

Springer-Verlag
Berlin · Heidelberg · New York 1971

AMS Subject Classifications (1970): 32B20, 32C40, 57D45, 58C25

ISBN 3-540-05402-2 Springer-Verlag Berlin · Heidelberg · New York
ISBN 0-387-05402-2 Springer-Verlag New York · Heidelberg · Berlin

Offsetdruck: Julius Beltz, Hemsbach üb. Weinheim/Bergstr.

INTRODUCTION

The papers in this volume, and a second one to appear shortly, were submitted by participants of the Symposium on Singularities of smooth manifolds and maps, held at the Department of Pure Mathematics at the University of Liverpool from September 1969 - August 1970, and supported by the Science Research Council. They include also the texts of the courses of lectures given at a Summer School held in July 1970 in connection with the Symposium. I have added some further notes to bring out the relations of the different papers in an attempt to make this volume as complete an exposition of the present state of knowledge in the subject as is possible within its scope. For this reason also we have reprinted the classic notes by Levine of lectures by Thom, which had become unavailable, and which represent the source of many of the ideas further developed in these papers.

Volume 1 contains the papers closely related to the problem of classification of singularities of smooth maps and is thus fairly coherent. Volume 2 will represent applications of these ideas (and others) to other branches of pure and applied mathematics.

Thanks are due to the S.R.C. for financial support; to Mrs.Evelyn Quayle (nee Hastwell) and Mrs.Eileen Bratt for doing the typing; to members of the department for proofreading it; and of course to the participants in the Symposium for creating the body of mathematics here presented.

Liverpool, December 1970.

C.T.C. Wall

C O N T E N T S

SINGULARITIES OF DIFFERENTIABLE MAPPINGS

H.I. Levine

ACKNOWLEDGEMENTS

I wish to express my warmest thanks to Prof. Thom for introducing me to this subject and for the stimulation I received from conversations with him as well as from his lectures, to Dr. Peter Dombrowski for many valuable suggestions and for his help with the proof that the "Strong Conjecture" is false, and to Frl. D. von Viereck formerly with the Mathematical Institute at Bonn who did the typing.

I would also like to thank the Fulbright Commission and the Forschungsgemeinschaft under whose financial support I did this work.

H.I. Levine

Note to the present edition.

Prof. Levine's notes of Prof. Thom's lectures, previously published in mimeographed form by the University of Bonn, 1960, are here republished with the full approval of Prof. Levine, Prof. Thom and Prof. Hirzebruch. The text is unaltered save for some corrections to the statement and proof of the Weak Transversality Theorem in §6, several necessary corrections in the Lemmas of 7.1, and the addition of a few references.

INTRODUCTION

L'étude des singularités des applications différentiables remonte à l'article de Whitney sur les applications plan sur plan. Il a ainsi créé toute une direction de recherches, dont le présent cours - donné à Bonn au printemps 1959 - vise à donner un apercu assez complet. Deux aspects principaux dominent la théorie: l'aspect local, qui cherch à caractériser les singularités "génériques"; l'aspect global, qui s'intéresse aux propriétés globales des ensembles critiques, et s'efforce, - dans la mesure du possible - à généraliser aux applications les résultats bien connus de la théorie de M. Morse sur les fonctions. Il n'est que just de remarquer qu'en aucun de ces aspects la théorie actuelle n'est satisfaisante; le seul progrès notable réside dans l'utilisation du "lemme de transversalité", dont on trouvera ici la demonstration dans tous ses details. Le problème global le plus intéressant est la "conjecture faible"; il s'agit de démontrer que dans l'espace fonctionnel $L(R^n, R^p, \infty)$ "presque toute" application est topologiquement équivalente aux applications suffisamment voisines - au moins localement. De même on peut conjecturer que "presque partout", une application C^∞ est topologiquement équivalente à son développement de Taylor - poussé suffisamment loin. Ces questions, encore ouvertes, méritent de nouvelles recherches.

Outre les résultats fondamentaux de Whitney, il convient également de citer les travaux de Haefliger sur les singularités de self-intersection. Le présent exposé, rédigé par M. Levine, n'a avec la substance de mon cours qu'un rapport relativement lointain: il en constitue une présentation beaucoup plus développé et très améliorée; bien des démonstrations qui ne se trouvaient qu'esquissées dans mon cours - ou encore complaisamment laissées dans une ombre propice - ont été courageusement et totalement explicitées par M. Levine. Je lui sais particulièrement gré d'avoir relevé dans ma démonstration du lemme de transversalité une lacune dans l'emploi du théorème des fonctions implicites, et je lui suis encore plus reconnaissant de l'avoir complée. Enfin, M. Levine a apporté quelques contributions personnelles très précieuses, par exemple dans l'étude des singularités de 2e ordre $S_h(S_k)$. De ce point de vue il m'est agréable de relever les services que m'a rendus mon auditoire de Bonn par ses remarques et observations, notamment sur la question des orbites d'un groupe algébrique; je lui en exprime ici tout ma reconnaissance.

R. Thom

CONTENTS

I JETS

1. Definitions and Preliminaries

1.1 Let $\mathcal{E}^r(n, p)$ be the set of all r-times continuously differentiable mappings, $f : R^n \to R^p$ such that $f(0) = 0$; hereafter we will refer to R^n as the source and R^p as the target. We introduce the following equivalence relation : $f, g \in \mathcal{E}^r(n, p)$ are equivalent of order r at 0, if at 0, their partial derivatives of orders less than or equal to r are all equal.

DEFINITION. A jet of order r (or an r-jet) is an equivalence class in $\mathcal{E}^r(n, p)$ under the above equivalence relation. For the class of f, we write $f^{(r)}$.

Taking the values of the partial derivatives as the coordinates of a jet, the space of r-jets becomes a Euclidean space which we denote by $J^r(n, p)$. (By an abuse of language we may refer to the coordinate of a jet as partial derivatives, and also speak of the Jacobian etc. of a jet.)

For $r \geq s$, we have a natural projection $\pi_{r,s} : J^r(n, p) \to J^s(n, p)$ where $\pi_{r,s}(f^{(r)}) = f^{(s)}$.

By means of the composition of maps $f : R^n \to R^p$, $g : R^p \to R^q$ we can define a canonical algebraic map :

$$\begin{array}{ccccc} J^r(p, q) & \times & J^r(n, p) & \to & J^r(n, q) \\ g^{(r)} & \times & f^{(r)} & \to & (g \circ f)^{(r)} = g^{(r)}.f^{(r)} \end{array}$$

That this map is well defined and algebraic follows from the chain rule which expresses the partials of $g \circ f$ as polynomials in the partials of g and those of f. For $r = 1$, this product is merely the usual product of the Jacobian matrices. Further, for $r \geq s$, the diagram :

$$\begin{array}{ccc} J^r(p, q) \times J^r(n, p) & \rightarrow & J^r(n, q) \\ \pi_{r,s}\downarrow \qquad \downarrow\pi_{r,s} & & \downarrow\pi_{r,s} \\ J^s(p, q) \times J^s(n, p) & \rightarrow & J^s(n, q) \end{array}$$

is commutative.

1.2 In the special case $n = p = q$, this composition map defines a product in $J^r(n, n)$. Relative to this product, $(1_n)^{(r)}$ is a left and right identity where 1_n is the identity map of R^n onto itself. Let $L^r(n)$ denote the subset of $J^r(n, n)$ of invertible elements, i.e. $L^r(n)$ is the set of $f^{(r)} \in J^r(n, n)$ such that there exists a $g^{(r)} \in J^r(n, n)$ such that $f^{(r)} \cdot g^{(r)} = (1_n)^r$. By the implicit function theorem we know that a map is invertible in a neighbourhood of 0 if and only if its Jacobian is non-singular. Thus an r-jet, $f^{(r)}$, is invertible if and only if $\pi_{r,1}(f^{(r)}) = f^{(1)}$ is in $L^1(n)$. Since $f^{(1)}$ is just the Jacobian matrix of f at 0, we have :

PROPOSITION. $L^1(n) = GL(n, R)$ and $L^r(n) = (\pi_{r,1})^{-1}(L^1(n))$. $L^r(n)$ <u>is a Lie group which is obtained from</u> $GL(n, R)$ <u>by successive extensions with vector groups</u>.

PROOF : The first statement is already proved. That $L^r(n)$ is a Lie group is clear since it is an open subset of Euclidean space, diffeomorphic to $GL(n, R) \times R^k$, where $k = \sum_i n\binom{n+i-1}{i}$; $i = 1, \ldots, r$, whose group law is given by means of polynomials in the coordinates. Associativity of the product follows from associativity of composition of maps.

We define $K^j(n)$ by the exactness of :

$$0 \rightarrow K^j \rightarrow L^j \overset{\pi}{\rightarrow} L^{j-1} \rightarrow 0 , \quad \text{for } j = 2, \ldots, r .$$

To prove the second statement of the proposition, it must be shown that K^j is a vector group. Since the identity in L^{j-1} is the jet whose Jacobian is the identity matrix and all other partials vanish we see that K^j is the subset of L^j whose Jacobian matrix is the identity and whose partials of orders between and including 2 and $(j - 1)$ all vanish. On K^j the group law is easily seen to be addition of the corresponding j-order partials.

2. Singularities

2.1 In the space of r-jets, $J^r(n, p)$ we can define the action of $L^r(p) \times L^r(n)$ as a transformation group by means of composition of maps :

$$R^n \overset{b}{\to} R^n \overset{f}{\to} R^p \overset{a}{\to} R^p$$

Letting $L^r(p) \times L^r(n) = L^r(n, p)$ we obtain :

$$L^r(n, p) \times J^r(n, p) \to J^r(n, p)$$

$$((a^{(r)}, b^{(r)}), f^{(r)}) \to (a \circ f \circ b)^{(r)} = a^{(r)}.f^{(r)}.b^{(r)}$$

The product in $L^r(n, p)$ is defined as follows : (a, b) and (a', b') are elements of $L^r(n, p)$ then $(a', b') \cdot (a, b) = (a' \cdot a, b \cdot b')$.

DEFINITION. For any pair of positive integers (n, p), a singularity of order r is an orbit, $L^r(n, p)(z) \subset J^r(n, p)$, for $z \in J^r(n, p)$.

2.2 By manifold we will always mean differentiable manifold, and maps will always be differentiable.

If V and M are manifolds, and if there is a map $i : V \to M$ such that i is $1:1$ and such that i_* is also $1:1$ then we say that (V, i) is a submanifold of M. If V is a subset of M, and if the inclusion map satisfies the above two conditions then we say simply that V is a submanifold of M.

Note : We do not demand that i be a homeomorphism into.

When i is a homeomorphism into we call V a regular submanifold of M.

PROPOSITION 1. <u>The singularities of order r are submanifolds of</u> $J^r(n, p)$.

This is a special case of :

PROPOSITION 2. <u>If G is a Lie group of transformations on a manifold M , then the orbits are submanifolds of M.</u>

PROOF : The action of the group is given by a differentiable mapping, $f : G \times M \to M : (g, x) \to f(g, x)$. For a fixed $x \in M$, let $\underline{x}$ denote the mapping $\underline{x}\,(g) = f(g, x)$. Let $\underline{x}^{-1}(x) = H$; this is the closed subgroup of G which leaves x fixed. Since H is closed it is again a Lie group and G/H is an analytic manifold. The map $\underline{x}$ induces a map $\bar{\underline{x}} : G/H \to M : (g, H) \to f(g, x)$. This mapping is well defined since H leaves the point x fixed and the diagram

$$\begin{array}{ccc} G & \xrightarrow{\underline{x}} & M \\ \downarrow\pi & \nearrow \bar{\underline{x}} & \\ G/H & & \end{array}$$

commutes (π is the usual projection) .

Since the map $\bar{\underline{x}}$ is $1:1$, it suffices to prove that $\bar{\underline{x}}_*$, is $1:1$ at the identity coset (e) . Suppose $\bar{\underline{x}}_*$ is not $1:1$ at (e) then there is a vector $X_{(e)} \neq 0$ such that $\bar{\underline{x}}_*(X_{(e)}) = 0$. In a sufficiently small neighbourhood U of (e) we have a section $\sigma : U \to G$ by means of which we lift $X_{(e)}$ to the vector $Y_e = \sigma_*(X_{(e)})$ $Y_e \neq 0$ since $\pi_*(Y_e) = \pi_*\sigma_*(X_{(e)}) = X_{(e)} \neq 0$. However by the commutativity of the diagram above $\underline{x}_*(Y_e) = \bar{\underline{x}}_*\pi_*(Y_e) = \bar{\underline{x}}_*(X_{(e)}) = 0$. Thus we have a one parameter subgroup C with tangent vector at e equal to Y_e . We will show that $C \subset H$ which would imply that $\pi_*(Y_e) = X_{(e)} = 0$. Contradiction.
The tangent vector field along C is the restriction to C of a left invariant vector field Y on all of G. Y is obtained from Y_e by left translation by all the elements of G . For any $g \in G$ let L_g denote left translation by g. Then $Y_g = (L_g)_*(Y_e)$ and $\underline{x}_*(Y_g) = \underline{x}_*(L_g)_*(Y_e) = \underline{g}_*\underline{x}_*(Y_e) = 0$, where $\underline{g} : M \to M$, $\underline{g}(x) = f(g, g)$. That $\underline{x}_*(L_g)_* = \underline{g}_*\underline{x}_*$ follows from the commutativity of

$$\begin{array}{ccc} G & \xrightarrow{L_{\underline{g}}} & G \\ \downarrow{\underline{x}} & & \downarrow{\underline{x}} \\ M & \xrightarrow{\underline{g}} & M \end{array}$$

which follows from the fact that G acts as a transformation group. i.e. $f(gh, x) = f(g, f(h, x))$

Thus since $\underline{x}_*$ of the tangent vectors to C are all 0, $\underline{x}(C)$ = one point = $\underline{x}(e) = x$, which is equivalent to $C \subset H$.

DEFINITION 1. If S and S' are singularities, <u>S' is incident to S</u> if $S' \subset \bar{S}$.

If S' is incident to S and $S' \neq S$, then $S' \subset (\bar{S} - S)$.

2.3 If the rank at 0 of $f \in \mathcal{E}^r(n, p)$ is equal to $\min(n, p)$ we say that $f^{(r)}$ is <u>regular</u>, and we denote the subset of regular r-jets by ${}^{\rho}J^r(n, p)$. By choosing suitable coordinate systems in a neighbourhood of 0 in R^n and R^p, say $(x_1, \ldots, x_n)$ and $(X_1, \ldots, X_p)$ respectively, a regular map has the form :

$$X_1 = x_1, \ldots, X_p = x_p \qquad \text{for } n \geq p, \text{ and}$$

$$X_1 = x_1, \ldots, X_n = x_n, X_{n+1} = 0, \ldots, X_p = 0 \quad \text{for } n \leq p .$$

Thus :

PROPOSITION 1. <u>$L^r(n, p)$ acts transitively on</u> ${}^{\rho}J^r(n, p)$, i.e. ${}^{\rho}J^r(n, p)$ <u>is the orbit of the jet of the mapping described above.</u>

This is one of the few known instances where the r-jet of a mapping at a point determines, not only the s-jet of the mapping for all s, but actually the germ of the mapping at the point.

PROPOSITION 2. ${}^{\rho}J^r(n, p)$ <u>is everywhere dense in</u> $J^r(n, p)$

PROOF : Since the condition of regularity is only a condition on the rank of the Jacobian it suffices to see that ${}^{\rho}J^1(n, p)$ is everywhere dense in $J^1(n, p)$. This amounts to the fact that the $(n \times p)$ matrices of rank $= \min(n, p)$ are dense in all $(n \times p)$ matrices, which is trivial.

COROLLARY 1. <u>Every singularity of order r is incident to the regular orbit, ${}^{\rho}J^r$.</u>

2.4 We can identify $J^1(n, p)$ with the set of all real $(n \times p)$ matrices. Let $S_k(n, p)$ be the subset of $J^1(n, p)$, with rank equal to $(q - k)$, where $q = \min(n, p)$ and $0 \leq k \leq q$. If $f^{(1)} \in S_k(n, p)$, we can always choose coordinates in a neighbourhood of 0,

$(Y_1, \ldots, Y_{q-k}, X_1, \ldots, X_{p-q+k})$ and $(y_1, \ldots, y_{q-k}, x_1, \ldots, x_{n-q+k})$, in R^p and R^n respectively so that f has the form

$$Y_i = y_i \quad ; \quad i = 1, \ldots, q-k$$

$$X_\alpha = \phi_\alpha(x,y); \; \alpha = 1, \ldots, p-q+k$$

where all first partials of ϕ_α vanish at 0. In other words in the L^1 orbit of $f^{(1)}$, the 1-jet appears which looks like

$$\underbrace{\begin{pmatrix} I_{q-k} & & 0 \\ & & \\ 0 & & 0 \end{pmatrix}}_{p}\Big\} n \quad , \quad \text{where } I_{q-k} \text{ is the } (q-k \times q-k) \text{ identity matrix.}$$

In the future we will write simply S_k, J^r, and L^r for $S_k(n, p)$, $J^r(n, p)$, and $L^r(n, p)$ respectively, where no confusion arises.

PROPOSITION 1. $J^1 = S_o \cup S_1 \cup \ldots \cup S_q$, <u>and</u> $S_o = {}^{\rho}J^1$.
<u>This is a decomposition of</u> $J^1(n, p)$ <u>into disjoint orbits with</u>
$\bar{S}_k = \bigcup_i S_{k+i}$, $i = 0, \ldots, q - k$.

If $z \in S_k(n, p)$, then $(n - q + k)$ = <u>corank in the source</u> and $(p-q+k)$ = <u>corank in the target</u>.

PROPOSITION 2. S_k *is a regular submanifold of* $J^1(n, p)$ *and*

codim $S_k = (n-q+k)(p-q+k)$ = product of the coranks.

PROOF : It suffices to show that in a neighbourhood W of a point $z \in S_k$, $S_k \cap W$ is the set of zeros of $(n-q+k)(p-q+k)$ functions, which can be completed to a coordinate system in the neighbourhood of the point. In other words, it suffices to give a mapping $\Psi : W \to R^{(n-q+k)(p-q+k)}$ of maximal rank at z such that $S_k \cap W = \Psi^{-1}(0)$.

Since S_k is the orbit of a point, it suffices to show this for a single point. In particular if it has been shown for a point $z \in S_k$, suppose $z' \in S_k$, then there is a $g \in L^1$ such that $gz = z'$. Since L^1 is a transformation group on J^1, gW is a neighbourhood of z' and $\Psi \circ g^{-1}(gW) \to R^{(n-q+k)(p-q+k)}$ is the desired map. Thus we work in a neighbourhood of $\begin{pmatrix} I_{q-k} & 0 \\ 0 & 0 \end{pmatrix} = z$. We take our neighbourhood W of z so small that the upper left hand $(q-k) \times (q-k)$ corner is non-singular. Thus W is a subset of $(n \times p)$ matrices, $x = \begin{pmatrix} A & B \\ C & D \end{pmatrix}$, where A, a $(q-k) \times (q-k)$ matrix is nonsingular. $S_k \cap W$ is the subset of matrices in W of rank exactly $q - k$, thus there is a matrix E such that $B = AE$ and $D = CE$. Solving for E we obtain $D = CA^{-1}B$ as the equations of S_k in W, or $S_k \cap W = \{x \in W \mid D - CA^{-1}B = 0\}$. The number of functions here is $(n-q+k)(p-q+k)$ and they clearly define a mapping of rank $(n-q+k)(p-q+k)$ since each is linear in a different coordinate in $J^1(n, p)$.

PROPOSITION 3. $\bar{S}_k$ *is an algebraic set in* $J^1(n, p)$ *and the locus of singularities of* $\bar{S}_k$ *is* $\bar{S}_{k+1}$, $k = 1, \ldots, q - 1$

PROOF : Since $\bar{S}_k = \bigcup_i S_{k+i}$, $i = 0, \ldots, q-k$, we know that $\bar{S}_k$ is the set of all $(n \times p)$ matrices such that all $(q - k + 1)$ determinants vanish. So $\bar{S}_k$ is an algebraic set. We show that a particular point $x \in S_{k+1}$ is singular point of S_k. Thus since the group L^1 maps the singular locus of $\bar{S}_k$ into itself, this yields

$S_{k+1} \subset$ (singular locus of $\bar{S}_k$) and since the singular locus is closed we have $\bar{S}_{k+1} \subset$ (singular locus of $\bar{S}_k$). However $\bar{S}_{k+1} = (\bar{S}_k - S_k) \supset$ (singular locus of S_k). Thus $\bar{S}_{k+1} =$ (singular locus of $\bar{S}_k$).

We take as x, the jet $\begin{pmatrix} I_{q-k-1} & 0 \\ 0 & 0 \end{pmatrix}$, let e_{ij} be the $(n \times p)$ matrix with all entries zero except at the $(i, j)^{th}$ place where there is a 1. Consider the curves $(x + e_{ij}t)$ where $t \in R$. These are all curves in $\bar{S}_k$ which pass through the point x for $t = 0$. The vectors at x tangent to these curves are clearly all independent, and there are np of them. However if x were a regular point of $\bar{S}_k$ we could obtain at most $(np - (n-q+k)(p-q+k))$ independent vectors at x in this fashion. Thus for $k \neq 0$, x is a singular point of $\bar{S}_k$.

2.5 We can give another description of these singularities which will be useful later on. Let $(v_1, \dots, v_n)$ be a fixed n-frame in R^n. To each element of $J^1(n,p)$ corresponds a linear transformation $T : R^n \to R^p$. Consider now the n vectors $(v_1 \times T(v_1), \dots, v_n \times T(v_n))$ in $R^n \times R^p$. Since the n vectors are linearly independent, they span an n-dimensional subspace of R^{n+p}. We thus obtain a mapping $\sigma : J^1(n, p) \to G(n, p)$, where $G(n, p)$ is the Grassman manifold of n-planes through 0 in $(n+p)$-space. We may choose coordinates in R^n so that the matrix whose rows are the coordinates of the vectors $v_i \times T(v_i)$ has the form $(I_n\, T)$. The mapping σ is clearly $1:1$ and onto a neighbourhood of the horizontal plane, the plane $(I_n, 0)$. The image under σ of $J_1(n, p)$ is the set of all planes whose first Plücker coordinate is different from zero.

This is the set of all n-planes in R^{p+n} whose projection onto the horizontal plane is n-dimensional. In order to describe this image and the images of the singularities we need the notion of Schubert variety.

DEFINITION. In R^{n+p} choose a sequence of linear spaces through 0, $E^1 \subset E^2 \subset \dots \subset E^{n+p}$, where $\dim(E^k) = k$. For any sequence $0 \leqslant a_1 \leqslant \dots \leqslant a_n \leqslant p$, the set of elements $X \in G(n, p)$, i.e. the n-dimensional linear spaces through 0, satisfying

$$\dim(X \cap E^{i+a_i}) \geqslant i \; ; \; i = 1, \dots, n$$

is called a Schubert variety and is denoted by the Schubert symbol $(a_1, \dots, a_n)$.

The definition of the Schubert varieties depends on the choice of the sequence E^k, but the varieties defined by a given symbol relative to different sequences are carried into one another under the group of motions $0^+(n + p)$ in R^{n+p}. If $a_i = a_{i+1}$, then $(\dim(X \cap E^{i+1+a_{i+1}}) \geqslant i + 1)$ implies $(\dim(X \cap E^{i+a_i}) \geqslant i)$.

PROPOSITION 1. $\sigma : J^1(n, p) \to G(n, p)$ is a diffeomorphism onto $G(n, p) - Q$, where Q is the Schubert variety with symbol $(p-1, p, \dots, p)$.

PROOF : By the preliminary remarks we know that σ is a diffeomorphism onto the set of all n-planes whose projection onto the horizontal plane is n-dimensional. Let $Q = \{X \in G(n, p) \mid \text{projection of } X \text{ into } R^n \text{ has dimension less than } n\}$ $= \{X \in G(n, p) \mid \dim(X \cap R^p) \geqslant 1\}$. Thus $Q = (p-1, p, \dots, p)$, for if we choose a sequence of spaces E^i such that $E^p = R^p$, then $\dim(X \cap E^{i+p-1}) = \dim(X \cap R^p) \geqslant 1$ and $\dim(X \cap R^{n+p}) \geqslant n$ is automatic.

Identifying $J^1(n, p)$ with $\sigma(J^1(n, p))$ we have :

PROPOSITION 2. $\bar{S}_k = (\underbrace{q-k, \dots, q-k}_{(n-q+k)}, p, \dots, p) \cap (G(n,p) - Q)$, where $q = \min(n, p)$.

PROOF : Suppose $X \in G(n, p) - Q$, then $X \in \bar{S}_k$ if and only if its projection into R^p has dimension less than or equal to $q - k$. Thus $S_k = \{X \in G(n, p) - Q \mid \dim(X \cap R^n) \geqslant (n-q+k)\}$. If we choose a sequence E^i, such that $E^n = R^n$, then the statement of the proposition is merely that

$$\dim(X \cap E^{(n-q+k)+(q-k)}) = \dim(X \cap R^n) \geqslant n-q+k \quad .$$

2.6 We consider now the simplest 2nd order singularities, $(\pi_{2,1})^{-1} S_q$, and show that for $p = 1$, J^2 is again a finite union of singularities but for certain (n, p) with $p > 1$ we get uncountably many distinct singularities. We let $(\pi_{2,1})^{-1} S_q = S_q^2$.

Consider first the case where $p = 1$. A 2-jet in S_q^2 is merely an $(n \times n)$ symmetric matrix, A , the 2-jet of a function, $f(x) = {}^t x A x$ where ${}^t x = (x_1, \ldots x_n)$. The action of the group $L^2(n, 1)$ on $f^{(2)}$ is described as follows. An element of $L^2(1)$ is the jet of a function $g(y) = uy + wy^2$ where $u \neq 0$ is constant. $(g \circ f(x) = u({}^t x A x) +$ (higher order terms). An element of $L^2(n)$ is the jet of a mapping $h : R^n \to R^n : z \to C z +$ (higher order terms) where ${}^t z = (z_1, \ldots, z_n)$ and $C \in GL(n, R)$. Thus $(g \circ f \circ h)(x) = {}^t x(u \,.\, {}^t C A C) x +$ higher order terms. Thus $(g \circ f \circ h)^{(2)} = (0, u\, {}^t C A C)$ and the orbit of the 2-jet $(0, A)$ under $L^2(n, 1)$ is the set of all $(0, u\, {}^t C A C)$ with $u \neq 0$ and $C \in GL(n, R)$. Thus the classification of orbits reduces to the classification of quadratic forms under the projective group. For this classification we have two integral invariants : the rank, k , and $j = |k - 2i|$ where i is the index of A , (the number of negative squares in ${}^t C A C$ in diagonal form.)

Let H_{kj} be the set of symmetric $n \times n$ matrices with rank k and index $i = \frac{1}{2}(k \pm j)$.

PROPOSITION 1. $J^2(n, 1) = {}^{\rho}J^2(n, 1) \cup \bigcup_{k,j} H_{kj}$ is a disjoint union of singularities

(The union is taken over all $0 \leq j \leq k \leq n$).

REMARK : $L^m(n, p)$ for $m \leq r$ also acts on $J^r(n, p)$, since any m-jets can be considered as an r-jet with the partials between m and r all equal to zero.

For the investigation of the singularities in S_q^2 for $p > 1$ we use the following lemma about the action of L^r on $\Sigma_k \subset J^r$, the set of all r-jets all of whose partials of orders $\leq k$ vanish.

LEMMA 1.

$$\begin{array}{ccccc} L^r & \times & \Sigma_k & \rightarrow & \Sigma_k \\ \pi_{r,r-k}\downarrow & & \downarrow \text{ id} & \nearrow & \\ L^{r-k} & \times & \Sigma_k & & \end{array}$$

<u>commutes, where the horizontal and slant maps are given by the action of the group on J^r</u>.

PROOF : This is easily seen by noting that an sth partial $\partial^s(f \circ g)$ is the sum of terms of the form $\partial^m f \, \partial^{i_1} g \ldots \partial^{i_m} g$ where $m \leq s$ and $\sum_{j=1}^{m} i_j = s$.

For the action of $L^r(p)$, suppose $\partial^i f_o = 0$ for $0 \leq i \leq k$ and suppose that in the term $\partial^m f \, \partial^{i_1} g \ldots \partial^{i_m} g$ in $\partial^s(f, g)$, some $i_j \geq r - k + 1$ we see that $s = \sum_j i_j \geq (r - k + 1) + m - 1$. Since $s \leq r$ we obtain $m \leq k$ so the term vanishes since $\partial^m f = 0$ for $m \leq k$.

For the action of $L^r(n)$ we suppose that $\partial^i g = 0$ for $0 \leq i \leq k$ and suppose that in the sth partial, $\partial^s(f \circ g)$, we have a term with $m \geq r - k + 1$, $s = \sum_j i_j \geq m(k+1) \geq (r - k + 1)(k + 1)$. Since s is smaller than r we obtain $k^2 > (k^2 - 1) \geq rk$, $k > r$.

Consider now the general case $J^2(n, p)$ for $p > 1$. We show that in this case unlike in the preceding we have, frequently, a continuous set of orbits. Here $S^2_q = \Sigma_1$ is the set of jets all whose first partials vanish; each such jet can be denoted by $(0, A_1, A_2, \ldots, A_p)$ where A_i are the $n \times n$ symmetric matrices of the 2nd partials of the ith coordinate of the jet. The action of $L^2(p)$ on $(0, A_1, A_2, \ldots, A_p)$ is given as follows :

suppose $g : R^p \rightarrow R^p$, $g^i(y) = \sum_j u_{ij} y_j$, where $(u_{ij}) = U \in GL(p, R)$. The image of $(0, A_1, A_2, \ldots, A_p)$ under $(g)^{(2)}$ is $(0, \sum_j u_{1j}A_j, \sum_j u_{2j}A_j, \ldots, \sum_j u_{pj}A_j)$.

Under an element $(h)^{(2)}$ $L^2(n)$ where $h_i(x) = \sum_j c_{ij} x_j$, where $(c_{ij}) = C \in GL(n,R)$,

$(0, A_1, \ldots, A_p)$ is mapped to $(0, {}^tC A_1 C, \ldots, {}^tC A_p C)$. Thus the orbit of $(0, A_1, \ldots, A_p)$ under the group $L^2(n, p)$ is the set of all jets $(0, \sum_j u_{1j} {}^tC A_j C, \ldots, \sum_j u_{pj} {}^tC A_j C)$ where $C \in GL(n, R)$ and $U \in GL(p, R)$.

Since S_q^2 is the p-fold cartesian product space of symmetric $(n \times n)$ matrices, $\dim(S_q^2) = \frac{n(n+1)p}{2}$. The dimension of an orbit in S_q^2 is smaller than or equal to $\dim(GL(n, R) \times GL(p, R)) = n^2 + p^2$. Thus there are certainly infinitely many orbits in S_q^2 if $\max \dim(\text{orbit}) < \dim S_q^2$.

PROPOSITION 2. <u>For all</u> (n, p) <u>such that</u> $(n^2 + p^2) < \frac{np(n+1)}{2}$, $J^2(n, p)$ <u>has infinitely many orbits</u>.

3. The closure of a real orbit may not be a real algebraic set.

3.1 Let A be a subset in R^m. We denote by A^*, the closure of A in the <u>Zariski topology</u>. That is :

$$A^* = \{x \in R^n \mid (P(A) = 0) \text{ implies } (P(x) = 0), \text{ for all real polynomials } P\}.$$

Since A^* is closed in the ordinary topology it contains $\bar{A}$.

DEFINITION 1. A closed set F in R^m is a <u>real algebraic set</u> iff $F^* = F$.

DEFINITION 2. A <u>critical manifold</u> in $J^r(n, p)$ is a regular submanifold of $J^r(n, p)$ invariant under the group $L^r(n, p)$, whose topological closure is a real algebraic set.

The singularities S_k are all critical manifolds but the H_{jk} of 2.6 are not, since $H_{jk} \neq H_{jk}^*$. Each of these submanifolds is the orbit of a single point in J^r under L^r, and although the action of L^r is algebraic we see that we cannot conclude that the topological closure of an orbit of a point is a real algebraic set in J^r. The following theorem due to Seidenberg and Tarski ([4], p. 370) indicates what could happen.

THEOREM. Let F be any finite set of simultaneous polynomial equations and inequalities in the unknowns $x_1, \ldots, x_n$ and the parameters $a_1, \ldots, a_m$ where all the polynomials in question are in $Q[a_1, \ldots, a_m; x_1, \ldots, x_n]$ where Q = the field of rational numbers. One can then construct (in a finite number of steps) sets $G_1, \ldots, G_s$ of polynomials equations and inequalities involving only the parameters $a_1, \ldots, a_m$, such that for any real-closed field K and any values a_i of the a_i in K, the following two statements are equivalent :

(a) F has at least one solution x_i in K.

(b) for at least one i, $i = 1, \ldots, s$, all the equations and inequalities of G_i are satisfied.

PROPOSITION 1 : The theorem of Seidenberg-Tarski implies the following : Let $\emptyset$ be an algebraic map from R^n in R^m and suppose $K \subset R^n$ is given by a finite number of polynomial equations and inequalities, then $\emptyset(K)$ is the union of a finite number of sets each of which is given by a finite number of polynomial equations and inequalities.

PROOF : The proof of theorem itself we will not give, (see [4]) we will merely prove this implication. Suppose $a_1, \ldots, a_m$ are the coordinates in R^m, and $x_1, \ldots, x_n$ those of R^n. Since $\emptyset$ is algebraic, the equations of its graph in $R^n \times R^m$ are polynomial equations. We take these equations together with the equations and inequalities defining K as our system F. Since the coefficients in these expressions are not all necessarily rational we might have to throw in a few more parameters in our defining equations and inequalities. The theorem states that it is possible to find a finite set of sets in R^m, each of which is defined by polynomial equations and inequalities in the a_1, the union of which is exactly the image of the set K under $\emptyset$. These defining polynomials may have some irrational coefficients since we have at our disposal the real numbers which entered into the system F. Since the closure of such a set is again of the same type the implication is proven.

In our special case of orbits of a point under an algebraic group, the set K would be given by polynomial equations only, but the Seidenberg-Tarski result does not guarantee that inequalities won't enter into the discription of the closure of the image.

4. Constructing critical manifolds from real orbits

4.1 We show in this section that we can choose certain real orbits, the union of which is the smallest critical manifold containing any point of the set. To do this we need the complex analog to the real jets, which we will denote consistently by $\sim$ over the corresponding real symbol. For instance $\tilde{J}^r(n, p)$ is the set of all complex r-jets of holomorphic maps $f : C^n \to C^p$ such that $f(0) = 0$, and $\tilde{L}^r(n, p)$ is the complex group which acts on $\tilde{J}^r(n, p)$. We consider $J^r(n, p)$ as canonically imbedded in $\tilde{J}^r(n, p)$, and $\tilde{L}^r(n, p)$ is just the complexification of $L^r(n, p)$. Again we will omit (n, p) when the omission is not confusing, and A^* will again mean Zariski closure of A - whether complex or real will be clear from the context.

PROPOSITION 1. $\tilde{L}^r(z)$ is Zariski-open in $(\tilde{L}^r(z))^*$ for $z \in J^r$.

Before proving this we draw some conclusions. Suppose $z \in J^r$, then since $\tilde{L}^r$ is the complexification of L^r we have :

$$(L^r(z))^* = (\tilde{L}^r(z))^* \cap J^r) = (\tilde{L}^r(z) \cap J^r)^* = \overline{(\tilde{L}^r(z) \cap J^r)} .$$

The last equality requires a trivial argument about the Zariski topology.
Thus if we knew that $\tilde{L}^r(z) \cap J^r$ was a regular submanifold in J^r we would have:

PROPOSITION 2. ([3], p.7 - 05) For $z \in J^r$, $\tilde{L}^r(z) \cap J^r$ <u>is a minimal critical manifold containing</u> $L^r(z)$.

PROOF : By Proposition 1, and the fact that $\tilde{L}^r$ is the complexification of L^r, we know that $L^r(z)$ is open in $(L^r(z))^*$. Thus since L^r maps the singular set of $(L^r(z))^*$ into itself and dimension of the singular set in $(L^r(z))^*$ is lower dimensional and L^r acts transitively on $L^r(z)$ we have $L^r(z)$ is contained in the regular points of $(L^r(z))^*$. Exactly the same argument can be used to show that if $x \in J^r \cap \tilde{L}^r(z)$, then $L^r(x)$ is open in the set of regular points of $(L^r(z))^*$. Thus $J^r \cap \tilde{L}^r(z)$ is an open subset of the regular points of $(L^r(z))^*$. Note : We cannot conclude that $J^r \cap \tilde{L}^r(z)$ is the set of regular points of $(L^r(z))^*$ - counter example : ${}^{\rho}J^r$.

4.2 Proposition 1 of 4.1 is a special case of the following : (everything is complex and the topology referred to is the Zariski topology).

LEMMA 1. ([1] , exposé 5, Lemma 4, p.13) <u>Let</u> G <u>be an algebraic group of transformations of an algebraic set</u> A . <u>Then the orbit under</u> G <u>of a point in</u> A <u>is open in its closure</u> .

This in turn is a consequence of :

LEMMA 2. ([2], exposé 7, Théorème 3, p.9) <u>Let</u> f <u>be a regular map of an algebraic set</u> G <u>in another</u> E . <u>Then the interior of</u> f(G) <u>in</u> $(f(G))^*$ <u>is dense in</u> f(G).

PROOF : We show first that Lemma 2 implies Lemma 1 and then sketch the proof of Lemma 2.

By Lemma 2, the interior of f(G) is non empty. Thus the interior of an orbit in its closure is non-empty. Since the interior of an orbit relative to its closure is mapped into itself by the group and since the group is transitive on its orbit we see that the whole orbit is open in its closure.

For the proof of Lemma 2, it suffices to show that $f(G)$ contains a non-empty open subset of $f(G)^*$, and to restrict to the case of G irreducible. Suppose $G \subset C^n$ and $E \subset C^m$. We consider the map $f : G \to f(G)^*$.

Let A be the quotient of the ring of polynomials in C^n modulo the ideal of polynomials vanishing on G, and B, the corresponding quotient for $f(G)^*$. Since $f(G)$ is dense in $f(G)^*$, the induced mapping $f^* : B \to A$ is a monomorphism. Thus we can regard B as a subring of A. Since we assumed that G was irreducible, A is an integral domain and we can apply the following which we state without proof.

LEMMA 3. ([2], exposé 3, Lemma, p.1) _Let A be an integral domain and B a subring such that A is generated by B and a finite number of elements. Let $0 \neq v \in A$. Then there exists a $0 \neq u \in B$, such that any homomorphism h of B into an algebraically closed field, K, such that $h(u) \neq 0$ can be extended to a homomorphism g of A into K with $g(v) \neq 0$._

Applying Lemma 3 to our case we choose $v = 1$ and let $u \in B$, given by the Lemma. Let U be the open subset of $f(G)^*$ where u doesn't vanish. This is non-empty since $u \neq 0$. We show $U \subset f(G)$. Let $y \in U$ and define the homomorphism $h : B \to C$ by taking each $w \in B$, to its value $w(y)$. This makes sense since two elements in the equivalence class of w differ by a polynomial that vanishes on $f(G)^*$. Since $u(y) \neq 0$, we can extend this homomorphism to $g : A \to C$, such that $g(1) \neq 0$. The kernel, N, of g is a proper ideal of A since $1 \notin N$, and there is a point $x \in G$ such that $N(x) = 0$. But the kernel of h, also annihilates $f(x)$. Thus $f(x) = y$.

REFERENCES

[1] Chevalley, C. Séminaire 1956-1958, Classification des Groupes de Lie Algebrique, vol. 1.

[2] Chevalley, C. - Cartan, H. Séminaire 1955-1956, Géométrie Algébrique.

[3] Haefliger, A. Séminaire H. Cartan 1956-1957.

[4] Seidenberg, A. A New Decision Method for Elementary Algebra Annals of Math., vol. 60, 1954, pp. 365-374.

II SINGULARITIES OF MAPPINGS

5. General Definitions

5.1 Let V and M be manifolds of dimensions n and p respectively. We define the r-jet from V to M with source x and target y of a C^r - map $f : V \to M$ such that $f(x) = y$ as the equivalence class of all C^r - maps from V to M which take x into y, all of whose partials at x of orders $\leq r$ are equal to those of f. We denote the r-jet of f at x by $J^r(f)(x)$. The set of all r-jets from V to M, we denote by $J^r(V, M)$. By choosing local coordinate systems in V and M, it is clear that $J^r(V, M)$ is a fibre bundle over $V \times M$ with fibre $J^r(n, p)$ and group $L^r(n, p)$. When $V = R^n$ and $M = R^p$, $J^r(R^n, R^p)$ is the product bundle $R^n \times R^p \times J^r(n, p)$.

DEFINITION 1. Let $f : V \to M$ be of class at least C^r, the r-extension of f is defined by $J^r(f) : V \to J^r(V, M) : x \to J^r(f)(x)$.

In the special case where $V = R^n$, and $M = R^p$ we can further follow this mapping by the projection into the fibre, which yields :

$$f^{(r)} : R^n \to J^r(n, p) : x \to f^{(r)}(x) ,$$

where $f^{(r)}(x)$ is an r-jet in the sense of 1.1 where we've merely shifted the origin of R^n to x and shifted the origin of R^p to $f(x)$.

5.2 For a regular submanifold S of codimension q in a manifold M of dimension m, the following is always true : for each point $y \in S$ there exists a neighbourhood U of y and a mapping $g : U \to R^q$ of rank q such that $U \cap S = g^{-1}(0)$. We call g a defining mapping of S in M at y.

DEFINITION 1. Let N be a manifold of dimension n and S and M as in the preceding paragraph. Let $f : N \to M$ be a differentiable mapping. f is said to be transversal to the submanifold S at a point $x \in N$ if either :

i. $f(x) \notin S$, or

ii. $f(x) \in S$ and the image under f_* of the tangent space to N at x and the tangent space to S at $f(x) = y$ span the tangent space of M at y , or equivalently $g \circ f$ is of rank q at x , where g is a defining map of S in M at $y = f(x)$.

If f is transversal to S at every point $x \in N$, we say merely that f is transversal to S .

PROPOSITION 1. Let $f : N \to M$ be transversal to the regular submanifold S in M at every point of $x \in N$, then $f^{-1}(S)$ is a regular submanifold of N of codimension q in N , or void.

PROOF : If at a point $y = f(x)$ we have the defining mapping $g : U \to R^q$ for some neighbourhood U of y , we take as defining mapping at x of $f^{-1}(S)$ in N, $g \circ f : f^{-1}(U) \to R^q$. The q components of $g \circ f$ can be taken as q coordinate functions on $f^{-1}(U)$, so in this neighbourhood $f^{-1}(S)$ is defined by setting q coordinate functions equal to zero.

We give an equivalent definition of transversality (using the same notation as above). Suppose $y \in S \subset M$, we denote by $T(S)_y$ the tangent space to S at y and $T(M)_y$ the tangent space to M at y . Let $p : T(M)_y \to T(M)_y / T(S)_y$ be the obvious projection.

If $x \in N$ and $f : N \to M$, $f(x) = y$, then we have the induced map $f_* : T(N)_x \to T(M)_y$.

DEFINITION 1'. f is transversal to S at $x \in N$ iff either

i. $f(x) \notin S$

ii. $p \circ f_* : T(N)_x \to T(M)_y / T(S)_y$ is surjective, where $y = f(x)$.

The equivalence of the two definitions is trivial.

5.3 Let V and M be manifolds. Denote by $L(V, M, q)$ the set of all q-times continuously differentiable maps from V to M. On $L(V, M, q)$ we define the topology as follows. Let $J^q : L(V, M, q) \to L(V, J^q(V, M), 0)$, where $J^q(f)$ has the usual meaning (5.1). Taking the compact open topology on $L(V, J^q(V, M)\ 0)$ we define the topology on $L(V, M, q)$ to be the weakest in which J^q is continuous. In the case $V = R^n$, $M = R^p$ we will have occasion to use another topology on $L(R^n, R^p, q)$ - the topology of uniform convergence of all partials of orders $\leq q$ (including the oth). Unless however we state explicitly to the contrary $L(V, M, q)$ will be understood with the topology of "compact convergence of all partials of orders $\leq q$" defined in the preceding paragraph.

PROPOSITION 1. $L(V, M, q)$ <u>is a Baire space</u>

PROOF : Since this is a local question we may assume that V and M are R^n and R^p respectively. But locally $L(R^n, R^p, q)$ is a complete metric space, and locally complete metric implies Baire [1].

<u>Note</u> : $L(R^n, R^p, q)$ is a Baire space in either topology mentioned above.

6. Applications of Sard's Theorem

6.1 Suppose P and Q are surface differentiably embedded in R^3. We can represent the kinds of local intersections by :

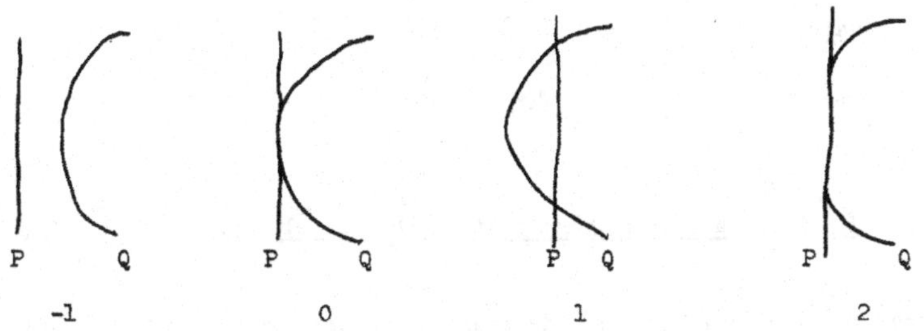

where the integers give the dimensions of the intersections. The figures would indicate that in the 0 - and 2 - dimensional intersection cases a small deformation of one of the surfaces would change the dimension to either -1 or $+1$, whereas small deformations would not change the dimensions in the -1 or $+1$ cases. The *Transversality Theorem* proves this intuition to be essentially correct.

Assume that the surface P is locally the plane $x = 0$, and that Q is the set of solutions of $f(x, y, z) = 0$. Thus locally the intersection is given by $f(0, y, z) = 0$. If at a point $(0, y, z)$, $\frac{\partial f}{\partial y} \neq 0$ (or $\frac{\partial f}{\partial z} \neq 0$), we can solve for y (or z) in a neighbourhood of the point and obtain an equation for the intersection $y = \phi(z)$ on P in a neighbourhood of the point. If $d(f|P) \neq 0$ then the intersection of P and Q consists of non-singular non intersecting curves. If we take any function g close to f in the sense that the values and the first derivatives of g and f are close on a compact set K, then the intersection of the solutions of g with $P \cap K^{o}$ would consist of non-singular curves.

PROPOSITION 1. *Let K be a compact subset of R^3 and P a surface differentiably embedded in R^3, then the set of $g \in L(R^3, R, 1)$ which have the property that the surface $g = 0$ intersects $P \cap K^{o}$ in non-singular non-intersecting curves is open.*

DEFINITION 1. Let f be a real valued function on R^m. $x \in R^n$ is called a *critical point* of f if $df(x) = 0$. P_f = the set of critical points (or the critical set) of f . $y \in R$ is called a <u>critical value</u> of f if $y \in f(P_f)$.

$$\Sigma_f = f(P_f) .$$

PROPOSITION 2. <u>If</u> f <u>is a proper function then</u> Σ_f <u>is closed.</u>

PROOF : Since f is proper, $f^{-1}(K)$ is compact for any compact $K \subset R$. Since P_f is closed in R^m , $P_f \cap f^{-1}(K)$ is compact. Hence $\Sigma_f \cap K$ is closed for every compact K. Thus Σ_f is closed. We know thus that $R - \Sigma_f$ is open for f a proper function on a manifold. It would be more interesting to know if it is everywhere dense. If it were then given any function $f \in L(R^3, R, 1)$ and P any surface in R^3 then the critical values of $f|P$ would be nowhere dense. So there would be values c arbitrarily close to zero such that if $x \in P$ is a solution of $f|P = c$ then $df|P(x) \neq 0$; thus the solution surface of $f - c = 0$ cuts P in non-singular curves. Thus under the assumption that the critical set is nowhere dense we obtain that the set of all $g \in L(R^3, R, 1)$ whose set of zeros intersect a given surface P in non-singular curves is everywhere dense in $L(R^3, R, 1)$. We will see that we will have to demand more differentiability than C^1 before the above statement is true. Clearly the above discussion can be modified for R^n instead of R^3 , but the differentiability of the functions and the surfaces will depend on n. The information we're after is contained in :

THEOREM 1. (A.P. Morse [2]) <u>If</u> $f \in L(R^n, R. r)$ <u>where</u> $r > (n - 1)$ <u>then</u> Σ_f <u>has outer measure zero (in particular is nowhere dense in</u> R).

We will also need information about maps in $L(R^n, R^p, s)$. To this end we adapt definition 1, and state the theorem of Sard [3].

DEFINITION 1'. Let $f \in L(R^n, R^p, 1)$; $x \in R^n$ is called a <u>critical point of</u> f if the rank of f at x is less than $\min(n, p)$.

Let P_f = the set of critical points of f.

A point $y \in R^p$ is called a <u>critical value</u> of f, if $y \in f(P_f)$. Let $\Sigma_f = f(P_f)$.

Proposition 2 above clearly goes over for proper mappings.

THEOREM 2. If $f \in L(R^n, R^p, s)$ and $s > \max(n - p, 0)$, then Σ_f has outer measure zero.

6.2 Throughout this section V, M, N will denote differentiable manifolds of dimensions $n, p, p - q$ respectively all of whose structures are at least s-differentiable and N is a regular submanifold of M.

WEAK TRANSVERSALITY THEOREM. Suppose $s > \max(n - q, 0)$ and K is a compact subset of V. The set of maps in $L(V, M, s)$ which are transversal to N on K is a countable intersection of open dense sets. If N is closed this set is open.

PROOF : Let $S_K = \{F \in L(V, M, s) \mid f$ is not transversal to N on $K\}$, and let $T_K = L(V, M, s) - S_K$.

LEMMA 1. If N is closed, S_K is closed in $L(V, M, s)$ (T_K is open).

PROOF : Let $g \in (L(V, M, s) - S_K)$, then g is transversal to N on K. Cover K with a finite number of compact sets K_i each of which is contained in a coordinate neighbourhood and such that $g(K_i) \subset \mathcal{M}_i$, coordinate neighbourhoods in M. Let W_i be the neighbourhood of g defined by $W_i = \{h \in L(V, M, s) \mid h(K_i) \subset \mathcal{M}_i\}$ and let $W = \cap_i W_i$.

Let L_i be an open neighbourhood in K_i of $g^{-1}(N) \cap K_i$ such that the composite $p \circ g_* : T(V)_x \to T(M)_{gx}/T(N)_{gx}$ is surjective for all x in the closure $\bar{L}_i$ of L_i in K_i. Choose a neighbourhood W' of g such that, for $h \in W'$,

$h(K_i - L_i) \cap N = \emptyset$ and $p \circ h_* : T(V)_x \to T(M)_{hx}/T(N)_{hx}$ is surjective for $x \in \bar{L}_i$. Since $K_i - L_i$ and $\bar{L}_i$ are compact and N is closed, W' is an open neighbourhood of g consisting of maps transversal to N on K_i.

LEMMA 2. <u>If</u> N <u>is not closed</u> S_K <u>is a countable union of closed sets</u> (T_K <u>is a countable intersection of open sets</u>).

PROOF : Cover N by countably many closed neighbourhoods N_i and define $f \in L(V, M, s)$ to be transversal to N_i at x if either $f(x) \notin N_i$, or $f(x) \in N_i$ and f is transversal to N at x. The argument in Lemma 1 gives $\{g \in L(V, M, s) \mid g$ is transversal to N_i on $K\}$ is open. Thus $\{g \in L(V, M, s) \mid g$ is transversal to N on $K\}$ $= \bigcap_i \{g \in L(V, M, s) \mid g$ is transversal to N_i on $K\}$, as required.

Since this theorem is a special case of the general transversality theorem which we prove later, we merely sketch the proof that S_K is nowhere dense in $L(V, M, s)$.

To prove S_K is nowhere dense it suffices to show that for each $f \in L(V, M, s)$ there exists a neighbourhood W_f of f such that $S_K \cap W_f$ is nowhere dense in W_f. Let $\mathcal{M}$ be a coordinate neighbourhood of a point of N such that $N \cap \mathcal{M}$ is defined by an s-differentiable map $\phi : \mathcal{M} \to R^q$ of rank q, i.e. $N \cap \mathcal{M} = \phi^{-1}(0)$. We may assume $f(K) \subset \mathcal{M}$ and that K is contained in a coordinate neighbourhood. (Otherwise cover K with a <u>finite</u> number of compact K_i and find $\mathcal{M}_i$ such that K_i is contained in a coordinate neighbourhood; $\mathcal{M}_i$ is a coordinate neighbourhood in which $N \cap \mathcal{M}_i$ is defined by ϕ_i and such that $f(K_i) \subset \mathcal{M}_i$. As indicated below we will prove there is a W_f^i such that $S_{K_i} \cap W_f^i$ is nowhere dense in W_f^i. Thus if $W_f = \cap W_f^i$, then $(\bigcup_i S_{K_i}) \cap W_f$ is nowhere dense in W_f. Thus $\bigcup_i S_{K_i} = S_K$ is nowhere dense.)

Let $W_f = \{g \in L(V, M, s) \mid g(K) \subset \mathcal{M}\}$.

A map $g \in W_f$ is transversal to N on K if $\phi \circ g : g^{-1}(\mathcal{M}) \to R^q$ has rank q on $K \cap g^{-1} \circ \phi^{-1}(0)$. Thus if $\mathcal{V}$ is a coordinate neighbourhood containing K, then transversality of g on K is implied by the transversality of g on $\mathcal{V} \cap g^{-1}(\mathcal{M})$. Thus g is transversal to N on $\mathcal{V} \cap g^{-1}(\mathcal{M})$ if and only if the map $\phi \circ g : \mathcal{V} \cap g^{-1}(\mathcal{M}) \to R^q$ has 0 as a non-critical value.

Suppose $\phi \circ g : \mathcal{V} \cap g^{-1}(\mathcal{M}) \to R^q$ has 0 as a critical value. We must find an $h \in W_f$, arbitrarily close to g such that $\phi \circ h : \mathcal{V} \cap h^{-1}(\mathcal{M}) \to R^q$ has 0 as a non-critical value. Since $s > \max(n - q, 0)$ by assumption, theorem 2 of 6.1 (Sard's theorem) tells us that it is possible to find a value $c \in R^q$ arbitrarily close to 0 which is a regular value of $\phi \circ g | \mathcal{V} \cap g^{-1}(\mathcal{M})$. If c is in a coordinate ball B_ρ about 0, we can find a diffeomorphism of B_ρ into itself which maps c to 0 which is the identity map on $B_\rho - B_{\rho'}$, for some ρ such that $|c| < \rho' < \rho$; this map can be extended to a diffeomorphism, j_c, of R^q which is the identity outside $B_{\rho'}$. Thus $j_c \circ \phi \circ g | \mathcal{V} \cap g^{-1}(\mathcal{M})$ has 0 as a non-critical value, i.e. g is transversal to $(j_c \circ \phi)^{-1}(0)$ on $\mathcal{V} \cap g^{-1}(\mathcal{M})$. Since ϕ is of rank q we can find a diffeomorphism $\bar{j}_c$ of $\mathcal{M}$ onto itself such that $\phi \circ \bar{j}_c | \mathcal{M} = j_c \circ \phi | \mathcal{M}$. It is reasonably clear that j_c can be extended to a diffeomorphism of M which is the identity outside a neighbourhood. $\mathcal{M}' \supset \mathcal{M}$. Introduce a coordinate system in $\mathcal{M}'$ whose first q functions are $\phi_1, \ldots, \phi_q$. In this coordinate system $N \cap \mathcal{M}'$ is just the plane defined by $\phi_i = 0, (i = 1, \ldots, q)$; $\bar{j}_c$ just slides the part of this plane $\phi_i = c_i$ in $\mathcal{M}$ down to $\phi_i = 0$ plane, and above and below a certain level $\bar{j}_c$ is the identity in $\bar{\mathcal{M}}$. All we must do is slow down this sliding process until when we get to the boundary of $\mathcal{M}'$ we have the identity. Everything can be described precisely if the neighbourhoods $\mathcal{M}$ and $\mathcal{M}'$ were chosen as coordinate balls as shown in the figure ...

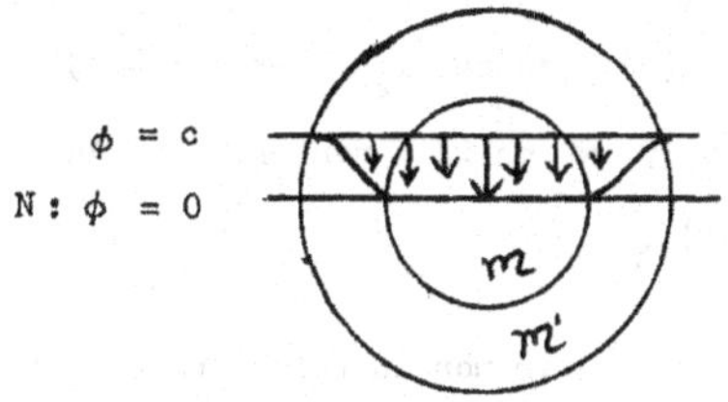

Define $h = \bar{j}_c \circ g$, h is clearly in W_f for c close enough to 0 and is transversal to N on K. Thus $(L(V, M, s) - S_K) \cap W_f$ is dense in W_f and so $S_K \cap W_f$ is nowhere dense in W_f.

6.3 Let V, M be compact, oriented, connected manifolds both of dimension n. Suppose $f \in L(V, M, 2)$. Let y be a non-critical value of f, then $f^{-1}(y) = x_i$, an isolated set of points in V. At each of these points f is either orientation preserving or reversing. We define $\operatorname{sgn}(x_i) = +1$ if f is orientation preserving and -1 if it is orientation reversing at x_i.

DEFINITION 1. $$\gamma_y(f) = \sum_{f(x_i)=y} \operatorname{sgn}(x_i),$$ the degree of f at y.

PROPOSITION 1. a) $\gamma_y(f)$ is independent of the choice of regular value y.

b) $\gamma(f) = \gamma(g)$ if f is homotopic to g, and $g \in L(V, M, 2)$.

PROOF : We first prove given a non-critical value y_1 of f there is a non-critical value y of g such that $\gamma_{y_1}(f) = \gamma_y(g)$ if f and g are homotopic. Since V is compact we know that f is proper thus the set of non-critical values is open. Thus there is a neighbourhood $\mathcal{M}$ of y_1 consisting only of regular points of f. $\mathcal{M}$ can be taken small enough so that for $y' \in \mathcal{M}$, $\gamma_{y'}(f) = \gamma_{y_1}(f)$. By an argument similar to that given in Steenrod [4] for approximating a continuous section by a differentiable one, we can assume that our homotopy between f and g is given by a C^2 mapping $F : V \times I_\epsilon \to M$ such that $F(x \times 0) = f(x)$, $F(x \times 1) = g(x)$, where $I_\epsilon = (-\epsilon, 1 + \epsilon)$. In $\mathcal{M}$ we can, by Sard's theorem, find a point y which is a non-critical value for f, g and F. We show that $\gamma_y(g) = \gamma_y(f)$.

Let $Z = F^{-1}(y)$. Z consists of non-singular non-intersecting curves since y is a non-critical point of F. $Z \cap (V \times 0) = f^{-1}(y) \times 0$ and $Z \cap (V \times 1) = g^{-1}(y) \times 1$.

Since y is non-critical for f and g these curves cannot be tangent to $V \times 0$ or $V \times 1$. Thus $Z \cap (V \times I_0)$ consists of simple arcs joining pairs of points in $f^{-1}(y) \times 0$, $g^{-1}(y) \times 1$ and some closed curves also. We consider just the arcs having endpoints among the $\{x_i, x_j'\}$ where $\{x_i\} = f^{-1}(y) \times 0$ and $\{x_j'\} = g^{-1}(y) \times 1$. We orient these arcs as follows : Let C be an arc joining p and q, two points of the set $\{x_i, x_j'\}$. At each point of C choose an $(n+1)$ frame whose first vector is along C in a given fixed direction, the remaining vectors of the frame are so ordered so that the vectors of the frame have a positive order. Suppose the direction given to C by the first vector is from p to q. At each point of C, the last n-vectors of the frame in order are mapped by F into an ordered n-tuple of vectors in M. This image n-tuple of vectors is ordered either positively or negatively. It is easy to see that at each point of C the image frame are all ordered the same - either positively or negatively. If positively, we take as orientation on C that given by the vector of the frame which is tangent to C; if negatively the orientation on C is taken opposite to that given by the first vector of the frame. The boundary of the chain $Z \cap (V \times I_0)$ is $\Sigma \epsilon_i x_i + \Sigma \delta_j x_j'$. ($\epsilon_i, \delta_i$ are ± 1.) Thus we know that $\Sigma \delta_j = -\Sigma \epsilon_i$. It remains to show that $\epsilon_i = -\mathrm{sgn}(x_i)$ and $\delta_j = \mathrm{sgn}(x_j')$. Suppose an arc joins x_i and x_j' then $\mathrm{sgn}(x_i) = \mathrm{sgn}(x_j')$ and if $\mathrm{sgn}(x_i) = +1$, the arc goes from x_i to x_j'. Thus the boundary of the arc is $x_j' - x_i$. If an arc joins x_{i_1} and x_{i_2} (or x_{j_1}', x_{j_2}') then $\mathrm{sgn}(x_{i_1}) = -\mathrm{sgn}(x_{i_2})$ (or $\mathrm{sgn}(x_{j_1}') = -\mathrm{sgn}(x_{j_2}')$) since the direction of the vector along C starts by pointing up, increasing $t \in I_\epsilon$, (down) and ends pointing down (up). Thus if $\mathrm{sgn}(x_{i_1}) = +1$, the arc goes from x_{i_1} to x_{i_2} (if $\mathrm{sgn}(x_{j_1}') = +1$, the arc goes from x_{j_2}' to x_{j_1}'). The rule is simply that the proper orientation of the arcs is in the direction of increasing values of $t \in I_\epsilon$ as it passes through a point with sgn $+1$. This proves that $\gamma_y(f) = \gamma_y(g)$.

We now show that $\gamma_y(f)$ is independent of the choice of non-critical value y. Let y' be another non-critical value of f and let C be any curve joining y to y'. Without loss of generality we can assume that there is a sequence of points

$y = y_0, y_1, \ldots, y_{k-1}, y_k = y'$ such that y_j are all non-critical values of f and such that y_j, y_{j+1} are both interior to a coordinate ball B_j. We show that $\gamma_{y_j}(f) = \gamma_{y_{j+1}}(f)$. We can find a homotopy of class C^2, $F_j: B_j \times I \to B_j$ such that $F_j(y, 0) = y$ and $F_j(y_{j+1}, 1) = y_j$ with the following further additional property: suppose y_j, y_{j+1} are contained in a strictly smaller concentric ball $B'_j \subset B_j$ then $F_j(y, t) = y$ for $y \in B_j - B'_j$. Further $F_j(\cdot, t)$ is a diffeomorphism of B_j onto itself. We can extend F_j to be a C^2-homotopy $G_j : M \times I \to M$ as follows: $G_j(y, t) = y$ for $y \notin B_j$; $G_j(y, t) = F_j(y, t)$ for $y \in B_j$. Consider now the mapping $G_j \circ (f \times i) : V \times I_\epsilon \to M$, where i is the identity map on I_ϵ. This is a C^2-homotopy of f (since $G_j(f(x) \times 0) = f(x)$) with g, where $g(x) = G_j(f \times i)(x \times 1)$. y_j is a non-critical value of f. It is also a non-critical value of g since $g^{-1}(y_j) = \{x | G_j(f(x) \times 1) = y_j\} = \{x | f(x) = y_{j+1}\}$. Since G_j is a diffeomorphism for each t, we see that y_j is a non-critical value of g and further $\gamma_{y_j}(g) = \gamma_{y_{j+1}}(f)$. But by the first part of the proof $\gamma_{y_j}(f) = \gamma_{y_j}(g)$.

7. Transversality Theorem

7.1 Assume $s > r \geqslant 0$ and let N be an $(s-r)$ - differentiable regular submanifold of $J^r(V, M)$ where V and M at least s differentiable paracompact manifolds of dimension n and p respectively. Suppose the codimension of N in $J^r(V, M)$ is equal to q.

THEOREM 1 (Transversality Theorem of Thom). *The set of maps in $L(V, M, s)$ whose r-extensions are transversal to N on V is everywhere dense if $(s - r) > \max(n - q, 0)$.*

PROOF : The theorem is proved by means of a sequence of reductions until we have finally just a lemma about R^n and R^p to prove which contains the central difficulty.

Let $S_A = \{g \in L(V, M, s) | J^r(g)$ is not transversal to N on $A \subset V\}$. (The dependence of S_A on r, s, V, M, N will be suppressed). We write S for S_V. Equivalent to the transversality theorem is :

THEOREM 1'. If $(s-r) > \max(n-q, 0)$, then S is nowhere dense in $L(V, M, s)$

To prove theorem 1' it suffices to prove :

LEMMA 1 : For K any compact set of V, S_K is a countable union of closed nowhere dense sets in $L(V, M, s)$ if $(s-r) > \max(n-q, 0)$.

Lemma 1 implies theorem 1' as follows. Let K_i be a countable collection of compact subsets of V whose union is V. S_{K_i} are all countable unions of closed nowhere dense sets in $L(V, M, s)$ under the assumption $(s-r) > \max(n-q, 0)$. Thus since $L(V, M, s)$ is a Baire space (see 5.3.), $\bigcup_i S_{K_i} = S_K$ is again nowhere dense in $L(V, M, s)$, by definition of Baire space.

(We will use the notation for Theorem x, (V_1, M_1) Theorem x means Theorem x for the case $V = V_1$ and $M = M_1$ and similarly for Lemma x.)

LEMMA 2. The (R^n, R^p) Theorem 1' implies the (V, M) Lemma 1.

PROOF : a) If U is an open solid ball in R^n, the (U, R^p) Theorem 1' is a trivial consequence of the (R^n, R^p) Theorem 1'.

b) Assume $(s-r) > \max(n-q, 0)$ and let K be a compact subset of V. We will prove given any $f \in L(V, M, s)$ there exists a neighbourhood W_f of f such that $S_K \cap W_f$ is nowhere dense in W_f. Cover K with a finite number of coordinate neighbourhoods $\mathcal{V}_i$ whose closures $\bar{\mathcal{V}}_i$ are compact and contained in open coordinate balls U_i such that $f(U_i) \subset \mathcal{M}_i$ where $\mathcal{M}_i$ are open coordinate neighbourhoods in M. Let $W_i = \{g \in L(V, M, s) | g(U_i) \subset \mathcal{M}_i\}$. Since $J^r(U_i, \mathcal{M}_i)$ is an open subset of $J^r(V, M)$, $N_i = J^r(U_i, \mathcal{M}_i) \cap N$ is a regular submanifold of both $J^r(V, M)$ and $J^r(U_i, \mathcal{M}_i)$. By means of coordinate maps the sets $\mathcal{U}_i$, U_i (and $\mathcal{M}_i$) can be considered as subsets of R^n (and R^p). We denote by * superscript the corresponding sets. For example U_i^* is the image in R^n of the set U_i under the coordinate map of U_i.

Since U_i^* is an open solid ball in R^n, we have the (U_i^*, R^p) Theorem 1' for the submanifold N_i^*. Here $N_i^* \subset J^r(U_i^*, \mathcal{M}_i^*) \subset J^r(U_i^*, R^p)$ is the submanifold whose coordinates are exactly those of N_i. Let $S_{U_i^*} = \{g \in L(U_i^*, R^p, s) | g$ is not transversal to $N_i^*\}$. Thus $S_{U_i^*}$ is nowhere dense in $L(U_i^*, R^p, s)$. Let $W_i^* = \{g \in L(U_i^*, R^p, s) | g(\bar{\mathcal{U}}_i^*) \subset \mathcal{M}_i^*\}$. Thus $S_{U_i^*} \cap W_i^*$ is nowhere dense in W_i^*.

Let $\rho_i : W_i \to W_i^* : g \to g^* | U_i^*$. This mapping is clearly an open mapping onto an open set. Thus $\rho_i^{-1}(S_{U_i^*} \cap W_i^*)$ is nowhere dense in W_i. But $\rho_i^{-1}(S_{U_i^*} \cap W_i^*) = \{g \in W_i | g|U_i$ is not transversal to $N_i\} \supset S_{\bar{\mathcal{U}}_i} \cap W_i$. Thus $S_{\bar{\mathcal{U}}_i} \cap W_i$ is nowhere dense in W_i. Let $\bigcap_i W_i = W_f$, then $S_{\bar{\mathcal{U}}_i} \cap W_f$ is nowhere dense in W_f. Thus also $\bigcup_i (S_{\bar{\mathcal{U}}_i}) \cap W_f$ is nowhere dense in W_f. Thus since $\bigcup_i S_{\bar{\mathcal{U}}_i} \supset S_K$, we conclude that S_K is nowhere dense in $L(V, M, s)$.

The proof that S_K is a countable union of closed sets is the same as that given in 6.2, Lemma 2, with M replaced by $J^r(M, V)$ and s replaced by $(s-r) > 0$.

7.2 Now all that remains to prove is the (R^n, R^p) Lemma 1. Since as remarked at the end of the preceding section the proof that S_K is a countable union of closed sets is easily seen directly we confine our attention to showing that it is nowhere dense.

We reduce the proof to the following :

LOCAL LEMMA 1. <u>Suppose</u> $f \in L(R^n, R^p, s)$ <u>and</u> $N \subset J^r(R^n, R^p)$ <u>is an</u> $(s - r)$ <u>differentiable submanifold of codimension</u> q. <u>If</u> $(s-r) > \max(n-q, 0)$ <u>then for each pair</u> (x, u) <u>where</u> $x \in R^n$ <u>and</u> $J^r(f)(x) = u \in N$ <u>we can find</u> :

1) <u>A neighbourhood</u> V_u <u>of</u> u <u>in</u> $J^r(R^n, R^p)$

2) <u>A neighbourhood</u> W_f <u>of</u> f <u>in</u> $L(R^n, R^p, s)$

3) <u>A compact neighbourhood</u> U_x <u>of</u> x <u>in</u> R^n <u>such that</u> :

a) For each $g \in W_f$, $J^r(g)(U_x) \subset V_u$

b) For each $h \in W_f$, there is a g arbitrarily close to h in W_f such that $J^r(g)|U_x$ is transversal to N.

That Local Lemma 1 implies the (R^n, R^p) Lemma 1 is easy.

Suppose we are given a compact set $K \subset R^n$ and a map $f \in L(R^n, R^p, s)$. By Local Lemma 1 we can find a finite set of triples of neighbourhoods, U_{x_i}, V_{u_i}, W_{i_f} which satisfy a) and b) such that $K = \bigcup_i U_{x_i}$. Let $W_f = \bigcap_i W_{i_f}$, and $K_i = U_{x_i} \cap K$, then b) yields $S_{K_i} \cap W_f$ is nowhere dense in W_f. Thus $S_K \cap W_f$ is nowhere dense in W_f, and so S_K is nowhere dense in $L(R^n, R^p, s)$.

We will only prove a special case of Local Lemma 1 and for its statement we need some new notation. We let $T = R^{p'} \times J^r(n, p')$ which can be identified with a subspace of $J^r(R^n, R^{p'}) = R^n \times R^{p'} \times J^r(n, p')$. If $g \in L(R^n, R^{p'}, s)$, we denote by the corresponding capital letter G, the map $J^r(g)$ followed by projection on T.

i.e. $\quad G : R^n \to T : x \to (g(x), g^{(r)}(x))$.

Equivalently, $G = g \times g^{(r)}$ (see 5.1.).

LOCAL LEMMA 2 [5]. Suppose $f \in L(R^n, R^{p'}, s)$ and $N' \subset T$ is an $(s-r)$ differentiable regular submanifold of codimension q. If $(s-r) > \max(n-q, 0)$ then for each $x \in R^n$ and each $u \in N' \subset T$ such that $F(x) = u$ we can find :

1) A neighbourhood V_u of u in T such that $N' \cap V_u$ is defined by $\phi : V_u \to R^q$.

2) A neighbourhood W_f of f in $L(R^n, R^{p'}, s)$

3) A compact neighbourhood U_x of x in R^n such that

a) For each $g \in W_f$, $G(U_x) \subset V_u$

b) For each $h \in W_f$, there exists a $g \in W_f$ arbitrarily close to h such that $G|U_x$ is transversal to N'.

In order to prove that Local Lemma 2 implies Local Lemma 1, we use a stronger conclusion b) in a special case. In particular if $R^n = R^n$ and $R^{p'} = R^n \times R^p$ and $i : R^n \to R^n$ is the identity map and $N' = P_1^{-1}(N)$ where N is a regular submanifold of $J^r(R^n, R^p)$ and P_1 is defined as follows :

Let $\pi_2 : R^n \times R^p \to R^p$ be the projection and $P : L(R^n, R^{n+p}, s) \to L(R^n, R^p, s)$, $P(g) = \pi_2 \circ g$. The map P induces a projection on the jets $P' : J^r(n, n+p) \to J^r(n, p)$ and this in turn induces the map ,

$$P_1 : (R^n \times R^p) \times J^r(n, n+p) \to J^r(R^n, R^p) : (x \times y) \times z \to x \times y \times P'(z) .$$

The strengthened conclusion b) which we use in this case is :

c) For each $(i \times h) \in W_f$, where $h \in L(R^n, R^p, s)$ there is a map $g \in W_f$ arbitrarily close to $i \times h$ such that $G|U_x$ is transversal to N' and such that on U_x, $g(x) = (x + \lambda, g'(x))$ where λ is a constant vector and $g' \in L(R^n, R^p, s)$.

Conclusion c) is contained in the proof of Local Lemma 2. We show that (Local Lemma 2 + c)) implies Local Lemma 1.

Note first that in Local Lemma 2, condition b) that $G|U_x$ is transversal to N' is equivalent with $J^r(g)|U_x$ is transversal to $R^n \times N'$.

We have the following natural injections :

$Q : L(R^n, R^p, s) \to L(R^n, R^{n+p}, s) : g \to i \times g$, where i is the identity mapping on R^n. Q induces $Q' : J^r(n, p) \to J^r(n, n+p)$ and this induces the map $Q_1 : J^r(R^n, R^p) \to J^r(R^n, R^{n+p})$.

Let $f \in L(R^n, R^p, s)$ and let N be a regular submanifold of $J^r(R^n, R^p)$ of codimension q . Then $i \times f = Q(f) \in L(R^n, R^{n+p}, s)$. $P_1^{-1}(N) = N'$ is a regular submanifold of $(R^{n+p} \times J^r(n, n+p))$, of codimension q in this space. Since

P_1 is a projection of one Euclidean space onto another, we can pull back the coordinates of $J^r(R^n, R^p)$ to obtain a partial coordinate system on $R^{n+p} \times J^r(n, n+p)$ in terms of which the defining local equations of N' are the same as those of N, in particular if in a neighbourhood V of a point $u \in J^r(R^n, R^p)$, N is defined by $\psi : V \to R^q$, then defining mapping for N' in $P_1^{-1}(V)$ is $\psi \circ P_1 : P_1^{-1}(V) \to R^q$.

We apply (Local Lemma 2 + c)) to the map $Q(f) \in L(R^n, R^n \times R^p, s)$ and the submanifold $N' \subset (R^{n+p} \times J^r(n, n+p))$. We can find for each $x \in R^n$ and each $(x, u) = J^r(i \times f)(x)$ such that $u \in N'$, neighbourhoods $U_x \subset R^n$, $V_u \subset R^{n+p} \times J^r(n, n+p)$, $W_{Q(f)} \subset L(R^n, R^{n+p}, s)$ which satisfy conditions a), b) and c) of Local Lemma 2.

Consider triples of neighbourhoods U'_x, $P_1(V_u)$, $Q^{-1}(W_{Q(f)})$, where $U'_x \subset U_x$ will be determined later. We show that we can find a $U'_x \subset U_x$ so that the triple satisfies conditions a) and b) of Local Lemma 1.

a) Let $g \in Q^{-1}(W_{Q(f)})$, then $J^r(Q(g))(U_x) \subset R^n \times V_u$. However $J^r(Q(g))(x') = (Q_1 \circ J^r(g))(x')$. Thus $Q_1(J^r(g)(U_x)) \subset Q_1(P_1(V_u))$ and so $J^r(g)(U_x) \subset P_1(V_u)$.

b) Let $h \in Q^{-1}(W_{Q(f)})$. Conclusion b) of Local Lemma 2 says that we can find a $\tilde{g} = g_1 \times g_2 \in L(R^n, R^{n+p}, s)$, arbitrarily close to $i \times h$ such that $J^r(g_1 \times g_2)|U_x$ is transversal to $R^n \times P_1^{-1}(N)$. The additional conclusion c) that we've attached to Local Lemma 2 says that in this case g_1 in U_x is an arbitrarily small translation, $g_1(x) = x + \lambda_1$. Fix $\rho > 0$ small enough so that $x + \lambda \in U_x$ for all vectors λ with $\|\lambda\| < \rho$, and define :

$$U'_x = \{x' \in U_x \mid \text{ for all } \lambda \text{ with } \|\lambda\| < \rho,\ x' + \lambda \in U_x\} .$$

We may assume that $\|\lambda_1\| < \rho$.

In V_u, $P_1^{-1}(N)$ is defined by $\phi : V_u \to R^q$. However, as remarked above, ϕ can be written as $\psi \circ P_1$ where ψ defines N in $P_1(V_u)$. The condition that $\tilde{G} = \tilde{g} \times \tilde{g}^{(r)}$ restricted to U_x is transversal to $P_1^{-1}(N)$ is that $\psi \circ P_1 \circ G$ has rank q at each $x' \in U_x$ such that $\tilde{G}(x') \in P_1^{-1}(N)$ or whenever $P_1 \circ \tilde{G}(x') \in N$. Since $g_1(x') = x' + \lambda$ for all $x' \in U_x$, define $g \in L(R^n, R^p, s)$ by $g(x') = g_2(x' - \lambda)$. Thus $(i \times g)(x' + \lambda) = (x' + \lambda), g_2(x') = \tilde{g}(x')$. Thus $P_1 \circ \tilde{G}(x') = J^r(g)(x' + \lambda)$. Thus since $\tilde{G}|U_x$ is transversal to $P_1^{-1}(N)$ on $U_x' \subset U_x$ we obtain $J^r(g)|U_x'$ is transversal to N. Clearly since (g_1, g_2) can be taken arbitrarily close to $(i \times h)$, g can be found in this way arbitrarily close to h, and of course in $Q^{-1}(W_{Q(f)})$.

7.3 We come at last to the proof of Local Lemma 2 + c.
(In the statement of Local Lemma 2 replace $R^{p'}$ by R^p and N' by N)

Notation : Let $\mathcal{N}$ be the set of all ordered n-tuples of positive integers $(i_1, \ldots, i_n)$. For $\omega, \omega' \in \mathcal{N}$ we write $\omega \geq \omega'$ iff $i_j \geq i_j'$ where $\omega = (i_1, \ldots, i_n)$ and $\omega' = (i_1', \ldots, i_n')$. If $\omega \geq \omega'$, $\omega - \omega' = (i_1 - i_1', \ldots, i_n - i_n')$ and $\omega! = (i_1!)(i_2!) \ldots (i_n!)$. We write x_ω for $x_1^{i_1} . x_2^{i_2} . . x_n^{i_n}$ and ∂_ω for $\dfrac{\partial^{|\omega|}}{\partial x_1^{i_1} \ldots \partial x_n^{i_n}}$ where $|\omega| = \sum_{j=1}^{n} i_j$.

Using this notation we find for $\omega \geq \omega'$, $x_\omega = x_{(\omega - \omega')} x_{\omega'}$ and

$$\partial_{\omega'} x_\omega = \frac{\omega!}{(\omega - \omega')!} x_{(\omega - \omega')}. \text{ If } \omega' > \omega \text{ then } \partial_{\omega'} x_\omega = 0 .$$

Suppose $(x_1, \ldots, x_n)$, $(y_1, \ldots, y_p)$ and $(u_1, \ldots, u_m)$ are coordinates in R^n, R^p, and T respectively. Since $T = R^p \times J^r(n, p)$, p of the u's are y's but rather than complicate the notation we allow the same functions to have two names. We make the following conventions on the indices $(i = 1, \ldots, n)$, $(j = 1, \ldots, p)$ $(k = 1, \ldots m)$ $(\alpha, \beta, \gamma = 1, \ldots, q)$.

Let $f \in L(R^n, R^p, s)$ and $N \subset T$ be a regular submanifold of codimension q. Suppose $x \in R^n$ and $F(x) = u \in N$. To simplify the notation, we assume that $x = 0$, the origin of R^n. Suppose that in a neighbourhood of u, say $V \subset T$, N is defined by $\phi : V \to R^q$, i.e. $V \cap N = \{u' \in V | \phi(u') = 0\}$. ϕ is a $C^{(s-r)}$

mapping of rank q at u.

Since ϕ has rank q at u, there is a $(q \times q)$ matrix, say $\left(\frac{\partial \phi_\alpha}{\partial u_\beta}\right)$ which is non-singular at u. Let J be the determinant of this matrix. If $|J(u)| = 2A > 0$, choose $V_u \subset V$ such that $\bar{V}_u$ is compact and $|J| \geq A > 0$ on V_u. Thus for $g \in L(R^n, R^p, s)$, $J \circ G$ is a function defined on $G^{-1}(V_u) \subset R^n$ and $|(J \circ G|G^{-1}(V_u))| \geq A > 0$.

Suppose $u_1 = \partial_{\omega_1}(y_{j_1}), \ldots, u_q = \partial_{\omega_q}(y_{j_q})$; here denoting the coordinates in $T = R^p \times J^r(n, p)$ as partial derivatives is the usual abuse of lánguage (see 1.1.). The u_β listed here are exactly those occuring in $J = \det\left(\frac{\partial \phi_\alpha}{\partial u_\beta}\right)$. Let $x_{\omega_1}, \ldots, x_{\omega_q}$ be the corresponding monomials. It is of course possible that not all of the y_j are distinct.

Let $\pi_j = \{\alpha | y_j = y_{j_\alpha}\}$.

Suppose $g \in L(R^n, R^p, s)$ and is defined by $y_j = g_j(x)$. We define a deformation, $\bar{g}$, of g which depends on q real parameters, λ_β, as follows :

$$\bar{g}_j(x, \lambda) = g_j(x) + \Big(\sum_{\beta \in \pi_j} \lambda_\beta \frac{x_{\omega_\beta}}{(\omega_\beta)!}\Big) \phi(x)$$

where $\phi = 1$ on a compact neighbourhood U' of 0 and vanishes outside a neighbourhood containing U'.

(Note : It will be proved later that the $\bar{g}$ thus defined for arbitrarily small values of λ is the sought "transversal approximation" to g. If $u_{\beta'} = y_j$ for some β' and if $\pi_j = \{\beta'\}$, then for $x \in U'$, $\bar{g}_j(x, \lambda) = g_j(x) + \lambda_{\beta'}$. From this fact, conclusion c) follows, since in the special situation of conclusion c), $N' = P_1^{-1}(N) \subset R^{n+p} \times J^r(n, n+p)$. R^{n+p} here plays the role of R^p and so among the coordinates $u_1, \ldots, u_m$ in $R^{n+p} \times J^r(n, n+p)$ will appear $x_1, \ldots, x_n$. Since however the defining mapping of N' in a neighbourhood V in $R^{n+p} \times J^r(n, n+p)$ has the form $\psi \circ P_1 : V \to R^q$ we see that it is possible in the non-singular matrix that $\left(\frac{\partial \psi_\alpha \circ P_1}{\partial x_i}\right)$ occurs, as a column (i.e. for some fixed x_i all α. However $\psi_\alpha \circ P_1$ is independent of the coordinates in $J^r(n, n+p)$ which we write symbolically $\partial_\omega x_i$, $1 \leq |\omega| \leq r$. Thus for a map $(i \times g) \in L(R^n, R^p, s)$, the deformation $(\overline{i \times g})$ relative to a submanifold $N' = P_1^{-1}(N)$ has the form :

$$(\overline{i \times g})(x') = (x' + \lambda, \ h(x, \lambda). \quad).$$

The following computations are made inside U', so we may suppress the ϕ.

Computing : $$\frac{\partial(\phi_\alpha \circ \bar{G})}{\partial \lambda_\gamma} = \sum_k \left(\frac{\partial \phi_\alpha}{\partial u_k} \circ \bar{G}\right) \frac{\partial(u_k \circ \bar{G})}{\partial \lambda_\gamma}$$

$$u_k(\bar{G}) = \partial_{\omega_k}(\bar{g}_{j_k}) = \partial_{\omega_k}(g_{j_k}) + \partial_{\omega_k}\Big(\sum_{\beta \in \pi_{j_k}} \lambda_\beta \frac{x^{\omega_\beta}}{(\omega_\beta)!}\Big).$$

If $k \leq q$, then $k \in \pi_{j_k}$, and :

$$u_\alpha(\bar{G}) = \partial_{\omega_\alpha}(g_{j_\alpha}) + \lambda_\alpha + \sum \lambda_\beta \frac{x^{(\omega_\beta - \omega_\alpha)}}{(\omega_\beta - \omega_\alpha)!},$$

where the sum is taken over all $\beta \in \pi_{j_\alpha}$ such that $\beta \neq \alpha$ and $\omega_\beta > \omega_\alpha$.

Thus , $$\frac{\partial(u_k \circ \bar{G})}{\partial \lambda_\gamma} = \begin{cases} \dfrac{x^{(\omega_\gamma - \omega_k)}}{(\omega_\gamma - \omega_k)!}, & \text{if } \gamma \in \pi_{j_k} \text{ and } \omega_\gamma \geq \omega_k \\ 0, & \text{otherwise} \end{cases}$$

Thus , $$\frac{\partial(\phi_\alpha \circ G)}{\partial \lambda_\gamma} = \sum_k \left(\frac{\partial \phi_\alpha}{\partial u_k} \circ \bar{G}\right) . \frac{x^{(\omega_\gamma - \omega_k)}}{(\omega_\gamma - \omega_k)!} + \frac{\partial \phi_\alpha}{\partial u_\gamma} \circ \bar{G},$$

where the sum is taken over all $k \neq \gamma$ such that $\gamma \in \pi_{j_k}$ and $\omega_\gamma > \omega_k$.

At $x = 0$, $\dfrac{\partial(\phi_\alpha \circ \bar{G})}{\partial \lambda_\gamma} = \dfrac{\partial \phi_\alpha}{\partial u_\gamma} \circ \bar{G}$.

We may suppose that $F(U') \subset V_u$ and define W_f as the set of all $g \in L(R^n, R^p, s)$ such that $G(U') \subset V_u$.

Define $H_{\gamma\alpha} : U' \times V_u \to R$ by :

$$H_{\gamma\alpha}(x, u) = \frac{\partial \phi_\alpha}{\partial u_\gamma}(u) + \sum \left(\frac{\partial \phi_\alpha}{\partial u_k}(u) \right) \cdot \frac{x^{(\omega_\gamma - \omega_k)}}{(\omega_\gamma - \omega_k)!},$$

where the sum is taken over all $k \neq \gamma$ such that $\gamma \in \pi_{j_k}$ and $\omega_\gamma > \omega_k$.

$$H_{\gamma\alpha} \circ (\text{id} \times G) = \left(\frac{\partial \phi_\alpha}{\partial u_\gamma} \circ G\right) + \sum \left(\frac{\partial \phi_\alpha}{\partial u_k} \circ G\right) \cdot \frac{x^{(\omega_\gamma - \omega_k)}}{(\omega_\gamma - \omega_k)!}$$

where id is the identity map on U'.

Let $H(G) = \det(H_{\gamma\alpha} \circ (\text{id} \times G))$. Note that at $x = 0$, $H(G) = J \circ G$. The neighbourhood V_u was so chosen that on it $|J| \geq A > 0$. Now choose a compact neighbourhood $U \subset U'$, so small that for all $g \in W_f$, $|H(G) - J \circ G| < A/2$ for all $x \in U$. This is possible since we know that for all $g \in W_f$, $G(U') \subset V_u$, so that the partials of g are all uniformly bounded. Thus by choosing U small enough we can bring $H(G)$ uniformly, for all $g \in W_f$, as close as we please to $J \circ G$.

Claim, V_u, W_f, U, satisfy the conditions a) and b). The first condition, a), follows by definition. For b), suppose $g \in W_f$. For sufficiently small $\lambda \in R^q$, $\bar{g}(\cdot, \lambda) = \bar{g}_\lambda \in W_f$. Let Λ be a cube in R^q of such sufficiently small λ's. From now on when we write $\bar{g}$ we will mean the deformation of g with λ values taken from Λ.

On $U \times \Lambda$, $|J \circ \bar{G}| \geq A > 0$ and $|J \circ \bar{G} - H(\bar{G})| < A/2$, thus $|H(\bar{G})| \geq A/2$.

Let $E = \bar{G}^{-1}(N) \cap (U \times \Lambda)$. This is the solution set in $U \times \Lambda$ of the equations

(*) $\qquad \phi_1 \circ \bar{G} = 0, \ldots, \phi_q \circ \bar{G} = 0$.

Suppose that there exists an $x_o \in U$ such that $(x_o, 0) \in E$. If this were not the case, then $G|U$ would already be transversal to N - trivially by not intersecting N at all.

$H(\bar{G}) = \det\left(\frac{\partial \phi_\alpha \circ \bar{G}}{\partial \lambda_\gamma}\right) \neq 0$ on $U \times \Lambda$, and so in particular at each point $(x, \lambda) \in E$. We apply the implicit function theorem to the system (*) at the point (x, λ). This gives neighbourhoods, $a(x)$, $b(\lambda)$ and a map $\psi_{x\lambda} : a(x) \to b(\lambda) \subset R^q$ of class $C^{(s-r)}$ such that $\psi_{x\lambda}(x) = \lambda$ and $(x', \psi_{x\lambda}(x'))$ are the only solutions of (*) in $a(x) \times b(\lambda)$ i.e. graph $(\psi_{x\lambda}) = (a(x) \times b(\lambda)) \cap E$. $\Sigma_{x\lambda}$, the set of critical values of $\psi_{x\lambda}$, is of outer measure zero in R^q, since by assumption $(s-r) > \max(n-q, 0)$, (Sard's theorem, see 6.1). Suppose $\lambda' \in (b(\lambda) - \Sigma_{x\lambda})$, then $\bar{G}_{\lambda'}$, is transversal to N on $a(x)$. For if there is an $x' \in a(x)$ such that $(x', \lambda') \in E$ then $\psi(x') = \lambda'$. (We write now ψ rather than $\psi_{x\lambda}$).
Since λ' is a non-critical value of ψ, $d\psi_1 \wedge \ldots \wedge d\psi_q \neq 0$ at x'.
Thus

$$0 = d(\phi_\alpha \circ \bar{G} \circ (i \times \psi))_{x'} = d(\phi_\alpha \circ \bar{G}_{\lambda'})_{x'} + \sum_\beta \left(\frac{\partial(\phi_\alpha \circ \bar{G})}{\partial \lambda_\beta}\right)_{(x',\lambda')} \cdot (d\psi_\beta)_{x'}$$

Thus $$d(\phi_\alpha \circ \bar{G}_{\lambda'})_{x'} = - \sum_\beta \left(\frac{\partial(\phi_\alpha \circ \bar{G})}{\partial \lambda_\beta}\right)_{(x',\lambda')} \cdot (d\psi_\beta)_{x'} ,$$

and finally by exterior multiplication :

$$\bigwedge_\alpha d(\phi_\alpha \circ \bar{G}_{\lambda'})_{x'} = (-1)^q \, H(\bar{G})_{(x',\lambda')} \bigwedge_\beta (d\psi_\beta)_{x'}$$

The right side is different from zero since $(x', \lambda') \in a(x) \times b(\lambda) \subset U \times \Lambda$ and λ' is a non-critical value of ψ.

$E \subset U \times \Lambda$ can be covered by a finite number of such neighbourhoods, $a(x_\nu) \times b(\lambda_\nu)$. Since $\bigcup_\nu \Sigma(x_\nu, \lambda_\nu) = \Sigma$, is again of outer measure zero in Λ, we can find a $\lambda^* \in \Lambda$, arbitrarily close to zero, such that $\bar{G}_{\lambda^*}$ is transversal to N on U. For suppose $\bar{G}_{\lambda^*}(x) \in N$ for $x \in U$, then $(x, \lambda^*) \in E$ and therefore for some ν $(x, \lambda^*) \in a(x_\nu) \times b(\lambda_\nu)$ and $\lambda^* = \psi_{(x_\nu \lambda_\nu)}(x)$. The preceding argument yields that $\bar{G}_{\lambda}*$ is transversal to N at x.

7.4 COROLLARY 1. If $(s-r) > \max(n-q, 0)$ and N_i is a countable collection of $(s-r)$ differentiable regular submanifolds of $J^r(V, M)$ where the codimension of N_i is greater than or equal to q, $\dim V = n$, $\dim M = p$. Then the set of maps in $L(V, M, s)$ whose r-extensions are transversal to all N_i is everywhere dense in $L(V, M, s)$.

PROOF : Let K_j be a compact, countable covering of V. Let $S^i_{K_j}$ be the subset of $L(V, M, s)$ of mappings which are not transversal to N_i on K_j. Lemma 1 of 7.1 says that the sets $S^i_{K_j}$ are closed nowhere dense sets in $L(V, M, s)$. Thus the union over all i and j is again nowhere dense since $L(V, M, s)$ is Baire.

8. Singularities of Mappings - General

8.1. We extend the notion of a singularity of order r (see 2.1)

DEFINITION 1. For any pair of positive integers (n, p) a singularity manifold of order r (or an r-singularity), is a regular submanifold of $J^r(n,p)$ which is invariant under the group $L^r(n, p)$.

Given a singularity manifold of order r, $S \subset J^r(n, p)$ we denote by $S(V, M)$ the subset of $J^r(V, M)$ ($\dim V = n$, $\dim M = p$) defined as follows : $J^r(V, M)$ is a fibre bundle over $V \times M$ with fibre $J^r(n, p)$ and group $L^r(n, p)$. Suppose $\pi : J^r(V, M) \to V \times M$ is the bundle projection. There is a covering of $V \times M$ by neighbourhoods $\mathcal{V} \times \mathcal{M}$ such that $\pi^{-1}(\mathcal{V} \times \mathcal{M})$ is diffeomorphic to $\mathcal{V} \times \mathcal{M} \times J^r(n,p)$. In each $\pi^{-1}(\mathcal{V} \times \mathcal{M})$, let $S(V, M)$ be the inverse image of $\mathcal{V} \times \mathcal{M} \times S$ by this diffeomorphism. Since S is invariant under the group $L^r(n, p)$ this set is well defined and is a regular submanifold in $J^r(V, M)$.

Note : The codimension of S in $J^r(n, p)$ equals the codimension of $S(V, M)$ in $J^r(V, M)$.

We will often write simply S for $S(V, M)$ when it is clear from the context what is meant.

Suppose $f : V \to M$ is of class C^r and $S \subset J^r(n, p)$ is an r-singularity. We let $S(f) = (J^r(f))^{-1} (S(V, M))$. A point $x \in S(f)$ is called a singular point of type S of f. If $J^r(f)$ is transversal to S and $S(f) \neq \emptyset$, then we say f displays the singularity S, transversally.

If the codimension of S is q and $n \geq q$ and $J^r(f)$ is transversal to $S(V, M)$ then either $S(f)$ is empty or is a regular submanifold of V of codimension q in V. (see 5.2). If $n < q$ and $J^r(f)$ is transversal to $S(V, M)$ then $S(f)$ is empty. Thus the Transversality Theorem says : given any map $g \in L(V, M, s)$ and any r-singularity $S \subset J^r(n, p)$ such that codim $S > n$, then it is possible by an arbitrarily small deformation of g in $L(V, M, s)$, (for s sufficiently large, in this case $s \geq r + 1$) to find a map f in $L(V, M, s)$ such that $S(f) = \emptyset$. If, on the other hand codim $S = q \leq n$, then it is possible by an arbitrarily small deformation of g in $L(V, M, s)$ (for s sufficiently large, in this case $s \geq r + n - q + 1$) to find a map $f \in L(V, M, s)$ such that either $S(f) = \emptyset$ or $S(f)$ is a regular submanifold of V of dimension $(n - q)$. We see that by varying mappings a little we can arrange it so their singularity sets are regular submanifolds simultaneously for a countable number of singularities.

8.2 One may ask if it is possible, by small variations to get rid of singular points that a mapping displays transversally. The answer is :

PROPOSITION 1. Let S be an r-singularity in $J^r(n, p)$ of codimension $q \leq n$. Suppose $f \in L(V, M, r + 1)$ displays the singularity S transversally at x. Then given any neighbourhood U of x there is a

<u>neighbourhood</u> W_f <u>of</u> f <u>in</u> $L(V, M, r+1)$ <u>such that if</u> $g \in W_f$, g <u>displays the singularity</u> S <u>transversally at some point of</u> U .

PROOF : Since this is a local proposition it suffices to prove it for $V = R^n$ $M = R^p$. The reduction is simple : take a small enough compact coordinate neighbourhood W of x in U such that $f(W)$ is contained in a coordinate neighbourhood $\mathcal{M}$ in M and consider only maps $g \in L(V, M, r+1)$ such that $g(W) \subset \mathcal{M}$. This is a neighbourhood of f and it suffices to find for any neighbourhood U' in W a neighbourhood W_f among this set of maps for which the conclusion holds.

LEMMA 1. <u>Let</u> $f \in L(R^n, R^p, 1)$ <u>and let</u> S <u>be a submanifold in</u> R^p <u>of codimension</u> $q \leqslant n$ <u>such that at some point</u> $x \in R^n$, $f(x) \in S$ <u>and</u> f <u>is transversal to</u> S <u>at</u> x . <u>Then for any neighbourhood</u> U <u>of</u> x <u>there is a neighbourhood</u> W_f <u>of</u> f <u>in</u> $L(R^n, R^p, 1)$ <u>such that for any</u> $g \in W_f$ <u>there is a</u> $x' \in U$ <u>such that</u> $g(x') \in S$ <u>and</u> g <u>is transversal to</u> S <u>at</u> x'.

PROOF : Suppose in a neighbourhood $\mathcal{W}$ of $f(x)$, S is defined by $\psi : \mathcal{W} \to R^q$. Since f is transversal to S at x , $\psi \circ f$ has rank q at x and therefore in a small closed n-ball B contained in U . By (5.2) we know that $f^{-1}(S) \cap B$ is a submanifold P in B of dimension $n-q$. In fact if the first q coordinates in B are taken as $\psi_i \circ f$, $i = 1,\dots, q$, P is just the $(n-q)$-plane in B through the origin obtained by setting the first q coordinates equal to zero , f then has the form $(x_1,\dots, x_q, f_{q+1},\dots, f_p)$. To simplify the notation assume x and $f(x)$ are origins in R^n and R^p respectively.

Let K be the q-ball in B defined by setting the vanishing last $m-q$ coordinates. $P \cap K = x$. The boundary of K, R, a$(q-1)$ sphere, is disjoint from P . Thus the distance between $f(R)$ and S is greater than zero i.e. $\rho(f(R), S) = \epsilon_0 > 0$. (ρ is Euclidean distance in R^p).

Let $\epsilon_1 > 0$, be chosen so small that $(I_q + (a_{ij}))$ is non singular if $|a_{ij}| < \epsilon_1$. $(i, j = 1,\ldots, q)$ and I_q is the $q \times q$ identity matrix. Let

$$W_f^1 = \{g \in L(R^n, R^p, 1) | \; \left|\frac{\partial g_i}{\partial x_j}(x') - \delta_j^i\right| < \epsilon_1 \; (i, j = 1,\ldots, q) \text{ for all } x' \in B\}$$

and

$$W_f^0 = \{g \in L(R^n, R^p, 1) \,|\, \rho(g(x'), f(x')) < \epsilon_0 \,, \text{ for all } x' \in B\} .$$

Let $W_f = W_f^0 \cap W_f^1$. The first condition assures us that for $g \in W_f$, $\psi \circ g$ has rank q on B. This together with the second condition yields that for $g \in W_f$ there is a point $x' \in K \subset B$ such that $g(x') \in S$ as follows :

Define a homotopy between f and g by $F_t = f + t(g-f)$, and consider $F: K \times I \to R^p$. $F(R \times I)$ is disjoint from S since

$$\inf \| f(x') + t(g(x') - f(x')) - y \| \geq \inf \| f(x') - y \| - \sup \| g(x') - f(x') \| > 0$$

for all $x \in R$, $y \in S$.

Let $I^1 = \{t \in I \,|\, F(K \times t) \cap S \neq 0\}$; clearly I^1 is closed and non-empty, $(0 \in I^1)$. Claim I^1 is also open in I. F is transversal to S on $K \times I$, since $g \in W_f^1$; thus $F^{-1}(N) \cap (K \times I)$ is a set of non-singular, non-intersecting curves. Suppose $t \in I_1$, then there exists an x' interior to K such that $F(x', t) \in S$. We know that x' is interior to K since $F(R \times I)$ is disjoint from S. Since F_t is transversal to N, we know that the curve of $F^{-1}(N)$ through (x', t) is not tangent to $K \times t$, thus there is a relatively open neighbourhood of t in I_1. Thus I_1 is relatively open, therefore $I_1 = I$.

If we let $J^r(n, p)$ play the role of R^p of the lemma we obtain the result by noting that $J^r : L(R^n, R^p, r+1) \to L(R^n, J^r(R^n, R^p), 1)$ is continuous by definition of the topology in $L(R^n, R^p, r+1)$.

In the above proposition we have demanded that the approximations g to f be good for all partials through the $(r+1)$ st. If we relaxed this and demanded that the approximations, g, be good only through some lower order partial derivative, it may be possible to find an arbitrarily good approximation g to f in this weaker sense which in some neighbourhood of x has no points of type S. An example of this type is given by Whitney ([6], p.403). In other words, the proposition is false if we take for the topology in $L(V, M, r+1)$ the topology induced by the inclusion of $L(V, M, r+1)$ in $L(V, M, k)$ for $k < r+1$.

9. Higher Order Singularities

9.1 Given an r-singularity $S \subset J^r(n, p)$, suppose codim $S = q \leq n$. Let $f \in L(V, M, s)$ where $s > r + \max(n-q, 0)$, be such that $J^r(f)$ is transversal to $S(V, M)$. $S(f)$ is then a regular submanifold of codimension q in V, or is empty. Restricting f to $S(f)$, we say that $x \in S(f)$ belongs to $S_k(S(f))$ if $x \in S_k(f|S(f))$ where $0 \leq k \leq \min(n-q, p)$. The points of $S_0(S(f))$ are called the regular points of $S(f)$. The $\bigcup_k S_k(S(f)) = S(f)$ and the $S_k(S(f))$ are disjoint and also $\bigcup_{i \geq k} S_i(S(f)) = \overline{S_k(S(f))} \cap S(f)$, exactly as in 2.4 for the S_k themselves. This subdivision of $S(f)$ has been introduced by Thom [5]. The conditions that determine whether or not a point x is in $S_k(S(f))$ are conditions on the $(r+1)$ jet of the mapping f at x.

Since the definition of $S_k(S(f))$ uses the mapping f and in fact requires that $J^r(f)$ be transversal to $S(V, M)$ it is not clear, a priori, how to iterate this process. If, however, we could define regular invariant submanifolds $(S_k S) \subset J^{r+1}(n, p)$ such that $(J^{r+1}(f))^{-1}((S_k S)(V, M)) = S_k(S(f))$, for $J^r(f)$ transversal to $S(V, M)$ then it would be clear how to define $S_k(S_{k_1}(S(f)))$, for f such that $J^r(f)$ is transversal to $S(V, M)$, and $J^{r+1}(f)$ is transversal to $(S_{k_1} S)(V, M)$. We would then merely repeat the same process with $(S_{k_1} S)$ playing the role of S.

If $\overline{S} \subset J^r(n, p)$ is a real agebraic set, is it possible to define regular invariant submanifolds, $(S_k S) \subset J^{r+1}(n, p)$, such that for $J^r(f)$ transversal to $S(V, M)$,

$(J^r(f))^{-1}((S_k S)(V, M)) = S_k(S(f))$ and such that $(\overline{S_k S})$ is a real algebraic set in $J^{r+1}(n, p)$, i.e. is it possible to define $(S_k S)$ so that they are critical manifolds ?

9.2 In the next four sections, we carry out the procedure of the last section for the case of the 1-singularities, S_k, $k = 1,\dots, q = \min(n, p)$ and find regular, invariant submanifolds $S_{k,h} \subset J^2(n, p)$ such that for $J^1(f)$ transversal to S_k, $(J^2(f))^{-1}(S_{k,h}) = S_h(S_k(f))$. We also determine the codimension of $S_{k,h}$ in $J^2(n, p)$. (Clearly $S_o(f)$ doesn't subdivide further since $S_o(f)$ is open and at $x \in S_o(f)$ the kernel of f_* at x is equal to the kernel of $(f|S_o(f))_*$ at x.) It suffices to consider maps $f : R^n \to R^p$.

We outline the procedure used. Let $S_k' = \pi_{21}^{-1}(S_k)$. We find a regular submanifold $V \subset S_k'$ such that if $S_{k,h} = L^2(n, p)(V)$, the orbit of V under $L^2(n, p)$; we obtain a regular invariant submanifold of S_k' such that for $J^1(f)$ transversal to $S_k(R^n, R^p)$, $S_{k,h}(f) = S_h(S_k(f))$. The proof that $S_{k,h}$ is a regular submanifold of S_k' and of $J^2(n, p)$ is made by showing that at any point $v \in V$ there is a neighbourhood W of v in $J^2(n, p)$ such that $S_{k,h} \cap W$ is defined by the vanishing of a number, d, of polynomials in the coordinates of $J^2(n, p)$. The mapping of W into R^d defined by these polynomials is of rank d in W. To obtain such a defining map at any other point of $S_{k,h}$, we just translate by an appropriate element of $L^2(n, p)$.

Let F equal the set of all 2-jets of maps $f : R^n \to R^p$ of the form :

$$(1)\quad \begin{cases} Y_i = y_i \\ X_\alpha = {}^t y\, B^\alpha x + {}^t x\, A^\alpha x \end{cases}$$

where $i = 1,\dots, s$; $\alpha = 1,\dots, p-s$, ${}^t x = (x_1,\dots, x_{n-s})$, and $s = \min(n,p) - k$. We will use these conventions throughout; also $m = 1,\dots, n-s$. Clearly $L^2 F = S_k'$.

Let T be the set of 2-jets in F such that if a map, $f : R^n \to R^p$, has the property that $f^{(2)}(0) \in T$, then $J^1(f)$ is transversal to $S_k(R^n, R^p)$ at 0.

LEMMA 1. T <u>consists of all 2-jets of the form</u> (1) <u>such that the rank of</u> $\binom{2A}{B}$ <u>equals</u> $(n-s)(p-s)$, <u>where</u> $B = (B^1, \dots B^{p-s})$ <u>and</u> $A = (A^1 \dots A^{p-s})$.

PROOF : Suppose $f : R^n \to R^p$ and $f^{(2)}(0) \in F$. Let W be a neighbourhood of $J^1(f)(0)$ such that $S_k(R^n, R^p) \cap W$ is defined by $\psi : W \to R^{(n-s)(p-s)}$. If U is a neighbourhood of 0 in R^n such that $J^1(f)(U) \subset W$, then the condition that $J^1(f)$ is transversal to $S_k(R^n, R^p)$ is that $\psi \circ J^1(f)$ has rank $(n-s)(p-s)$ at 0. Because f is assumed to have form (1) at 0, the composition is particularly easy.

$$\psi \circ J^1(f) : U \to R^{(n-s)(p-s)} : (x, y) \to \left(\frac{\partial X_\alpha}{\partial x_m}\right)(x, y)$$

Thus the Jacobian matrix is $\binom{2A}{B}$.

Note that the matrices A and B are obtained by placing the matrices A^α, resp. B^α side by side. Since A^α is $(n-s) \times (n-s)$ and B^α is $s \times (n-s)$, $\binom{2A}{B}$ is $n \times (p-s)(n-s)$.

COROLLARY 1. $T \neq \emptyset$ iff $n \geqslant (p-s)(n-s)$.

9.3 Suppose $f : R^n \to R^p$ and $f^{(2)}(x) \in S_k'$, then f_* annihilates an $(n-s)$ dimensional subspace, say K of the tangent space at x in R^n. Let $\vartheta : R^{(n-s)} \to R^n$, $\vartheta(0) = x$ and $\vartheta_*(T(R^{(n-s)})_0) = K$. As in the preceding section we assume that in a neighbourhood W of $J^1(f)(x)$ in $J^1(R^n, R^p)$ we have a defining map $\psi : W \to R^{(n-s)(p-s)}$ for $S_k \cap W$ and that U is a neighbourhood of x in R^n such that $J^1(f)(U) \subset W$. $\psi \circ J^1(f)$ is a defining mapping for $S_k(f) \cap U$. Motivated by 9.1. we have :

DEFINITION 1. (f, ψ, ϑ, x as above) The point $x \in S_{k,h}(f)$ iff the rank of $(\psi \circ J^1(f) \circ \vartheta)_*$ at 0 is $(n-s-h)$. $(0 \leq h \leq n-s)$.

Comparing this definition with that given in 9.1 for $S_h(S_k(f))$ we find that for f such that $J^1(f)$ is transversal to S_k, $S_{h,k}(f) = S_h(S_k(f))$. If $S_k(f) = \emptyset$, then both of these sets are also empty. By definition for such "transversal" f, $S_h(S_k(f)) = S_h(f|S_k)$. $S_h(S_k(f))$ is thus the set of points on $S_k(f)$ where the rank of $f|S_k(f)$ equals $\min(\dim S_k(f), p) - h$. Since $\dim S_k(f) = n-(n-s)(p-s) \leq p$ (assuming $S_k(f) \neq \emptyset$) we have

$$S_h(S_k(f)) = \{x \in S_k(f) \mid \text{rank } (f|S_k(f)) = \dim S_k(f) - h\}$$

Since the kernel of f at any point of $S_k(f)$ has dimension $(n-s)$ we see that $S_h(S_k(f)) = \emptyset$ for $h > (n-s)$. On the other hand, since for such an f, $(\psi \circ J^1(f))_*$ at $x \in S_k(f)$ annihilates only vectors tangent to $S_k(f)$ at x, we see that

$$S_{k,h}(f) = \{x \in S_k(f) \mid \dim(K \cap T(S_k(f))_x) = h\} .$$

In other words, for $h \leq \dim S_k(f)$

$$S_{k,h}(f) = \{x \in S_k(f) \mid \text{rank } (f|S_k(f)) = \dim S_k(f) - h\}$$

and otherwise $S_{k,h}(f) = \emptyset$.

Thus the two definitions coincide, whenever they both make sense, and for h such that $\min(\dim S_k(f), (n-s)) < h \leq \max(\dim S_k(f), n-s)$, when one of the two is not defined the other is always empty. Note also that since
$\dim K + \dim S_k(f) = (n-s) + n - (n-s)(p-s) \leq n$, it is possible that $K \cap T(S_k(f)) = \emptyset$.

The condition that $x \in S_{k,h}(f)$ is merely a condition on the 2-jet of f at x, independent of coordinates. Thus there is a subset $S_{k,h} \subset S'_k$ such that $S_{k,h}$ is invariant under $L^2(n, p)$ and $(J^2(f))^{-1}(S_{k,h}) = S_{k,h}(f)$, justifying the notation.

LEMMA 1. <u>Let</u> $N = S_{k,h} \cap F$, <u>then</u> $L^2N = S_{k,h}$ <u>and</u> N <u>consists of all 2-jets at 0 of mappings of the form</u> (1) <u>such that rank</u> $A = (n-s-h)$.

PROOF : $f : R^n \to R^p$ and $f^{(2)}(0) \in F$. In the coordinates (x, y) (see 9.2), f_* annihilates the vectors $\frac{\partial}{\partial x_m}$, thus we may take the mapping $\vartheta : R^{n-s} \to R^n : (x) \to (x, 0)$. We have then :

$$\psi \circ J^1(f) \circ \vartheta \ : \ \vartheta^{-1}(U) \to R^{(n-s)(p-s)} : x \to \frac{\partial({}^t x A x)}{\partial x_m} .$$

The Jacobian of this mapping is $2A$.

COROLLARY 1. $T \cap N \neq \emptyset$ iff $(n-h) \geq (n-s)(p-s)$

PROOF. If $T \cap N \neq \emptyset$ then rank $B \geq (n-s)(p-s) - (n-s-h)$. Thus $s \geq (n-s)(p-s) - (n-s-h)$ or $(n-h) \geq (n-s)(p-s)$. For the converse, just construct matrices A and B with the two rank properties.

9.4 In N we define a subset V, which will have the properties announced in 9.2. Let V be the set of all 2-jets at 0 of mappings of the form :

$$(2) \qquad \begin{cases} Y_i = y_i \\ X_1 = \sum_\mu \epsilon(\mu)\, x_\mu^2 + {}^t y B^1 x \\ X_\alpha = {}^t x'(A^\alpha)'\, x' + {}^t y B^\alpha x \end{cases} ,$$

where $\mu = 1, \ldots, n-s-h$, $\epsilon(\mu) = \pm 1$, $\alpha = 2, \ldots, p-s$ and ${}^t x' = (x_1, \ldots, x_{n-s-h})$. If we call the diagonal $(n-s-h) \times (n-s-h)$ matrix with diagonal entries $\epsilon(\mu)$, $(A^1)'$, we may let α run from $1, \ldots, p-s$.

V is clearly a regular submanifold in $J^2(n, p)$ and in S'_k since the only condition on $(A^\alpha)'$ and B^α is that $(A^\alpha)'$ is symmetric.

In order to show that $L^2V = S_{k,h}$ it suffices to show that $N \subset L^2V$. So suppose f has form (1) with rank $A = (n-s-h)$, then $f^{(2)}(0) \in N$. We choose maps $\pi : R^p \to R^p$ and $\nu : R^n \to R^n$ such that $(\pi^{(2)}(0) \times \nu^{(2)}(0)) \in L^2(n, p)$ and such that $(\pi \circ f \circ \nu)^{(2)}(0) \in V$.

Let π be given by $\begin{cases} \bar{Y}_i = Y_i \\ \bar{X}_\alpha = (CX)_\alpha \end{cases}$ and ν by $\begin{cases} y_i = \bar{y}_i \\ x_m = (c\bar{x})_m \end{cases}$

where C and c are non singular matrices. Dropping the bars we obtain for $(\pi \circ f \circ \nu)$:

$$\begin{cases} Y_i = y_i \\ X_\alpha = {}^t y\left(\sum_\beta C_{\alpha\beta} B^\beta c\right) x + {}^t x\left(\sum_\beta C_{\alpha\beta} {}^t c A^\beta c\right) x \end{cases}$$

where the summation is for $1 \leq \beta \leq p-s$. Showing that for appropriate choices of C and c , $(\pi \circ f \circ \nu)^{(2)}(0) \in V$ is equivalent to :

LEMMA 1. <u>If</u> $A = (A^1 \ldots A^{p-s})$ <u>has rank</u> $(n-s-h)$ <u>where</u> A^α <u>are</u> $(n-s) \times (n-s)$ <u>symmetric matrices, it is possible to choose an</u> $(n-s) \times (n-s)$ <u>non-singular matrix</u> c <u>and a</u> $(p-s) \times (p-s)$ <u>non-singular matrix</u> C <u>such that</u> $\sum_\beta C_{\alpha\beta} {}^t c A^\beta c$ <u>has nonzero entries only in the first</u> $(n-s-h)$ <u>rows and such that</u> $\sum_\beta C_{1\beta} {}^t c A^\beta c$ <u>has the form</u> $\begin{pmatrix} \epsilon & 0 \\ 0 & 0 \end{pmatrix}$ <u>where</u> ϵ <u>is a diagonal matrix with</u> ± 1's <u>on the diagonal, of rank,</u> $(n-s-h)$.

PROOF. We may assume A^1 has the maximal rank r , obtainable by such transformations and that A^1 is in diagonal form with an $(r \times r)$ non-singular matrix, with entries ± 1 , in the upper left hand corner. If $r = (n-s-h)$, the rank condition on A guarantees that there are only non-zero terms in the first $(n-s-h)$ rows of A and we're done.

Suppose $r < (n-s-h)$. Again by the rank condition on A we can assume that there is an A^α with non-zero entry in its $(r+1)$ th row. We may assume that A^2 has a non-zero entry in its $(r+1)$ st row. Consider $A^1 + uA^2$ with u indeterminate. If on the $(r+1)$ st row there is a non zero element in the $(r+j)$ th column $(j \geq 1)$ we let $H(n)$ be the $(r+1) \times (r+1)$ matrix indicated in the figure :

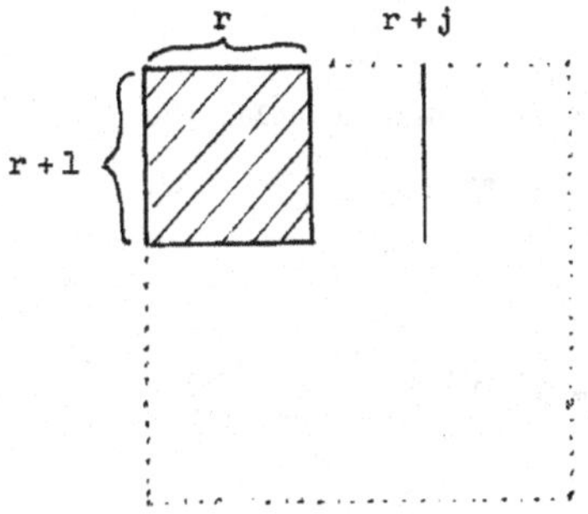

$A^1 + uA^2$

i.e. we take the upper left hand corner $(r+1) \times r$ matrix of $A^1 + uA^2$ and as $(r+1)$ st column we take the first $(r+1)$ row entries of the $(r+j)$ th column of $A^1 + uA^2$. In the determinant of $H(u)$ the coefficient of u is precisely the non zero element of A^2 in the $(r+1)$, $(r+j)$ place. Thus since this determinant doesn't vanish identically we may pick a value of u so that $A^1 + uA^2$ has rank at least $(r+1)$-Contradiction.

On the other hand if on the $(r+1)$ st row of A^2 all the terms in the $(r+j)$ th column vanish for $j \geq 1$, we follow the same procedure as above but take as $H(u)$ the upper left hand corner $(r+1) \times (r+1)$ matrix of $A^1 + uA^2$. Now the lower right hand term of $H(u)$ is zero so the coefficient of u^2 in the determinant of $H(u)$ is minus the sum of the squares of the entries in the $(r+1)$ st row of A^2 which by assumption doesn't vanish. Thus we have a contradiction again.

COROLLARY 1. $L^2V = S_{k,h}$.

9.5 We prove in this section that $L^2V = S_{k,h}$ is a regular submanifold. Let $v \in V$ be the 2-jet of a mapping of the form (2). We determine a neighbourhood W of v in $J^2(n, p)$ in which we give defining equations for $S_{k,h}$. A point $w \in W$ is the two jet of a mapping :

$$(3) \quad \begin{cases} Y_i = (Gy)_i + (Hx)_i + {}^t y C^i y + {}^t y E^i x + {}^t x D^i x \\ X_\alpha = (gy)_\alpha + (hx)_\alpha + {}^t y c^\alpha y + {}^t y e^\alpha x + {}^t x d^\alpha x \end{cases}$$

The first condition on W is that G is non-singular, and thus the first set of polynomial equations in W which are satisfied by $S_{k,h}$ is

$$\text{(I)} \qquad h - g\,G^{-1}H = 0\,.$$

These are just the equations of $S'_k \cap W$ and so if we assume these to hold, the second set of equations below define $(S_{k,h}) \cap S'_k \cap W$ in $S'_k \cap W$.

We apply the following transformations to the point w.

$$\begin{cases} \bar{Y} = G^{-1}\,Y \\ \bar{X} = -g\,G^{-1}Y + X \end{cases} \qquad \text{and} \qquad \begin{cases} y = \bar{y} + G^{-1}\,H\bar{x} \\ x = -\bar{x} \end{cases}$$

After the first transformation we obtain :

$$\bar{Y}_i = y_i + (G^{-1}\,H\,x)_i + \sum_j (G^{-1})_{ij}({}^t y C^j y + {}^t y\,E^j x + {}^t x\,D^j x)$$

$$\bar{X}_\alpha = ((h - g\,G^{-1}H)x)_\alpha - \sum_j (g\,G^{-1})_{\alpha j}({}^t y\,C^j y + {}^t y\,E^j x + {}^t x\,D^j x) + ({}^t y c^\alpha y + {}^t y\,e^\alpha x + {}^t x\,d^\alpha x)$$

We apply the second transformation but compute explicitly only the quadratic terms in $\bar{x}$ in $\bar{X}_\alpha$

$$\bar{Y}_i = \bar{y}_i + \ldots$$

$$\bar{X}_\alpha = ((g\,G^{-1}H - h)\bar{x})_\alpha + {}^t\bar{x}[{}^t(G^{-1}H)\,c^\alpha(G^{-1}H) - \tfrac{1}{2}({}^t(G^{-1}H)e^\alpha + {}^t e^\alpha G^{-1}H) + d^\alpha$$

$$- \sum_j (g\,G^{-1})_{\alpha j}({}^t(G^{-1}H)\,C^j(G^{-1}H) - \tfrac{1}{2}({}^t(G^{-1}H)\,E^j + {}^t E^j G^{-1}H)) + D^j)]\,\bar{x} + \ldots$$

Call the matrix inside the square brackets $\mathfrak{A}^\alpha$. The second condition on W is that the upper left hand $(n-s-h) \times (n-s-h)$ corner of $\mathfrak{A}^1$ should be non-singular. This can clearly be arranged by taking, g, H, C^i, E^i, D^i, c^α in very small neighbourhoods of zero and e^α and d^α in a small neighbourhood of B^α, A^α. (Here A^α refers to the $(n-s) \times (n-s)$ matrix obtained by putting $(A^\alpha)'$ of the jet of v in the upper left hand corner and zeros every place else.)

Thus a jet in this neighbourhood W of a mapping of the form (3) is in $S_{k,h}$ if and only if (I) holds and the matrix $\mathfrak{A} = (\mathfrak{A}^1\ \mathfrak{A}^2\ \ldots\ \mathfrak{A}^{p-s})$ has rank precisely $(n-s-h)$.

Suppose $\mathfrak{A}^{\alpha}$, is written as $\begin{pmatrix} u_\alpha & {}^t v_\alpha \\ v_\alpha & w_\alpha \end{pmatrix}$ where u_α is $(n-s-h) \times (n-s-h)$.

In W, u_1 is non-singular. Thus the second set of equations for $S_{k,h}$ in W is.

$$\text{(II)} \quad \begin{cases} w_1 - v_1 u_1^{-1}\, {}^t v_1 = 0 & \\ v_\alpha - v_1 u_1^{-1} u_\alpha = 0 & \\ w_\alpha - (v_1 u_1^{-1}) u_\alpha {}^t(v_1 u_1^{-1}) = 0 & \end{cases} \quad \left.\begin{matrix} \\ \\ \end{matrix}\right\} \alpha \neq 1$$

Equations (II) are precisely the equations that express the fact that the matrix $\mathfrak{A}$ has rank $(n-s-h)$ under the assumption that u_1 has rank $(n-s-h)$. Thus (I) and (II) define $S_{k,h} \cap W$ in W. Taking account of symmetry of these matrix equations we obtain :

$$d = (n-s)(p-s) + h\left(\frac{(p-s)(h+1)}{2} + (n-s-h)(p-s-1)\right)$$

equations. The d polynomials on the left hand sides of (I) and (II) define a map $\phi : W \to R^d$ such that $\phi^{-1}(0) = S_{k,h} \cap W$. That this mapping is of rank d on W is obvious from the definition.

PROPOSITION 1. $S_{k,h}$ <u>is a 2-singularity</u> $\subset S'_k$, <u>and in</u> $J^2(n, p)$ codim $S_{k,h}$ = codim $S'_k + h\left(\frac{(p-s)(h+1)}{2} + (x) + (n-s-h)(p-s-1)\right)$. <u>For</u> $f : V \to M$, <u>with</u> $J^1(f)$ <u>transversal to</u> $S_k(V, M)$, $S_{k,h}(f) = S_h(S_k(f))$.
(Note : $k \geq 1$ and $(n-s) \geq h$, and the last equality holds when both sides make sense (see 9.3.))

Translating this proposition into a statement about mappings and combining this with the fact that codim $S_k = (n-s)(p-s)$ we have :

1) f can display the singularity S_k transversally iff $n \geq k\ (|n-p| + k)$

2) If $k \geq 1$, $h \leq (n-s)$; f can display the singularity $S_{k,h}$ transversally iff :

$$\text{a)} \quad n \geq k\,(n-p+k) + h\left(k\,\frac{(h+1)}{2} + (n-p+k-h)(k-1)\right) \quad \text{for } n \geq p$$

b) $n \geq k(p - n + k) + h(p - n + k) \frac{(h+1)}{2} + (k-h)(p-n+k-1))$ for $p \geq n$.

For $k = h = 1$, we have

$$\text{codim } S_{1,1} = (n - p + 2) \text{ when } n \geq p \text{ and}$$

$$\text{codim } S_{1,1} = 2(p - n + 1) \text{ when } p \geq n \text{ which}$$

agrees with the values given by Whitney [7]. Also for $k = 1$, $h = 2$ and $n \geq p$, codim $S_{1,2} = (n - p + 4)$. Since $h \leq (n - s)$, in this case $2 \leq n - p + 1$, we see that $n \geq p + 1$. Thus $S_{1,2}$ can be displayed only when $p \geq 4$ and $n \geq p + 1$, which agrees with Whitney's statement [7] that $S_{1,2}$ occurs for the first time for $n = 5$, $p = 4$.

PROBLEM : For sequences of positive integers $k_1, \ldots, k_r$, define the analogous regular invariant submanifolds $S_{k_1, k_2, \ldots, k_r}$ and determine their codimensions in $J^r(n, p)$. Of course not all sequences of k_i will make sense. Prove that $S_{k_1, \ldots, k_r}$ are critical manifolds, that is that they are regular invariant submanifolds whose closures are real algebraic sets.

REFERENCES

[1] Bourbaki, N. *Topologie Générale*, Chapter IX, §5, p.75.

[2] Morse, A.P. *The behaviour of a function on its critical set*, Ann. of Math. 40 (1939) pp. 62-70.

[3] Sard, A. *Images of critical sets*, Ann. of Math. 68 (1958) pp.247-259.

[4] Steenrod, N. *Topology of Fibre Bundles*, p.27.

[5] Thom, R. *Un lemme sur les applications differentiables*, Bol. Soc. Mat. Mexicana, (1956) pp. 59-71.

[6] Whitney, H. *On singularities of mappings of Euclidean spaces, I, mappings of the plane into the plane*, Ann. of Math. 62 (1955) pp. 374-410.

[7] Whitney, H. *Singularities of mappings of Euclidean Spaces*, Sym. Int. de Top. Alg. Mexico (1958) pp. 285-301.

III EQUIVALENCE AND STABILITY

10. Equivalence of Mappings.

10.1 Throughout V and M are paracompact C^∞ - manifolds.

DEFINITION 1. f and g in $L(V, M, r)$ are said to be r-equivalent if there exist C^r diffeomorphisms h_1 and h_2 of V and M respectively so that

$$\begin{array}{ccc} V & \xrightarrow{\;f\;} & M \\ {\scriptstyle h_1}\downarrow & & \downarrow{\scriptstyle h_2} \\ V & \xrightarrow[\;g\;]{} & M \end{array} \qquad \text{commutes .}$$

In case $r = 0$ this is the usual topological equivalence of mappings.
The problem of classifying C^r maps under this global equivalence relation suggests the corresponding local problem, that is the classification of germs of C^r maps.

DEFINITION 2. f and g in $L(V, M, r)$; we say that f at x is r-equivalent to g at x' if there are neighbourhoods U of x and U' of x' and W of $f(x)$ and W' of $g(x')$ and C^r diffeomorphisms h_1 of U onto U' and h_2 of W onto W' such that ,

$$\begin{array}{ccc} U & \xrightarrow{\;f\;} & W \\ {\scriptstyle h_1}\downarrow & & \downarrow{\scriptstyle h_2} \\ U' & \xrightarrow[\;g\;]{} & W' \end{array} \qquad \text{commutes .}$$

For the classification of germs of C^r mappings it suffices to consider $\mathcal{E}^r(n, p)$, i.e. maps from R^n to R^p which take the origin into the origin and to consider the r-equivalence classes of germs of maps in $\mathcal{E}^r(n, p)$ at the origin. Denote the

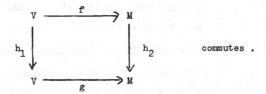

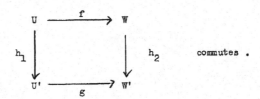

r-equivalence class of $f \in \mathcal{E}^r(n, p)$ by $[f]^r$.
In $[f]^r$, the r-jets of all $g \in [f]^r$ at the origin are r-equivalent in the sense that they all lie on the orbit under $L^r(n, p)$ of $f^{(r)}(0)$.

PROBLEM : We restrict our attention to ∞-equivalence of germs of maps in $C^\infty(n, p)$. When is it true that the s-jet of a mapping determines its germ equivalence class, i.e. for which s-jets is it true that $g \in [f]^\infty$ if and only if $g^{(s)}(o) \in L^s(n,p)$ $f^{(s)}(0)$?

There are some examples of s-jets which determine the ∞-equivalence class of germs. These are cases in which any mapping in $\mathcal{E}^\infty(n, p)$ whose s-jet belongs to a given orbit can be brought to a particularly simple form, "a normal form", by an infinitely differentiable coordinate change.

1) If $f^{(1)}(0) \in J^1(n, p) = S_0$, then the germ at f at 0 has the form given in 2.3, after a suitable change of coordinates.

2) If $f : R^n \to R$ and $J^1(f)$ is transversal to S_1 at 0 and $0 \in S_1(f)$, then f at 0 is equivalent to one of the mappings $g(x_1, \ldots, x_n) = \sum_1^n \epsilon_i x_i^2$, where $\epsilon_i = \pm 1$, in a neighbourhood of 0. In this case the 2-jet of f at 0 determines the germ equivalence class. This is merely a reformulation (see 9.2 Lemma 1) of the Theorem of Morse [1] which states that a function with an isolated, non-degenerate singular point at 0, can be put into such a form by a change of coordinates.

3) For maps $f : R^2 \to R^2$, only two types of singularities can appear transversally, namely S_1 and $S_{1,1}$. Whitney [4] has given local normal forms for mappings displaying such singularities transversally.
The 2-jet at a point of $S_1(f)$ and the 3-jet at a point of $S_{1,1}(f)$ determine the germ ∞-equivalence class. The results are as follows and we state them without assuming that f is infinitely differentiable.

PROPOSITION 1. ([4] Theorem 15A) <u>Let</u> $f : R^2 \to R^2$ <u>be of class</u> $s \geq 3$ <u>and suppose</u> $J^1(f)$ <u>is transversal to</u> S_1 <u>at</u> 0 <u>and</u> $0 \in S_{1,0}(f)$, <u>then we can find</u> $(s-3)$-<u>differentiable coordinate systems</u> (X, Y) <u>and</u> (x, y) <u>such that in a neighbourhood of</u> 0, f <u>has the form</u> :

$$\begin{cases} Y = y \\ X = x^2 \end{cases}$$

Whitney calls such points of $S_{1,0}(f)$, *fold points*, and $S_{1,0}(f)$ consists of the *fold curves*.

PROPOSITION 2. ([4] Theorem 16 A) *Let* $f : R^2 \to R^2$ *be of class* $s \geq 12$ *and suppose* $J^1(f)$ *is transversal to* S_1 *at* 0 *and* $J^2(f)$ *is transversal to* $S_{1,1}$ *at* 0 . *Let* $0 \in S_{1,1}(f)$ (see 9.1). *Then we can find* $([\frac{s}{2}] - 5)$- *differentiable coordinate systems* (X, Y) *and* (x, y) *in the neighbourhoods of the origins such that* f *has the form* :

$$\begin{cases} Y = y \\ X = xy - x^3 \end{cases}$$

Whitney calls such a point a *cusp point*.

By considering the map $f : R^2 \to R^2 : (y, x) \to (y, z)$, where $z = xy - x^3$ we can visualize such a cusp point with the two folds which end at it. The set of points of $S_1(f)$ is given by $y = 3x^2$ and the image of $S_1(f)$ by f is given by $27z^2 - 4y^3 = 0$.

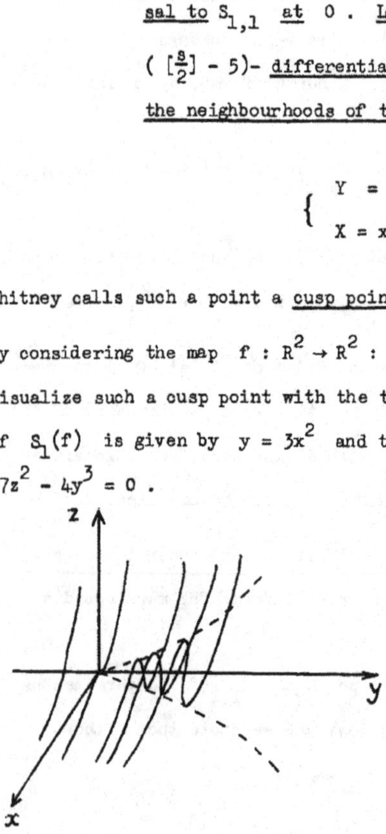

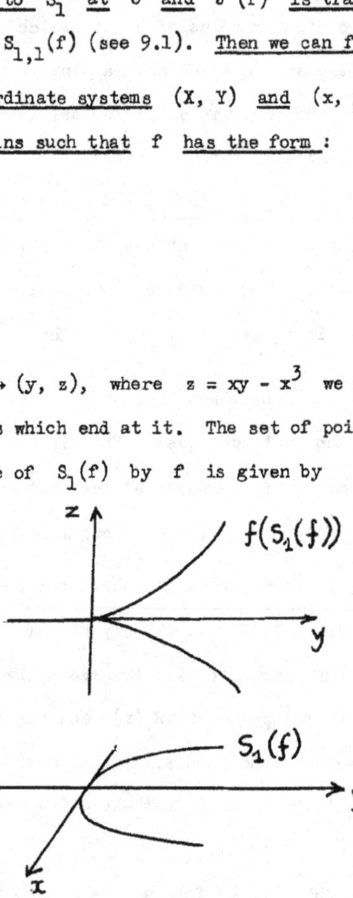

4) By the same method as is used in the proof of the theorem of Morse, (2) above, we can find a normal form for any map $f \in \mathcal{E}^\infty(n, p)$ for $n \geq p$ at 0 if $J^1(f)$ is

transversal to S_1 at 0 and $0 \in S_{1,0}(f)$. In this case by a suitable coordinate change, in a neighbourhood of 0, f has the form :

$$\begin{cases} Y_i = y_i \ , & i = 1,\dots, p-1 \\ X = \Sigma\, \epsilon_j x_j^2 \ , & j = 1,\dots, n-p+1 \ , \text{ and } \epsilon_j = \pm 1 \ . \end{cases}$$

Here again the 2-jet determines the germ equivalence class.

5) For $f \in \mathcal{E}^\infty(n, 2n-1)$ only one kind of singularity can occur transversally namely S_1 and this at isolated points. Whitney [3] has given a normal form for such singularities also. Again we state the precise result without assuming infinite differentiability of f :

PROPOSITION 3. ([3] Theorem 1, p.164) Let $f : R^n \to R^{2n-1}$ of class $s = 4r + 8$, $r \geq 1$, and suppose $J^1(f)$ is transversal to S_1 and $0 \in S_1(f)$, then there are C^r coordinate systems $(x_1,\dots, x_n)$ and $(Y_1,\dots, Y_{2n-1})$ such that in a neighbourhood of 0, f has the form :

$$\begin{cases} Y_1 = x_1^2 & \\ Y_i = x_i & , i = 2,\dots, n \\ Y_{n+i-1} = x_1 x_i & , i = 2,\dots, n \end{cases}$$

11. Stability of Mappings.

11.1 DEFINITION 1. A C^s map $f : V \to M$ is called r-stable $(s > r)$ if there is a neighbourhood W of f in $L(V, M, r+1)$ such that for any $g \in W$, g is r-equivalent to f.

Here of course also there is the corresponding local concept.

DEFINITION 2. A C^s map $f : V \to M$ is called r-stable at x, $(s > r)$ if for every neighbourhood U of x there is a neighbourhood W of f in

$L(V, M, r+1)$ such that for any g in W, there is a point $x' \in U$ such that f at x is r-equivalent to g at x'.

If we restrict our attention to C^∞ maps, there are two conjectures about stability that have been made.

STRONG CONJECTURE : Almost every map $f \in L(V, M, \infty)$ is ∞-stable. (Here almost means except for a countable union of closed sets without interior.)

WEAK CONJECTURE : Almost every map $f \in L(V, M, \infty)$ is 0-stable.

In the next paragraph we show that the Strong Conjecture is false. The counter example shows that if we substitute for ∞-stable, 2-stable, the statement is still false. Thus the stability conjectures that remain open* are the Weak Conjecture above and :

FEEBLE CONJECTURE : Almost every map $f \in L(V, M, \infty)$ is 1-stable.

* The "Weak Conjecture" has been proved by John Mather, following an outline proposed by Thom. The proof will appear in his book, copies of Chapter 1 of which were distributed at the symposium.

11.2 The counter example that we construct to the Strong Conjecture of the preceding paragraph will be for C^∞ mappings from R^{n^2} to R^{n^2}. As remarked in 5.3, there are two possible topologies on $L(R^n, R^p, s)$, the topology of uniform convergence, (U C), and the topology of compact convergence, (C C), of all partials of orders less than or equal to s. We topologize $L(R^n, R^p, \infty)$ by the weakest topology such that all the injections into $L(R^n, R^p, s)$ are continuous. We thus obtain two topologies on $L(R^n, R^p, \infty)$. In either the (C C) or (U C) topology, $L(R^n, R^p, \infty)$ is a Baire space. Note also that the Transversality Theorem (see 7.1) remains true if we take (U C) as the topology on $L(R^n, R^p, \infty)$.

The proof that the strong conjecture is false holds for both topologies and the slight differences will be noted.

Assume that the Strong Conjecture were true. Since $L(R^n, R^p, \infty)$ is a Baire space, the assertion implies that there exists a dense set $\mathcal{J}$ in $L(R^n, R^p, \infty)$ such that given any $f \in \mathcal{J}$ there is a neighbourhood W of f such that any $g \in W$ is equivalent to f.

Suppose S is any r-singularity in $J^r(n, p)$. In 8.2 Proposition 1 we proved that the set of mappings in $L(R^n, R^p, r+1)$ which display singularity of type S, transversally at at least one point, is open. This was proved in (C C) topology but this obviously implies the same proposition in the (U C) topology. Also by definition of the topology in $L(R^n, R^p, \infty)$ we see that the set of mappings which display the singularity of type S, transversally at at least one point, is open in $L(R^n, R^p, \infty)$ in either topology. Call this set of mappings $\mathcal{A}_S$. Assuming $\mathcal{A}_S$ is non-empty $\mathcal{A}_S \cap \mathcal{J}$ is non-empty since $\mathcal{J}$ is assumed dense.

Let $f \in \mathcal{A}_S \cap \mathcal{J}$. Since $f \in \mathcal{J}$ there is a neighbourhood W of f in $L(R^n, R^p, \infty)$ such that for all $g \in W$, $g = h \circ f \circ h'$ where h are h' are C^∞ homeomorphisms of R^n and R^p respectively. Thus we can assume that $J^r(f)$ is transversal to $S(R^n, R^p)$ at all points of R^n. For otherwise there would exist a $g \in W$ by the transversality theorem such that $J^r(g)$ is transversal to $S(R^n, R^p)$ on R^n. But this would give a contradiction since g and f are C^∞ equivalent and thus the set of orbits of the $(r+1)$ - jets of g and f must be equal.

For $f \in L(R^n, R^p, r)$, we let $f^{(r)}(R^n)$ be the set of all r-jets of f at all points of R^n, and for a set $W \subset L(R^n, R^p, r)$ we let $W^{(r)}(R^n)$ be the set of all r-jets of all $g \in W$ at all points in R^n. (see 5.1)

LEMMA 1. <u>Let W be a neighbourhood of</u> $f \in L(R^n, R^p, \infty)$, <u>then</u> $W^{(r)}(R^n)$ <u>contains an open neighbourhood of</u> $f^{(r)}(R^n)$.

We give the proof of this lemma later.

Now we restrict to the special case of $n = p = n^2$ and $S = S_n \subset J^1(n^2, n^2)$. Since the codimension of S_n in $J^1(n^2, n^2)$ is n^2, we know that $\mathcal{A}_{S_n}$ is not empty and $f \in \mathcal{A}_{S_n} \cap \mathcal{J}$ displays the singularity S_n in at most a <u>countable</u> number of isolated points of R^{n^2}.

Let $z = \pi_{21}^{-1}(S_n) = S'_n \subset J^2(n^2, n^2)$; $L^2 z$ denotes the orbit of z under $L^2(n^2, n^2)$.

LEMMA 2. $\dim L^2 z < \dim S'_n$ for $n \geq 4$.

We postpone the proof of this lemma also.

Let $S_n^T \subset S'_n$ be the set of 2-jets in S'_n which are the 2-jets at 0 of mappings, g, from R^{n^2} to R^{n^2} such that $J^1(g)$ is transversal to $S_n(R^{n^2}, R^{n^2})$ at 0 and $0 \in S_n(g)$. S_n^T is open in S'_n, thus $\dim L^2 z < \dim S_n^T$ for $n \geq 4$. Thus in S_n^T there are uncountably many orbits, so the jets which are inequivalent to the jets of f are dense in S_n^T.

Suppose W is the neighbourhood of $f \in \mathcal{a}_{S_n} \cap \mathcal{J}$ such that $g \in W$ implies g is equivalent to f.

By <u>Lemma 1</u> we know that $W^{(2)}(R^{n^2})$ is an open neighbourhood of $f^{(2)}(R^{n^2})$ and thus $W^{(2)}(R^{n^2}) \cap S_n^T$ is a non-empty open subset of S_n^T.

By the remarks following Lemma 2, above, it follows that there are 2-jets in $W^{(2)}(R^{n^2}) \cap S_n^T$ which are not equivalent to any 2-jet in $f^{(2)}(R^{n^2})$. But this is impossible since all maps $g \in W$ are equivalent to f which implies that $g^{(2)}(R^{n^2})$ consists only of jets equivalent to jets in $f^{(2)}(R^{n^2})$- contradiction.

Note : The proof has only used the fact that all maps $g \in W$ were 2-equivalent to f so that we could know that $W^{(2)}(R^n)$ is contained in the union of L^2 orbits of jets of $f^{(2)}(R^n)$.

11.3 In the next 2 paragraphs we give the proofs of Lemmas 1 and 2.

LEMMA 1. <u>Let</u> W <u>be a neighbourhood of</u> f <u>in</u> $L(R^n, R^p, \infty)$, <u>then</u> $W^{(r)}(R^n)$ <u>contains an open neighbourhood of</u> $f^{(r)}(R^n)$, <u>in</u> $J^r(n, p)$.

PROOF : We prove the (U C) version of the lemma since it implies the (C C) version. It suffices to prove the lemma for the inverse image of an element of the sub-base of the (U C) topology in $L(R^n, R^p, s)$. Such a neighbourhood is given by a sequence of $s+1$ positive numbers, $\epsilon_0, \ldots, \epsilon_s$, i.e.

$$W = \{g \in L(R^n, R^p, \infty) \mid |\partial_\omega g^i - \partial_\omega f^i| < \epsilon_{|\omega|}, \quad i = 1, \ldots, p, \quad 0 \leq |\omega| \leq s\}$$

(For the ω-notation see 7.3).

Letting $W_k = \{g \in L(R^n, R^p, \infty) \mid |\partial_\omega g^i - \partial_\omega f^i| < \epsilon_{|\omega|} \ ; \ i = 1, \ldots, p, \ |\omega| = k\}$

we see that $W = \cap W_k$ and we show that $(W_k)^{(r)}(R^n)$ contains an open neighbourhood of $f^{(r)}(R^n)$.

If $k > r$, if we add any polynomial of degree r to f we remain in W_k so that $(W_k)^{(r)}(R^n)$ contains an open neighbourhood of $f^{(r)}(R^n)$.

In fact, it suffices to assume $k \leq r$ and to show that there exists a δ, $0 \leq \delta \leq \epsilon_k$ such that for any set of $p\binom{n+k-1}{k}$ real numbers, a^i_ω, $(i = 1, \ldots, p \ ; \ |\omega| = k)$, such that $|a^i_\omega - \partial_\omega f^i(x_o)| < \delta$ for all $|\omega| = k$, and some $x_o \in R^n$, there exists a mapping $g \in W_k$ such that $\partial_\omega g(x_o) = a_\omega$. We assume x_o is the origin in R^n and proceed to determine δ.

LEMMA A. <u>For each integer</u> k, <u>there exists a</u> $d > 0$ <u>such that for all</u> $\delta > 0$ <u>and all homogeneous polynomials</u> h <u>of degree</u> k : <u>if</u> $|\partial_\omega h(0)| < \delta$ <u>for all</u> $|\omega| = k$, <u>then</u> $|\partial_{\omega'} h(x) - \partial_{\omega'} h(0)| < \delta$ <u>for all</u> $|x_i| < d$ <u>and</u> $|\omega'| \leq k$.

PROOF : Since $h(x) = \sum_{|\omega| = k} \frac{\partial_\omega h(0)}{\omega!} x_\omega$, we have

$$\partial_{\omega'} h(x) = \frac{\partial_\omega h(0)}{(\omega - \omega')!} x_{(\omega - \omega')} .$$

If $|\omega'| = k$, then $\partial_{\omega'} h(x) = \partial_{\omega'} h(0)$, otherwise $\partial_{\omega'} h(0) = 0$. Thus for

$$|\omega'| < k \ ; \ |\partial_{\omega'} h(x) - \partial_{\omega'} h(0)| = |\partial_{\omega'} h(x)| < \delta \sum_{|\omega| = k} \frac{|x_{(\omega - \omega')}|}{(\omega - \omega')!} .$$

Choosing $d = \binom{n+k-1}{k}^{-1}$ we have for $|x_i| \leq d$, $(i = 1, \ldots, n)$

$$\sum_{|\omega|=k} \frac{|x_{(\omega-\omega')}|}{(\omega-\omega')!} \leq d^{k-|\omega'|} \sum_{|\omega|=k} \frac{1}{(\omega-\omega')!} < d \sum_{|\omega|=k} (1) = 1 .$$

Thus $|\partial_{\omega'} h(x) - \partial_{\omega'} h(0)| < \delta$.

LEMMA B. For all $\eta > 0$ and $d > 0$ and all integers k , there exists a $\delta > 0$ and a C^∞ function ϕ in R^n such that

a) $0 \leq \phi \leq \delta$

b) $\phi(x) = \delta$ for $0 \leq |x_i| \leq \frac{d}{2}$ and $\phi(x) = 0$ if any $|x_i| \geq d$

c) $|\partial_{\omega'} \phi(x)| < \eta$ for all $|\omega'| \leq k$.

PROOF : This is trivial since it is well known that we can find a C^∞ function ψ satisfying a) and b) if we replace δ by 1 .
Since all of the $\partial_{\omega'} \psi$ are continuous on the set $|x_i| \leq d$ $(i = 1, \ldots, n)$ they are all bounded and for $|\omega'| \leq k$ there is only a finite number of functions so it is possible to choose a δ small enough so that $\phi = \delta \cdot \psi$ has the desired properties.

To complete the proof, choose d as in Lemma A for our given value of k. For the triple $(\eta = (\epsilon_k/2^{k+1}), d, k)$ we find by Lemma B a $\delta > 0$ and a C^∞ function ϕ .
Let $h \in L(R^n, R^p, \infty)$ be a p-tuple of homogeneous polynomials of degree k whose coefficients are determined by $\partial_\omega h^i(0) = a^i_\omega - \partial_\omega f^i(0)$ for all $|\omega| = k$, for any a^i_ω such that $|a^i_\omega - \partial_\omega f^i(0)| < \delta$ for all $|\omega| = k$. Let $\tilde{h} = (\phi \frac{h}{\delta})$. Thus $|\partial_\omega \tilde{h}(0)| = |\partial_\omega h(0)| < \delta$. Thus applying Lemmas A and B we have :

$$|\partial_\omega \tilde{h}^i| = |\partial_\omega (\phi \frac{h^i}{\delta})| \leq \sum_{\omega'+\omega''=\omega} |\partial_{\omega'} \cdot \phi| \cdot |\partial_{\omega''} \frac{h^i}{\delta}| < \eta\, 2^{k+1} = \epsilon_k .$$

Set $g = f + \tilde{h}$. Since $\partial_\omega (g - f) = \partial_\omega \tilde{h}$, we see that $g \in W_k$ and by definition of $\tilde{h}$, $\partial_\omega g^i(0) = a^i_\omega$.

11.4 LEMMA 2. $\dim L^2 z < \dim S'_n$ <u>for</u> $n \geq 4$ <u>where</u> $z \in \pi_{21}^{-1} S_n = S'_n \subset J^2(n^2, n^2)$.

PROOF : codim S_n in $J^1(n^2, n^2)$ = codim S'_n in $J^2(n^2, n^2) = n^2$.

$\dim J^2(n^2, n^2) = \dim L^2(n^2) = \frac{n^2}{2}(n^4 + 3n^2)$.

We write simply L^2 for $L^2(n^2, n^2) = L^2(n^2) \times L^2(n^2)$.

Let $z \in S'_n$ and let H be the subgroup of L^2 leaving z fixed, then :

$$\dim L^2 z = \dim(L^2/H) = \operatorname{codim} H \text{ in } L^2 .$$

Thus we must show that

$$\operatorname{codim} H \text{ in } L^2 < \frac{n^2}{2}(n^4 + 3n^2 - 2) .$$

Let $k = n^2 - n$; the coordinates in the source and target are respectively $(y, x) = (y_1, \ldots, y_k, x_1, \ldots, x_n)$ and $(Y, X) = (Y_1, \ldots, Y_k; X_1, \ldots, X_n)$. We make the convention on the indices that $1 \leq i, j \leq k$ and $1 \leq r, s \leq n$. We may assume that the coordinates have been so chosen that z is the 2-jet at zero of the map :

$$\mathfrak{z} : \begin{cases} Y_i = y_i \\ X_r = {}^t y B^r x + {}^t x A^r x \end{cases}$$

Let

$$\alpha : \begin{cases} Y'_i = (EY)_i + (FX)_i + {}^tYH^iY + {}^tYM^iX + {}^tXG^iX \\ X'_r = (CY)_r + (DX)_r + {}^tYP^rY + {}^tYQ^rX + {}^tXN^rX \end{cases}$$

$$\beta : \begin{cases} y_i = (ey')_i + (fx')_i + {}^ty'h^iy' + {}^ty'm^ix' + {}^tx'g^ix' \\ x_r = (cy')_r + (dx')_r + {}^ty'p^ry' + {}^ty'q^rx' + {}^tx'n^rx' \end{cases}$$

and we assume that $(\alpha^{(2)}(0) \times \beta^{(2)}(0)) \in L^2$. We determine the number of conditions on $\alpha^{(2)}(0)$ and $\beta^{(2)}(0)$ so that $(\alpha^{(2)}(0) \times \beta^{(2)}(0)) \in H$. The 1-jet invariance of z under $\alpha^{(2)}(0) \times \beta^{(2)}(0)$ is given by :

$$\begin{pmatrix} E & F \\ C & D \end{pmatrix} \begin{pmatrix} I & 0 \\ 0 & 0 \end{pmatrix} \begin{pmatrix} e & f \\ c & d \end{pmatrix} = \begin{pmatrix} E \cdot e & E \cdot f \\ C \cdot e & C \cdot f \end{pmatrix} = \begin{pmatrix} I & 0 \\ 0 & 0 \end{pmatrix}$$

Thus $E = e^{-1}$, $f = 0$, $C = 0$. This yields $(k^2 + 2nk)$ independent conditions on $\alpha \times \beta$. We now estimate the number of independent conditions arising from second order invariance. We set $E = e^{-1}$, $f = 0$, $C = 0$ and omitting terms of order higher than two we obtain :

$$\alpha \circ \mathfrak{z} : \begin{cases} Y'_i = (Ey)_i + \sum_r F_{ir} \left({}^t yB^r x + {}^t xA^r x\right) + {}^t yH^i y \\ X'_r = \sum_s D_{rs} \left({}^t yB^s x + {}^t xA^s x\right) + {}^t y\, P^r y \end{cases}$$

and finally suppressing all primes :

$$\alpha \circ \mathfrak{z} \circ \beta : \begin{cases} Y_i = y_i + \sum_j E_{ij}\left({}^t yh^j y + {}^t ym^j x + {}^t x\, g^t x\right) + \\ \qquad + \sum_r F_{ir}\left({}^t(ey)B^r(cy + dx) + {}^t(cy + dx)\, A^r(cy + dx)\right) \\ \qquad + {}^t(ey)\, H^i(ey) \\ X_r = \sum_s D_{rs}\left({}^t(ey)B^s(cy + dx) + {}^t(cy + dx)\, A^s(cy + dx)\right) \\ \qquad + {}^t(ey)\, P^r\, (ey) . \end{cases}$$

(Here $e = E^{-1}$).

We thus have $n^2(n^2 \frac{(n^2 + 1)}{2})$ equations in the coefficients if we set $(\alpha \circ \mathfrak{z} \circ \beta)^{(2)}(0) = z$.

In the expression for X_r ; D, c, d, E^{-1}, P^r appear as coefficients. The number of coefficients thus entering into these expressions is $2n^2 + nk + k^2 + nk\left(\frac{k+1}{2}\right) =$ $= \frac{n^2}{2}(n^3 + 3)$. Since the number of equations arising from setting

$${}^t xA^r x + {}^t yB^r x = X_r, \text{ (above) is } n(n^2 \frac{(n^2+1)}{2})$$

we see for $n \geq 3$; $\frac{1}{2} n^2(n^3 + 3) \leq \frac{1}{2} n^2(n^3 + n)$. Thus the number of independent equations arising from X_r , is at most $\frac{1}{2}n^2(n^3+3)$.

On the other hand the number of coefficients arising in the equations obtained by setting the above expression for Y_i equal to y_i is greater than the number of equations so we can only say that the number of independent equations is at most equal to the number of equations. Thus the total number of independent relations arising from second order invariance is less than or equal to

$$\frac{1}{2}n^2(n^3+3+k(n^2+1)) .$$

Adding the number of equations that we get from the 1-jet invariance we obtain :

$$\operatorname{codim} H \text{ in } L^2 \leq \frac{1}{2}n^2(n^3+3+k(n^2+1))+k^2+2nk$$

$$=\frac{1}{2}n^2(n^4+3n^2-n+1) \quad \text{for } n \geq 3$$

But $\frac{1}{2}n^2(n^4+3-n+1) < \frac{1}{2}n^2(n^4+3n^2-2) = \dim S'_n$ whenever $n \geq 4$.

11.5 Whitney [5] has considered a sequence of singularities starting with $S_k \subset J^1(n,p)$ and higher order singularities, obtained by considering the position of the kernel of the tangent map of a map f , to the limiting positions of the tangent spaces to all incident singularities, (see 9.3). The singularities $S_{1,k}$ are elements of this sequence, and $S_{k,h}$ for $k > 1$ are further subdivided by considering the various positions of the kernel K of f_* of mapping f at a point of $S_k(f)$ with respect to $T(S_k(f))$ at that point and also with respect to the limiting positions of the tangent spaces to $S_{k'}(f)$ at the point for all $k' < k$.

If such a sequence of singularities could be defined[†], then a mapping, f, such that the r-extension of f is transversal to all of the r-singularities of the sequence is called <u>generic</u>.

The above counter-example shows that it is not possible that the set of generic maps is dense and that the generic maps are also all 2-stable at all points. In particular, there are no 2-stable maps from R^{n^2} to R^{n^2} which display the singularity S_n, for $n \geq 4$.

[†] A definition was given by J.M. Boardman, in 'Singularities of Differentiable Maps', Publ. Math. I.H.E.S. 33(1967) 21-57.

12. Homotopic Stability

12.1 For a map, $F : X \times Y \to Z$ we will sometimes use the notation, $F_x : Y \to Z$, $F_y : X \to Z$ where $F_x(y) = F_y(x) = F(x, y)$. Further $F' : X \times Y \to Z \times Y$ is defined by $F'(x, y) = (F(x, y), y)$. As usual, $I = [0, 1)$, and $I_t = [0, t)$. By id_X, we will always mean the identity map of X onto itself, and we will suppress the X when it is clear what is meant. All maps and manifolds considered below will be C^∞.

DEFINITION 1. $f : V \to M$ is called homotopically stable if for every homotopy of f, $F : V \times I \to M$, there exists a $t_o > 0$ and homotopies of the respective identity maps $\phi : V \times I_{t_o} \to V$ and $\psi : M \times I_{t_o} \to M$ such that ϕ_t and ψ_t are homeomorphisms of V and M respectively for $t < t_o$ and $F | V \times I_{t_o} = \psi \circ f \times id_I \circ \phi'$, or equivalently $F_t = \psi_t \circ f \circ \phi_t$ for $t < t_o$.

In the following we will not be concerned with the relationship between homotopic stability and stability but will investigate the notion of homotopic stability itself, first giving a necessary and sufficient condition that a map be homotopically stable, and then applying this condition to determine some classes of homotopically stable maps. We will restrict ourselves to maps $f : V \to M$ where V is compact. This is prompted by the fact that if V were allowed to be non-compact even regular imbeddings would not be homotopically stable. For example $f : R \to R^2$ as shown in the figure where the two tails are asymptotic to one another. If F is the deformation suggested by the dotted line and arrows which merely pushes the upper tail down a little we see that for every $t > 0$, F_t will have self-intersections whereas f has none. Considering ordinary stability if W_f is a neighbourhood of f in the uniform topology the maps F_t of the above deformation will again be included in W_f for small enough t.

12.2 Suppose N is a differentiable manifold. Over $N \times I$ we let $\tilde{T}(N)$ and $\tilde{T}(I)$ be the bundles induced from the tangent bundles, $T(N)$ and $T(I)$, by the projections of $N \times I$ onto N and I respectively. $T(N \times I) \cong \tilde{T}(N) \oplus \tilde{T}(I)$. Let $P_1 : T(N \times I) \to \tilde{T}(N)$ be the projection via this isomorphism onto the first factor. In $\tilde{T}(I)$ let τ denote the vector field which is the "inverse image" of the vector field $\frac{\partial}{\partial t}$ in I (t is the coordinate in I). We will also have occasion to consider exactly the analogous bundles over $N \times R$, $\tilde{T}(N)$, $\tilde{T}(R)$ and the vector field τ in $\tilde{T}(R)$. We make no distinction in the notation since the context will make it clear which bundles are meant. When more than one copy of R or of I appears, we will index them and index the τ's correspondingly. (For example in the proof of the lemma below, τ_1 is the vector field along the parameter interval introduced in the integration, i.e. it is along R_1.) In the following diffeomorphism means homeomorphism which is differentiable in both directions.

LEMMA. Suppose $Z : N \times I \to \tilde{T}(N)$ is a differentiable vector field that vanishes outside a compact set, then there exists a differentiable mapping $g : N \times I \to N$ such that :

1) For each t, g_t is a diffeomorphism, $g_0 = id_N$

2) $Z = P_1 \circ g'_* \circ \tau \circ g'^{-1} = -g'_* \circ P_1 \circ g'^{-1}_* \circ \tau$ on $N \times I$.

PROOF : Since Z vanishes outside a compact set, there exists a $t_0 > 0$ such that Z vanishes for all $(x, t) \in N \times I$, $t \geq t_0$. Extend Z to a vector field on $N \times R$ by means of a partition of unity so that the extended vector field, W, agrees with Z on $N \times I$, but still vanishes outside a compact set in $N \times R$. Thus there exists $[-t_1, t_0]$ so that W vanishes outside $N \times [-t_1, t_0]$. Consider now the vector field $(W + \tau)$ on $N \times R$. This we can integrate locally and obtain local 1-parameter groups ([2] pp. 5-7) as follows. Suppose the support of W is contained in $K \times [-t_1 - \epsilon, t_0 + \epsilon]$, K compact in N. We can cover $N \times R$ with open neighbourhoods $\{U_\lambda\}$ such that $K \times [-t_1 - \epsilon, t_0 + \epsilon]$, is contained in a finite union of the U_λ and such that to each U_λ we have a constant $\delta_\lambda > 0$ and a differentiable map $G : U \times (-\delta_\lambda, \delta_\lambda) \to N \times R$ such that

1) for $|t| < \delta_\lambda$, $G_t : U_\lambda \to G_t(U_\lambda)$ is a diffeomorphism

2) for $|t|, |s|, |t+s| < \delta_\lambda$ and if $G_t(p) \in U_\lambda$, $p \in U_\lambda$ then $G_s(G_t(p)) = G_{s+t}(p)$

3) $G_* \circ \tau_1(p, 0) = (W + \tau)(p)$ for $p \in U_\lambda$

Because of the uniqueness of such maps G it is unnecessary to index them. Let $\delta_1 = \min \delta_\lambda$ for all U_λ such that $U_\lambda \cap K \times [-t_1 - \epsilon, t_o + \epsilon] \neq \emptyset$.

If $U_\lambda \cap K \times [-t_1 - \epsilon, t_o + \epsilon] = \emptyset$, then we can take $\delta_\lambda = \epsilon$ since on such a U_λ, $G_t(x, s) = (x, s + t)$ at least until $(x, s + t) \in K \times [-t_1, t_o]$ that is for all $|t| < \epsilon$. Thus letting $\delta = \min(\delta_1, \epsilon)$ we have a universal δ_λ for all U_λ so that 1), 2), 3) hold. Thus we can piece together all the local groups to give a global group of diffeomorphisms i.e. $G : (N \times R) \times R_1 \to N \times R$ such that :

1) $G_t : N \times R \to N \times R : (x, s) \mapsto G(x, s, t)$ is a diffeomorphism

2) $G_{s_1} \circ G_{s_2} = G_{s_1+s_2}$

3) $G_* \circ \tau_1 (x, s, 0) = (W + \tau)(x, s)$

We examine condition 2). This states that :

$$G(G(x, t, s_1) s_2) = G(x, t, s_1 + s_2), \text{ that is}$$

$G \circ (G \times id_{R_2}) = G \circ A : N \times R \times R_1 \times R_2 \to N \times R$, where

$A : N \times R \times R_1 \times R_2 \to N \times R \times R_1 : (x, t, s_1, s_2) \to (x, t, s_1 + s_2)$

Thus $(G \circ (G \times id_{R_2}))_* \circ \tau_2 = G_* \circ \tau_1 \circ G \times id_{R_2}$ and

$$(G \circ A)_* \circ \tau_2 = G_* \circ \tau_1 \circ A, \text{ that is } G_* \circ \tau_1 \circ G \times id_{R_2} = G_* \circ \tau_1 \circ A.$$

Let $i : N \times R \times R_1 \to N \times R \times R_1 \times R_2 : (x, t, s_1) \mapsto (x, t, s_1, 0)$.

Thus $G_* \circ \tau_1 \circ G \times id_{R_2} \circ i = G_* \circ \tau_1 \circ A \circ i$, which we can rewrite as :

(1) $$(W + \tau) \circ G = G_* \circ \tau_1 .$$

Writing $G = \eta \times \mathfrak{I}$, i.e. splitting G into its N and R components, we have : $\langle G_* \circ \tau_1, dt\rangle = \langle W + \tau, dt\rangle \circ G = 1$.

Thus $\frac{\partial}{\partial t_1}\mathfrak{J}(x, t, t_1) = 1$, and thus $\mathfrak{J}(x, t, t_1) = t_1 + h(x, t)$.

But since $\mathfrak{J}(x, t, 0) = t$ we see that $\mathfrak{J}(x, t, t_1) = t_1 + t$. Let $g(x,t) = \mathfrak{N}(x, 0, t)$;

thus $g' : N \times R \to N \times R$ is $G \circ j$,

where $j : N \times R \to N \times R \times R_1 : (x, t) \to (x, 0, t)$.

$g(x, 0) = \mathfrak{N}(x, 0, 0) = x$ so is the identity on N . For each t, g_t is a diffeomorphism, since for each t, G_t is a diffeomorphism of $N \times 0$ onto $N \times t$.

Also since $\tau_1 \circ j = j_* \circ \tau$ we obtain from (1) :

$$(W + \tau) = g'_* \circ \tau \circ (g')^{-1} .$$

Restricting both sides to $N \times I$ we obtain the desired result.

For the other expression for Z , we note that since $g' \circ (g')^{-1} = id_{N \times R}$

$$(g' \circ (g')^{-1})_* \circ \tau = \tau$$

$$g'_* \circ \tau \circ (g')^{-1} + g'_* \circ P_1 \circ (g')^{-1}_* \circ \tau = \tau$$

Thus
$$-g'_* \circ P_1 \circ (g')^{-1}_* \circ \tau = P_1 \circ g'_* \circ \tau \circ (g')^{-1} = W$$

or when restricted to $N \times I$, equals Z .

REMARK : $g : N \times I \to N$ then $(g')_* (\tilde{T}(N)) \subset \tilde{T}(N)$. For suppose $v \in \tilde{T}(N)$ then $\langle g'_*(v) , dt \rangle = \langle v , d(t \circ g') \rangle = \langle v, dt \rangle = 0$ which is equivalent with $g'_*(v) \in \tilde{T}(N)$.

12.3 THEOREM. $f : V \to M$, V <u>compact, is homotopically stable iff given any</u> $F : V \times I \to M$, <u>a homotopy of</u> f , <u>then there are vector fields</u> $X : V \times I \to \tilde{T}(V)$ <u>and</u> $Y : M \times I \to \tilde{T}(M)$ <u>such that</u>

$$F'_* \circ \tau = F'_* \circ X + Y \circ F' + \tau \circ F'$$

PROOF : Suppose f is homotopically stable, then any homotopy, F , of f can be written as follows : $F = \psi \circ f \times id \circ \phi'$ and

$$(F')_* \circ \tau = (\psi' \circ f \times id \circ \phi')_* \circ \tau = (\psi' \circ f \times id)_* \circ (\tau \circ \phi' + P_1 \circ \phi'_* \circ \tau)$$
$$= \psi'_* \circ \tau \circ f \times id \circ \phi' + (\psi' \circ f \times id)_* \circ P_1 \circ \phi'_* \circ \tau$$

$$= \tau \circ F' + P_1 \circ \psi'_* \circ \tau \circ (\psi')^{-1} \circ F' + F'_* \circ (\phi')^{-1}_* \circ P_1 \circ \phi'_* \circ \tau$$

$$= \tau \circ F' + Y \circ F' + F'_* \circ X .$$

Conversely, suppose $F'_* \circ \tau = F'_* \circ X + Y \circ F' + \tau \circ F'$.

Extend the vector field X to a vector field $A : V \times R \to \widetilde{T}(V)$ which agrees with X on $V \times I$ and which vanishes for $|t| > 2$. We can apply the preceding lemma to find a $\phi : V \times I \to V$ such that $X = (\phi')^{-1}_* \circ P_1 \circ \phi'_* \circ \tau$. Similarly we can replace Y by a vector field $B : M \times R \to \widetilde{T}(M)$, such that B agrees with Y on $F'(V \times \bar{I}_{t_o})$ but vanishes outside a compact set containing $F'(V \times \bar{I}_{t_o})$. We can then find a map $\psi : M \times I_{t_o} \to M$ such that $Y | M \times I_{t_o} = P_1 \circ \psi'_* \circ \tau \circ (\psi')^{-1}$. Consider now $H' = (\psi')^{-1} \circ F' \circ (\phi')^{-1}$ on $V \times I_{t_o}$. We show that H is independent of t.

$$H'_* \circ \tau = (\psi')^{-1}_* \circ F'_* \circ P_1 \circ (\phi')^{-1}_* \circ \tau + (\psi')^{-1}_* \circ (F'_* \circ X + Y \circ F') \circ (\phi')^{-1}$$
$$+ P_1 (\psi')^{-1}_* \circ \tau \circ F' \circ (\phi')^{-1} + \tau \circ H'$$
$$= (\psi')^{-1}_* \circ F'_* \circ (P_1 \circ (\phi')^{-1}_* \circ \tau \circ \phi' + X) \circ (\phi')^{-1}$$
$$+ (\psi')^{-1}_* \circ (Y + \psi'_* \circ P_1 \circ (\psi')^{-1}_* \circ \tau) \circ F' \circ (\phi')^{-1} + \tau \circ H' .$$

Thus using 2) of the preceding lemma we obtain :

$$H'_* \circ \tau = \tau \circ H'$$

Thus given any function λ on $M \times I$, $< H'_* \tau , \lambda > = < \tau, \lambda > \circ H'$,

that is $\dfrac{\partial(\lambda \circ H')}{\partial t} = \dfrac{\partial \lambda}{\partial t} \circ H'$. This means that H is independent of t. Thus $H(x, t) = H(x, 0) = f(x)$. So finally, on $V \times I_{t_o}$, $f \times id = (\psi')^{-1} \circ F' \circ (\phi')^{-1}$ or $F = \psi \circ f \times id \circ \phi'$.

COROLLARY . If $n \geq p$, and $f : V^n \to M^p$ has maximal rank everywhere then f is homotopically stable.

PROOF : Let F be a deformation of f ; $F : V \times I \to M$. The interval I is taken small enough so that for each $t \in I$, F_t also has maximal rank everywhere. By the theorem it suffices to find a vector field $X : V \times I \to \tilde{T}(V)$ such that $F_* \circ \tau = F_* \circ X$. Cover $V \times I$ and M with a locally finite system of coordinate neighbourhoods $\mathcal{W}$, $\mathcal{M}$ (we omit the index) such that $F(\mathcal{W}) \subset \mathcal{M}$ so small that if $y_1, \ldots, y_p$ is a coordinate system in $\mathcal{M}$, then $y_i \circ F_{t_o}$ is part of a coordinate system in $P_V(\mathcal{W})$ (P_V = proj on V of $V \times I$) for each $t_o \in P_I(\mathcal{W})$. Taking a partition of unity for $\{\mathcal{W}\}$ it suffices to define X on $\mathcal{W}$. If for $(x, t) \in \mathcal{W}$, $F_* \circ \tau(x, t) = \sum_{i=1}^{p} a_i(x, t) \frac{\partial}{\partial y_i} (F(x, t))$. Define

$$X(x, t) = \sum_{i=1}^{p} a_i(x, t)(\cdot\, ,t)_* \circ \frac{\partial}{\partial (y_i \circ F_t)} (x) .$$

(Here $(\cdot\, , t) : V \to V \times I : x \mapsto (x, t)$.) This is clearly differentiable.

$$F_* \circ X(x, t) = \sum_{i=1}^{p} a_i(x, t)\, F_* \circ (\,\cdot, t)_* \circ \frac{\partial}{\partial (y_i \circ F_t)} (x) .$$

But $< F_* \circ (\,\cdot\, ,t)_* \circ \frac{\partial}{\partial (y_i \circ F_t)} (x), dy_j > = \frac{\partial}{\partial (y_i \circ F_t)} (y_j \circ F \circ (\,\cdot\, ,t)(x) =$

$$= \frac{\partial (y_i \circ F_t)}{\partial (y_i \circ F_t)} (x) = \delta_i^j$$

Thus in particular all differentiable fibrings are homotopically stable.

12.4 A system, Σ , of planes $\{E_1, \ldots, E_k\}$, in R^p such that $\dim E_i = m_i$ is in <u>general position</u> if $\operatorname{codim} \bigcap_{i=1}^{r} E_{j_i} = \sum_{i=1}^{r} m_{j_i}$, where $\{j_1, \ldots, j_r\}$ is a subset of r distinct integers among $\{1, \ldots, k\}$.

DEFINITION. An immersion $f : V \to M$ is called an <u>immersion with normal crossings</u> if $y \in f(V)$ and $f^{-1}(y) = \{x_1, \ldots, x_s\}$, then $\{f_*(V_{x_1}), \ldots, f_*(V_{x_s})\}$ is a system in general position in M_y . (V_{x_i} is the tangent space to V at x_i , and M_y is the tangent space to M at y.)

As a non-trivial application of the theorem (12.3) we prove :

THEOREM . V, M _differentiable manifolds,_ V _compact; then immersions of_ V _in_ M _with normal crossings are homotopically stable._

COROLLARY. _Imbeddings of compact manifolds are homotopically stable._

In the course of the proof a few elementary facts are needed for linear spaces in general position and immersions with normal crossings which we isolate as lemmas.

12.5 If E and F are two linear spaces through 0 in R^p of codimensions m, n resp. then $\operatorname{codim} E \cap F \leq m + n$. This follows immediately from $\dim (E + F) = \dim E + \dim F - \dim E \cap F$. So E, F are in general position iff $E + F = R^p$.

Further if $\Sigma = \{E_i\}$ is a set of linear spaces through 0 in R^p such that $\operatorname{codim} E_i = m_i$, then $\operatorname{codim} \bigcap_{i=1}^{k} E_i = \sum_{i=1}^{k} m_i$ implies that Σ is in general position. For we know that $\operatorname{codim} \bigcap_{i=1}^{k} E_i \leq p$ since $0 \in \bigcap_i E_i$. Further suppose for some subset, say $E_1,\ldots,E_r$, $\operatorname{codim} \bigcap_{i=1}^{r} E_i < \sum_{i=1}^{r} m_i$. Then $\operatorname{codim} \bigcap_{i=1}^{k} E_i < \sum_{i=1}^{k} m_i$ - contradiction. Here clearly the assumption of $\bigcap_{i=1}^{k} E_i \neq \emptyset$ is essential.

LEMMA 1. _Suppose_ Σ _is in general position through_ 0 _in_ R^p, $\operatorname{codim} E_i = m_i$, _then there exists a basis_ $\{e_1,\ldots e_s;\ f_1^1,\ldots,\ f_{m_1}^1\ ;\ \ldots;f_1^k,\ldots,f_{m_k}^k\}$ _such that if_ $E = \text{span of } \{e_1,\ldots e_s\}$ _and_ $F^i = \text{span of } \{f_1^i,\ldots f_{m_i}^i\}$ $(s = p - \sum_{i=1}^{k} m_i)$ _then_ $E_i = E + \sum_{j \neq i} F^j$.

PROOF : For $k = 1$ the proposition is trivial. So assume $k \geq 2$. Let $E = \bigcap_{i=1}^{k} E_i$ and choose any basis $\{e_1, \ldots, e_s\}$ for E. Let $D^i = \bigcap_{j \neq i} E_j$. Since $E \subset D^i$ we may complete $\{e_1, \ldots, e_s\}$ to a basis for D^i. There will be exactly m_i new vectors since codim $D^i = \sum_{j \neq i} m_j = (\text{codim } E) - m_i$. Call the new vectors $f_1^i, \ldots, f_{m_i}^i$. Since $D^i \cap D^j = E$, for $i \neq j$, letting F^i = span of $\{f_1^i, \ldots, f_{m_i}^i\}$, we see that $F^i \cap F^j = 0$, $i \neq j$. Thus $\{e_1, \ldots, e_s ; f_1^1, \ldots f_{m_k}^k\}$ is a basis for R^p. Since $E \subset E_i$ and $F^j \subset E_i$, $i \neq j$; $E + \sum_{j \neq i} F^j \subset E_i$ and since the dimensions check $E_i = E + \sum_{j \neq i} F^j$. Thus given a system Σ in R^p in general position through 0, we can introduce an inner product so that $\{e_a, f_{b_i}^i\} (1 \leq a \leq s ; 1 \leq i \leq k ; 1 \leq b_i \leq m_i)$ is an orthonomal basis. Unless we state otherwise this is the metric used in the following and we will call such a metric "the usual one in R^p for the spaces E_i in general position".

A point of $G(p - m_1, p) \times \ldots \times G(p - m_k, p)$ is a set of k linear spaces through 0 in R^p, where $G(r, p)$ is the Grassmann manifold of r-planes in p-space.

LEMMA 2. <u>The subset of</u> $G(p - m_1, p) \times \ldots \times G(p - m_k, p)$ <u>of systems of</u> k <u>linear spaces in general position is open</u>.

PROOF : $G(p - m_i, p) \cong G(m_i, p)$; the mapping that implements this diffeomorphism takes a plane to its orthogonal complement, in some metric. Thus using this homeomorphism the lemma is equivalent to : the subset of elements of $G(m_1, p) \times \ldots \times G(m_k, p)$, whose component spaces span a space of dimension $\sum_{i=1}^{k} m_i$, is open. The proof of this is trivial. That it is equivalent to the statement of the lemma is also easy since if $\{E_1, \ldots, E_k\}$ is in $G(p - m_1, p) \times \ldots \times G(p - m_k, p)$ and $(E_i)^{\perp} = F^i$ then $(\bigcap_i E_i)^{\perp} = \sum_i F^i$. Thus codim $(\cap E_i) = \dim \sum F^i$.

Note : The following lemma will not be used in the proof of the theorem but since it is amusing in its own right we have included it anyway.

LEMMA 3. (Thom - Mexico speech) Let Σ be a system of linear spaces through 0 in R^p in general position. Suppose f is a function defined on Σ which is of class C^m on each element of Σ. Then there exists a C^m extension, ϕ, of f to R^p.

PROOF : By induction on the number, k, of linear spaces in Σ. For $k = 1$ the proposition is trivial. Letting P be the normal projection on E_1 define $\phi = f \circ P$.

Suppose the lemma is proved for systems of $k - 1$ spaces. Let $\Sigma = \{E_1, \ldots, E_k\}$ in general position through 0. Let $\Sigma_1 = \{E_2, \ldots, E_k\}$. By the induction hypothesis there is a function ϕ_1 on R^p such that $\phi_1 | \Sigma_1 = f$. On E_1 let $g = f - \phi_1 | E_1$. g vanishes on $E_1 \cap E_i$ for $i > 1$. Letting P be the normal projection to E_1 set $G = g \circ P$. G vanishes on all E_i, $i > 1$, since in the notation of lemma 1, $E_i = E + \sum_{j \neq i} F^j$, where all spaces are orthogonal, and since $(E_1)^\perp = F^1$, we see that $P(E_i) = E_i \cap E_1$. Finally set $\phi = G + \phi_1$. ϕ is clearly C^m and for $x \in E_i$, $i > 1$, $\phi(x) = \phi_1(x) = f(x)$, and for $x \in E_1$, $\phi(x) = f(x) - \phi_1(x) + \phi_1(x) = f(x)$.

REMARK : In lemma 3, general position was essential, since for example if Σ consisted of three lines in the (x, y) - plane, $x = 0$, $y = 0$, $x = y$, then any C^1 function F in R^2 satisfies $\frac{dF(t,t)}{dt}\Big|_{t=0} = \frac{dF(x,0)}{dx}\Big|_{x=0} + \frac{dF(0,y)}{dy}\Big|_{y=0}$; thus $F | \Sigma$ cannot be arbitrary.

LEMMA 4. Let $\Sigma = \{E_1, \ldots, E_k\}$ be a system of linear spaces through 0 in R^p in general position. If on each E_i in Σ there is given a C^m normal vector field n_i, then there exists a global vector field, m, such that $m|E_i = n_i + t_i$, where t_i is a vector field tangent to E_i.

PROOF : The metric referred to is still that given by the orthonomality of the basis in Lemma 1. We have as in Lemma 1 : $E_i = E + \sum_{j \neq i} F^j$ and $(E_i)^{\perp} = F^i$.

Let $P_j : R^p \to E_j$, be the orthogonal projections. Let $T_a : R^p \to R^p : y \mapsto y + a$ for a in R^p. Claim : $m(x) = \sum_j (T_{(x-P_j(x))})_* \circ n_j \circ P_j(x)$ defines a vector field with the desired properties. It is clearly as differentiable as the n_j's are and is globally defined. On E_i, $P_i(x) = x$, and there,

$m(x) = n_i(x) + \sum_{j \neq i} (T_{(x-P_j(x))})_* \circ n_j \circ P_j(x)$. Further, $n_j \circ P_j(x)$ is parallel to F^j since it is normal to E_j. Thus $(T_{(-P_j(x))})_* \circ n_j \circ P_j(x)$ is tangent to $F^j \subset E_i$. Thus $(T_{(x-P_j(x))})_* \circ n_j \circ P_j(x)$ is tangent to E_i. Thus so is the sum that appears on the right above.

The proof could have been carried out in two steps: First find such an m on Σ alone and then apply lemma 3 to extend it to all of R^p. Since however it required no extra work, both steps have been made together in lemma 4.

12.6 LEMMA 5. Let $G : X \to Y$ be an immersion with normal crossings, X compact, then at each point $y \in G(X)$ there is a coordinate neighbourhood W such that $G(X) \cap W$ is the union of a finite number of coordinate linear spaces which intersect at y. (dim X = n, dim Y = p).

PROOF : Let $y \in G(X)$ and let $G^{-1}(y) = \{x_1, \ldots, x_k\}$ (k distinct points). Separate the x_i with pairwise disjoint coordinate neighbourhoods, U_i'. Since on each U_i' we may assume that G is a homeomorphism into there exists a neighbourhood

W' of y such that $G^{-1}(W') \cap U_i' = U_i \subset \bar{U}_i \subset U_i'$.

We can in fact choose W' so small that $G^{-1}(W') \subset \bigcup_i U_i$. For suppose not, then we could find a sequence of points $y_n \in G(X) - \bigcup_i G(U_i)$ such that $y_n \to y$. Thus there is a sequence of points $z_n \in X - \bigcup_i U_i$ such that $G(z_n) = y_n$. Since $X - \bigcup_i U_i$ is compact, there is a $z \in X - \bigcup_i U_i$ such that $G(z) = y$ which is impossible since $G^{-1}(y) \bigcup_i U_i$. Thus we have W' , a neighbourhood of y , such that $W' \cap G(X) \subset \bigcup_i G(U_i)$.

Since each $G(U_i) \cap W'$ is a submanifold of W' through y of dimension n , there is a neighbourhood. $W_i \subset W'$ of y and a mapping $\phi_i : W_i \to R^{p-n}$ such that rank $\phi_i = (p-n)$, and such that $G(U_i) \cap W_i = \phi_i^{-1}(0)$. Let $W'' = \bigcap_i W_i$. Thus $\phi_i | W''$ defines $G(U_i) \cap W''$. We show that the map $\phi = \phi_1 \times \ldots \times \phi_k : W'' \to R^{k(p-n)}$ has rank $k(p-n)$ at y .

Choose a basis for the tangent space at y as in lemma 1 :
$\{e_1, \ldots, e_s ; f_1^1, \ldots, f_{p-n}^1; \ldots; f_1^k, \ldots, f_{p-n}^k\}$ where $s = p - k(p-n)$. (We agree that $1 \leqslant i, j \leqslant k$; $1 \leqslant a, b \leqslant p-n$; $1 \leqslant m \leqslant s$ in the following .)

The vectors $\{e_m, f_a^j\}$, $j \neq i$ span the tangent space to $G(U_i)$ at y . Since $\phi_i : W'' \to R^{p-n} : y \mapsto (\phi_i^1(y), \ldots, \phi_i^{p-n}(y))$ defines $G(U_i)$,

$< e_m, d\phi_i^a (y) > = < f_b^j , d\phi_i^a (y) > = 0$ for $j \neq i$. Further since $\bigwedge_a d\phi_i^a (y) \neq 0$,

$< \bigwedge_a f_a^i , \bigwedge_a d\phi_i^a (y) > \neq 0$, thus $< \bigwedge_{i,a} f_a^i , \bigwedge_{i,a} d\phi_i^a(y) > \neq 0$ and therefore

$\bigwedge_{i,a} d\phi_i^a (y) \neq 0$. Thus ϕ has rank $k(p-n)$ at y and therefore in a small neighbourhood of y, $W \subset W''$, and so the functions ϕ_i^a can be taken as coordinate functions there.

LEMMA 6. $f : V \to M$ *is an immersion with normal crossing,* V *compact, and* F *, a homotopy of* f *; then there exists a* $T > 0$ *such that* F_t *is an immersion with normal crossing for all* $0 \leqslant t < T$. (dim $V = n$, dim $M = p$).

PROOF : Since $F' : V \times I \to M \times I$ has a non-vanishing Jacobian on $V \times 0$ and V is compact there is a $t_1 > 0$ such that F' has non-vanishing Jacobian on $V \times I_{t_1}$. Let $A = \{x \in V \mid f^{-1}(f(x)) = \{x\}\}$. A is open. For suppose there existed a sequence $x_n \to x$ with $x_n \in V - A$ and $x \in A$, then for each x_n there would exist at least one $x_n' \in V - A$ such that $f(x_n) = f(x_n')$, $x_n \neq x_n'$. Since V is compact we may assume $x_n' \to x'$. Thus $f(x') = f(x)$, and since $x \in A$, $x = x'$. But in a neighbourhood of x, f is $1:1$, and we have a contradiction.

Let $B = f(V - A)$; B is compact. By lemma 5 for each point $y \in B$ there is a neighbourhood W' of y with compact closure such that $f(V) \cap W'$ is the union of coordinate linear n planes in general position all of which meet at y.

Let $W \subset \bar{W} \subset W'$, where W is again a neighbourhood of y having the same properties as W'. Since B is compact, a finite number of such W neighbourhoods suffices to cover B. Call the union of these W neighbourhoods, S.

<u>Note</u> : We have thus two coverings of B, one by W the other by the containing W'. $V - f^{-1}(S)$ is again compact and $f \mid V - f^{-1}(S)$ is $1:1$.

<u>Claim</u> : There is a $t_2 > 0$ such that $F_t \mid V - f^{-1}(S)$ is $1:1$ for all $0 \leq t < t_2$. Suppose not, then there exists a sequence of pairs x_n, x_n' in $V - f^{-1}(S)$ such that $F(x_n, \frac{1}{n}) = F(x_n', \frac{1}{n})$. Thus since $V - f^{-1}(S)$ is compact we may assume that $x_n \to x$ and $x_n' \to x'$. Thus $f(x) = f(x')$ and therefore $x = x'$. But this is impossible since F' is $1:1$ in a neighbourhood of $(x, 0)$. Thus if we let $t_3 = \min(t_1, t_2)$ we see that F_t is an immersion and is $1:1$ on $V - f^{-1}(S)$ for all $0 \leq t < t_3$. Thus all crossings of $F_t(V)$ occur in S. For each W_α in the covering of B, choose a $t_\alpha > 0$ such that $F_t(f^{-1}(\bar{W}_\alpha)) \subset W_\alpha'$ for all $0 \leq t < t_\alpha$ and let $t_4 = \min_\alpha (t_\alpha, t_3)$. It now suffices to find a t_5 so that $F_t | f^{-1}(W_\alpha)$ has only normal crossings for $0 \leq t < t_5$ for all α. We determine a value of t_5 for one such W_α - for the remainder of the argument we suppress the α subscript.

Suppose $W' \cap f(V) = \bigcup_{i=1}^{k} E_i'$; E_i' are coordinate linear spaces in general position, all intersecting, say, at y. $f^{-1}(E_i') = U_i'$ is a neighbourhood of one of the inverse images, x_i, of y and $U'_i \cap U'_j = \emptyset$ for $i \neq j$. Pulling back the coordinate system in W' to each of the U'_i we have that :

$J(f) : U'_i \to G(n, p) : x \mapsto$ (subspace of R^p spanned by the rows of the Jacobian of f at x, relative to these coordinate systems)

is constant.

Thus $J(f)^k : U'_1 \times \ldots \times U'_k \to G(n, p)^k$ is constant and the image Σ is a system of linear spaces in general position.

Let $H(F) : U'_1 \times I_{t_4} \to G(n, p) : (x, t) \mapsto J(F_t)(x)$;

$(H(F))^k : (U'_1 \times I_{t_4}) \times \ldots. \times (U'_k \times I_{t_4}) \to (G(n,p))^k$ is continuous. By lemma 2 we can find a neighbourhood $\mathcal{G}$ of Σ in $(G(n,p))^k$, each element of which is a system of k, n-planes in general position. Since $((H(F))^k)^{-1}(\mathcal{G})$ is a neighbourhood of $(U'_1 \times 0) \times \ldots \times (U'_k \times 0)$, there is a $t_5 > 0$ such that $(H(F))^k((U_1 \times I_{t_5}) \times \ldots \times (U_k \times I_{t_5})) \subset \mathcal{G}$, where $U_i = f^{-1}(W) \cap U'_i$.

Thus $F_t \mid (\bigcup_i U_i)$ has only normal crossings for $0 \leq t < t_5$.

LEMMA 7. $F : V \times I \to M$, <u>such that</u> $F_t : V \to M$ <u>is an immersion with normal crossings; then</u> $F' : V \times I \to M \times I$ <u>is also an immersion with normal crossings</u> ($\dim V = n$, $\dim M = p$)

PROOF : F' is an immersion since the Jacobian of F' at (x, t) equals the Jacobian of F_t at x. If $F'(x_1, t_1) = F'(x_2, t_2)$ then $t_1 = t_2$. Thus suppose for some t, $\{(x_1,t),\ldots, (x_k,t)\} = F'^{-1}(y,t)$. Let V_{x_i} be the tangent space to V at x_i, and $V_{x_i,t}$ be the tangent space to V in $V \times I$ at (x_i, t). Thus $V_{x_i,t} + \tau(x_i,t) = (V \times I)_{x_i,t}$ where τ is the vector field along I (see 12.2).

Let $E_i = (F_t)_*(V_{x_i}) = F_*(V_{x_i,t})$. $F'_*(V \times I)_{x_i,t} = \tilde{E}_i + F'_*(\tau(x_i,t))$, where $\tilde{E}_i = F'_*(V_{x_i,t}) = F'_{t*}(V_{x_i})$. Since the tangent space $(M \times I)_{y,t}$ is canonically isomorphic with $(M_y + I_t)$, $\tilde{E}_i$ can be identified with E_i, and $\tau(y,t)$ with $\tau(t)$.

Thus adopting the notation of lemma 1 , where E_i are in general position and $E_i = E + \sum_{j \neq i} F^j$, we have : $F'_*(V \times I)_{x_i,t} = E_i + (\phi^i + \tau(t))$, where $\phi^i \in F^i$, since $F'_*(\tau(x_i,t)) = \phi^i + \tau(t) +$ (a vector in E_i). Equivalently, $F'_*(V \times I)_{x_i,t}$ is spanned by E_i and $(\sum_{i=1}^{k} \phi^i + \tau(t))$. Thus $\bigcap_i F'_* (V \times I)_{x_i,t}$ is spanned by E and $(\Sigma \phi^i + \tau(t))$ and so has codimension $k((p+1) - (n+1))$ and so by the remark at the beginning of 12.5 the crossing at (y, t) is normal.

12.7 PROOF of Theorem 12.4.

Let $f : V \to M$ be an immersion with normal crossings, V compact, and $\dim V = n$ and $\dim M = p$. Let $F : V \times I \to M$ be any homotopy of f . By lemma 6 we may assume that F_t is also an immersion with normal crossing for $t \in I$. As usual we let $F' : V \times I \to M \times I : (x, t) \mapsto (F(x,t),t)$. Using the notation of 12.2 , $F'_* \circ \tau = P_1 \circ F'_* \circ \tau + \tau \circ F'$. Here of course τ has two meanings; in $\tau \circ F'$, $\tau : M \times I \to \widetilde{T}(I)$, whereas in the other two expressions $\tau : V \times I \to \widetilde{T}(I)$ (the two $\widetilde{T}(I)$ used here are also different, the first is the induced bundle over $M \times I$ and the second the induced bundle over $V \times I$). Since the context will make the domain of definition clear we will not distinguish between the "two τ's " .
In order to apply the theorem of 12.3, we must find vector fields $X : V \times I \to \widetilde{T}(V)$ and $Y : M \times I \to \widetilde{T}(M)$ such that

$$(*) \quad P_1 \circ F'_* \circ \tau = F'_* \circ X + Y \circ F' .$$

By lemma 7 we know that F' is again an immersion with normal crossings. Although lemma 5 was proved for immersions with normal crossings with compact domain of definition it is applicable to F'. (In fact, the only place in the proof of lemma 5 where compactness was used was in the first paragraph to show that we could take the sequence z_n convergent. However when we replace G by F' and X by $V \times I$ this sequence will look like $(z_n, t_n) \in V \times I$ where $t_n \to t \in I$. Thus the sequence can be taken convergent to (z, t) and the rest of the proof goes through without change.) Thus at each point (y', t') in $F'(V \times I)$ we can find a neighbourhood , W , of (y', t') such that $F'(V \times I) \cap W$ is a finite union of coordinate linear spaces of dimension $n + 1$, all intersecting at (y', t') . $F'^{-1}(W) = {}_1\cup^k U_i$ where U_i , $i = 1, \ldots, k$, are pairwise disjoint coordinate neighbourhoods of the inverse images of (y', t') . We proceed to construct vector fields, $X : \cup U_i \to \tilde{T}(V)|\cup U_i$ and $Y : W \to \tilde{T}(M)|W$, such that (*) holds when restricted to $\cup U_i$. We may assume that W is small enough so that there is a product coordinate system valid there, say, $y_1, \ldots, y_p, (t - t')$ with centre at (y', t') .

Let $E_i = F'(U_i)$. By lemma 5, E_i is given by the vanishing of $(p - n)$ functions ϕ_i^a , $a = 1, \ldots, p - n$, and $\bigwedge_{i,a} d\phi_i^a \neq 0$ in W . Since dt doesn't annihilate the tangent space to E_i (see proof of lemma 6), we may take W small enough so that $\{\phi_i^a , y_r, t - t'\}$, $i = 1, \ldots, k$; $a = 1, \ldots, p - n$; $r = 1, \ldots, p - k(p - n)$ is a coordinate system in W (we've perhaps had to renumber the y-coordinates). This coordinate system will in general not be a product coordinate system since the ϕ_i^a will not be independent of t .

(Notation : Given a set $A \subset B \times C$, we write $A(c)$ for $A \cap (B \times c)$ for $c \in C$.)

In W , we introduce a Riemann metric by declaring the vectors

$$\left\{\frac{\partial}{\partial(\phi_i^a)_{t_o}} , \frac{\partial}{\partial y_r} , \frac{\partial}{\partial(t - t')}\right\} \quad \text{to be orthonormal at all points of } W(t_o) .$$

These vectors are understood relative to the coordinate system $\{(\phi_i^a)_{t_o} , y_r, t - t'\}$.

<u>Note</u> : This does <u>not</u> give the usual metric in W for the space E_i . However it does induce the usual metric in $W(t)$ for the spaces $E_i(t)$ for each t .

Let $(x, t) \in U_i$, then $P_1 \circ F'_* \circ \tau\,(x, t) \in \tilde{T}(M) \mid W(t)$. The base point of this vector is in $E_i(t)$, thus we can resolve this vector into two components : $A(x, t) + B(x, t)$, where $A(x, t)$ is tangent to E_i (and by our choice of metric tangent to $E_i(t)$) and $B(x, t)$ is normal to $E_i(t)$. Since $F'|U_i$ is a diffeomorphism of U_i onto E_i , and $F'|U_i(t)$ is a diffeomorphism of $U_i(t)$ onto $E_i(t)$, we may write $A(x, t) = F'_* \circ C(x, t)$ and $B(x, t) = N_i \circ F'(x, t)$, where

$C : U_i \to \tilde{T}(V) \mid U_i$ and $N_i : E_i \to \tilde{T}(M) \mid E_i$. Thus ,

$$P_1 \circ F'_* \circ \tau \mid U_i = F'_* \circ C \mid U_i + N_i \circ F' \mid U_i \ .$$

It is unnecessary to index the C since the U_i are disjoint. Together they determine a vector field on $\cup U_i$. However since $\cap E_i \neq \emptyset$, we do not have one well defined vector field on $\cup E_i$. Assuming that W is a cubical neighbourhood relative to the coordinate system $\{\phi_i^a , y_r, (t - t')\}$, we apply lemma 3 to the N_i, slice by slice. That is in $W(t)$, the $E_i(t)$ are coordinate linear spaces in general position with the usual metric in $W(t)$. Thus there is a vector field Y_t on $W(t)$ such that $Y_t | E_i(t) = (N_i)_t + (D_i)_t$, where $(D_i)_t$ is a vector field on $E_i(t)$, tangent to $E_i(t)$. Piecing these vector fields together, we obtain $Y : W \to \tilde{T}(M) \mid W$ such that $Y | E_i = N_i + D_i$ where $Y(y, t) = Y_t(y, t)$ and $D_i(y,t) = (D_i)_t(y, t)$.

For a global definition of Y , we write in exact analog to lemma 3 :

$$Y(y, t) = \Sigma \, T_j(y, t)_* \circ N_j \circ P_j(y, t)$$

where $P_j : W \to E_j$ is the projection defined by means of the coordinates $\{\phi_i^a , y_r, (t - t')\}$ and $T_j(y, t)$ is the translation in R^{p+1} by the point whose coordinates are the difference of the coordinates of (y, t) and $P_j(y, t)$, symbollically $(y, t) - P_j(y, t)$.

Since this translation induces parallel displacement in the tangent space wherever it makes sense - and it always makes sense in the above equation - the argument at the end of the proof of lemma 3 still holds. By the same argument as applied above to A we can write $D_i \circ F' | U_i = F'_* \circ L | U_i$. The various $L \mid U_i$ pieced together yield a vector field $L : \cup U_i \to \tilde{T}(V) \mid U_i$. Thus ,

$$P_1 \circ F'_* \circ \tau = F'_* \circ (C - L) + Y \circ F' \quad \text{on } \cup U_i \ .$$

Thus we have obtained the decomposition (*) locally.

In order to pass from this local decomposition to a global one, we use, of course, a partition of unity. We first cover $M \times I$ with a locally finite covering such that if a neighbourhood of the covering meets $F'(V \times I)$, it has all the properties required in the above construction of the decomposition. This is clearly possible since such neighbourhoods form a base at any point of $F'(V \times I)$. The inverse images of the sets of the covering by F'^{-1} is a locally finite covering of $V \times I$, and the composition with F' of a partition of unity for the covering of $M \times I$ yields a partition of unity for the inverse image covering of $V \times I$ and it is with these two partitions of unity that we patch together Y and the $(C - L)$. By our decomposition construction we have only defined the Y-pieces in neighbourhood that meet $F'(V \times I)$. For the partition of unity construction define Y in any neighbourhood disjoint from $F'(V \times I)$ to vanish.

12.8 As a final application of the foregoing we prove that maps with particularly simple singularities are homotopically stable.

Recall that $J^1(V, M)$ is a bundle over $V \times M$ with fibre $J^1(n, p)$ and group $GL(n, R) \times GL(p, R)$ where $\dim V = n$ and $\dim M = p$. In this paragraph, we assume that $n \geq p$ and that V is compact. $J^1(V, M)$ is the bundle of all 1-jets of maps from V into M. Let $S_i(V, M)$ be the subset (actually sub-bundle) of $J^1(V, M)$ of 1-jets of corank i. (see 8.1) Given a map $f : V \to M$ we define $J^1(f): V \to J^1(V,M)$ and $S_i(f) = (J^1(f))^{-1}(S_i(V, M))$ = the set of points where f has corank i, in direct analog to the definitions in Euclidean space. Corank i means here rank $p - i$ since $n \geq p$.

THEOREM. Let $f : V \to M$ and suppose that $J^1(f)$ is transversal to $S_1(V, M)$ and that $S_i(f) = \emptyset$ for all $i > 1$. If $f|S_1(f)$ is an immersion with normal crossings then f is homotopically stable.

PROOF : (In the following I is to be interpreted as some half open interval, $[0, t)$; the interval will not be constant throughout the argument but will be shrunk a finite number of times. This convention merely saves us the trouble of indexing the I, the exact length of which is of little interest.) Let $F : V \times I \to M$ be a homotopy of f. Since V is compact, we may assume that $J^1(F_t)$ is transversal to $S_i(V, M)$ for all i, and that $S_i(F_t) = \emptyset$ for $i > 1$ and for $t \in I$. Thus $(J^1(F_t))^{-1}(S_1(V,M)) = (J^1(F_t))^{-1}(\overline{S_1(V, M)}) = S_1(F_t)$ is a closed, hence compact, submanifold of V, (see 5.2., Proposition 1). We first prove that $S_1(F') = \bigcup_t (S_1(F_t) \times t)$ is a submanifold of $V \times I$. By restricting the coordinate systems in $V \times I$ and $M \times I$ to be product coordinate systems with the same global coordinate, t, in both intervals I, we reduce the group of $J^1(V \times I, M \times I)$ to $GL(n, R) \times GL(p, R)$. If $v \subset V \times I$ and $m \subset M \times I$ are coordinate neighbourhoods with coordinates $\{x_i, t\}$ and $\{y_j, t\}$ respectively $J^1(V \times I, M \times I) | v \times m$ is isomorphic to $v \times m \times J^1(n+1, p+1)$. The subset of 1-jets of mappings $g \times id : R^n \times R \to R^p \times R$ in $J^1(n+1, p+1)$ is invariant under the action of $GL(n, R) \times GL(p, R)$ and is naturally homeomorphic to $J^1(n, p)$. We call the sub-bundle with fibre $J^1(n, p)$ and group $GL(n, R) \times GL(p, R)$, B. B can also be described in another way. Let $q : M \times I \to M$ be the projection. If $(x, t, y, t') \in V \times I \times M \times I$ then the points of B over this point are the 1-jets of all $q \circ G_t$ at (x, t) where $G : V \times I \to M \times I$ and $G(x, t) = (y, t')$. If we let $\mathcal{J}^1(V \times I, M \times I)$ be the bundle obtained from $J^1(V \times I, M \times I)$ by the above reduction of the group, we have the natural projection $Q : \mathcal{J}^1(V \times I, M \times I) \to B$, a fibre map. Define $b(F') = Q \circ J^1(F')$, and $s_1 \subset B$ to be the sub-bundle, which over $v \times m$ is $v \times m \times S_1(n, p)$ via the isomorphism between $B | v \times m$ and $v \times m \times J^1(n, p)$. Clearly, $S_1(F') = (b(F'))^{-1}(s_1)$. We show that $b(F')$ is transversal to s_1. Since this is a purely local problem we may restrict our attention to $b(F')|v$; we assume that $F'(v) \subset m$. Let $(x_o, t_o) \in v \cap S_1(F')$, and suppose $b(F')(x_o, t_o) = (x_o, t_o, F'(x_o, t_o), (F_{t_o})^{(1)}(x_o))$; here we've used the local representation of $B | v \times m$. If in a neighbourhood, W, of $(F_{t_o})^{(1)}(x_o) \in J^1(n,p)$, $S_1(n, p)$ is the inverse image of the origin by a map $h : W \to R^{n-p+1}$, we see that $h^j \circ (F_t)^{(1)}(x) = 0$, $j = 1, \dots, n-p+1$, are the defining equations for $v \cap S_1(F')$ in v. Since for each fixed t these are the defining equations for $S_1(F_t)$ in $v \cap (V \times t)$, the system of equations has maximal rank at (x_o, t_o). Thus $b(F')$ is transversal to s_1, and thus $S_1(F')$ is a submanifold in $V \times I$. The functions $\{h^j \circ (F_t)^1\}$ can be completed to a coordinate system in a neighbourhood, which we again call v, of (x_o, t_o), say $\{h^j \circ (F_t)^1, x_k, t\}$,

$(j = 1,\ldots, n-p+1\ , k = 1,\ldots, p-1)$.

Note : The functions $h^j \circ (F_t)^1$ are not considered for fixed t so the above coordinate system is not a product coordinate system. Relative to this coordinate system $\frac{\partial}{\partial t}$ defines a vector field in v , which when restricted to $S_1(F') \cap v$ is a tangent vector field to $S_1(F')$. Relative to the product coordinate system in v , $\{x_1,\ldots, x_n, t\}$ this vector field can be written as $\tau + X$, where $X : v \to \tilde{T}(V) \mid v$. Using a partition of unity we can patch together these local vector fields to a vector field on $V \times I$ which vanishes outside a neighbourhood of $S_1(F')$. We denote this globally defined vector field again by X . Note that $\tau + X : S_1(F') \to T(S_1(F'))$. We now apply lemma (12.2) to the vector field X, which we extend to vanish outside a compact set in $V \times R$ as in the proof of theorem (12.3). We obtain a map $\phi : V \times I \to V$ such that for each $t \in I$, ϕ_t is a diffeomorphism. By the uniqueness of the local 1-parameter group which is the 'integral' of the vector field $\tau + X$ on $S_1(F')$, and by our construction of ϕ , we see that $\phi_t | S_1(f) : S_1(f) \to S_1(F_t)$ is a diffeomorphism. Thus $S_1(F' \circ \phi') = S_1(f) \times I$. Let $H = F \circ \phi'$. H is again a homotopy of f , and $S_1(H_t) = S_1(f)$ for all $t \in I$, and $H_t | S_1(f)$ is an immersion with normal crossings. Applying theorem (12.4) to $H \mid S_1(f) \times I$ we obtain vector fields $X_1 : S_1(f) \times I \to \tilde{T}(S_1(f))$ and $Y_1 : M \times I \to \tilde{T}(M)$ such that $H'_* \circ \tau = H'_* \circ X_1 + Y_1 \circ H' + \tau \circ H'$. Extending the vector field X_1 to all of $V \times I$, we obtain, applying theorem (12.3), mappings $\phi_1 : V \times I \to V$ and $\psi_1 : M \times I \to M$ such that the $(\phi_1)_t$ and $(\psi_1)_t$ are diffeo - morphisms and such that $(\psi'_1)^{-1} \circ H' \circ (\phi'_1)^{-1} \mid S_1(f) \times I$ is equal to $(f \times id_I) \mid S_1(f) \times I$. Let $K' = (\psi'_1)^{-1} \circ H' \circ (\phi'_1)^{-1}$. $S_1(K_t) = S_1(f)$ and $K_t(S_1(K_1)) = f(S_1(f))$ for all $t \in I$, and K is again a homotopy of f. Using now the same procedure as in corollary (12.3), we can find a vector field, $W : (V - S_1(f)) \times I \to \tilde{T}(V)$ which we extend to $V \times I$ by defining W to vanish on $S_1(f) \times I$. By simple local considerations of the coefficients of $K'_* \circ \tau$, it is clear that the vector field W so defined is differentiable and that $P_1 \circ K'_* \circ \tau = K'_* \circ W$. Applying theorem (12.3) once more we obtain a family of diffeomorphisms , $\phi_2 : V \times I \to V$ such that $K' = f \times id_I \circ \phi_2'$, and so $F' = \psi_1' \circ (f \times id_I) \circ \phi_2' \circ \phi_1' \circ (\phi')^{-1}$. Thus f is homotopically stable.

REFERENCES

[1] Morse, M. The Calculus of Variations in the Large, A.M.S. Colloquium Publications, vol. XVIII, (1934) p.155 Theorem 4.2.

[2] Nomizu, K. Lie Groups and Differential Geometry, Publications of the Math. Soc. of Japan , 2 .

[3] Whitney, H. The general type of singularity of a set of 2n - 1 smooth functions of n variables, Duke Math. J. 10 (1943), pp. 161-172.

[4] Whitney, H. On singularities of mappings of Euclidean spaces, I, Mappings of the plane into the plane, Ann. of Math. 62 (1955) pp. 374-410 .

[5] Whitney, H. Singularities of Mappings of Euclidean Spaces, Symposium Internacional de Topologia Algebrica, Mexico 1958, pp. 285-301.

INTRODUCTION TO THE PREPARATION THEOREM

C.T.C. Wall

1. The complex analytic case

In about 1880, Weierstrass proved the following result [11]. Let

$$G(t, z) = G(t, z_1, \ldots, z_k)$$

be a power series in $(k + 1)$ complex variables, converging near 0, with $G(t, 0) \neq 0$. Let $G(t, 0)$ be of order s, i.e. s is the least integer for which $\partial^s G/\partial t^s$ is nonzero at 0. We call such a function G regular of order s.

Theorem. If $G(t, z)$ is regular of order s, then

$$G(t, z) = (t^s + H_1(z)\, t^{s-1} + \ldots + H_s(z))\, U(t, z) ,$$

where U and the H_i are also power series converging near 0. Moreover, U and the H_i are unique, and $U(0, 0) \neq 0$.

This result he called the preparation theorem (Vorbereitungssatz) because G is 'prepared' for the further study of the variety of its zeros. The result can be used inductively to reduce questions about convergent power series (in the complex domain) to ones about polynomials. For example, the ring of convergent power series in $z_1, \ldots, z_k$ is a noetherian unique factorization ring [8]. Note that for $s = 0$ the result is trivial $(U = G)$; for $s = 1$ we have $\partial G/\partial t \neq 0$, and it follows from the implicit function theorem.

This theorem was generalised by Späth [9] as follows. With G regular of order s, and $F = F(t, z)$ any power series converging near 0, we have

$$F = GQ + R ,$$

where $Q(t, z)$ is analytic at 0 and $R = \Sigma_1^s A_i(z) t^{s-i}$, with the A_i convergent power series in the z's. This is close to the division algorithm, so is called a division theorem. The result above follows on taking $F(t, z) = t^s$ (one can then take $H_i = -A_i$ and $U = 1/Q$: that $Q(0, 0) \neq 0$ follows on comparing coefficients of t^s). The result we call <u>the</u> division theorem is apparently more special.

Write $P_s(t, \lambda)$ for the generic monic polynomial in t of degree s,

$$P_s(t, \lambda) = t^s + \Sigma_{i=1}^{s} \lambda_i t^{s-i} .$$

<u>Division theorem</u> <u>For any</u> $F(t, z, \lambda)$ <u>analytic near</u> 0, <u>we have</u>

$$F(t, z, \lambda) = P_s(t, \lambda)\, Q(t, z, \lambda) + R(t, z, \lambda) ,$$

<u>with</u> $R(t, z, \lambda) = \Sigma_{i=1}^{s} A_i(z, \lambda)\, t^{s-i}$ <u>and</u> A_i, Q <u>analytic near</u> 0 . <u>Moreover,</u> Q <u>and the</u> A_i <u>are unique</u>.

To deduce the Weierstrass theorem, we let $z = 0$. If G is regular of order s - say $G(t, 0, 0) \sim c\, t^s$ - we easily find that

$$\frac{\partial A_i}{\partial \lambda_j}(0,0) = \begin{cases} 0 & \text{if } j < i \\ c & \text{if } j = i , \end{cases}$$

so the Jacobian is nonzero. By the implicit function theorem there are analytic functions $\lambda_j = H_j(z)$ near 0 such that $A_i(z, H_j(z)) = 0$ for all i. Thus

$$G(t, z) = P_s(t, H(z))\, Q(t, z, H(z))$$

which is in the required form since, as is easily seen, Q takes the value c at the origin.

As to the more general division theorem, we now simply write $F = P_s Q' + R$, with P_s as above, substitute $\lambda = H(z)$ and deduce $F = GQ'' + R$ with $Q'' = Q'/Q$.

The division theorem can be reduced to the case where F is independent of λ by taking the λ_i as extra parameters z, but this does not simplify the proof. The argument below (copied from [6]) is related to Weierstrass' original proof. For other proofs see the books [0],[1]: the former gives also a proof for the ring of formal power series.

<u>Proof</u>. Adding $P_s(t, \lambda)$ to both sides of the identity

$$P_s(u, \lambda) - P_s(t, \lambda) = (\Sigma_{i=1}^{s} P_{i-1}(u, \lambda)\, t^{s-i})\,(u - t) ,$$

and dividing through by $P_s(u, \lambda)(u - t)$, we obtain

$$\frac{1}{u-t} = \frac{P_s(t,\lambda)}{P_s(u,\lambda)(u-t)} + \sum_{i=1}^{s} \frac{P_{i-1}(u,\lambda)}{P_s(u,\lambda)} t^{s-i} .$$

Any simple loop γ in the u-plane containing the origin once will also contain t (for t near 0) and the roots of P_s (for λ near 0) in its interior. Thus by Cauchy's integral formula,

$$F(t, z, \lambda) = \frac{1}{2\pi i} \oint_\gamma \frac{F(u, z, \lambda)\, du}{u-t} .$$

Substituting from above for $(u-t)^{-1}$, we get an equation of the desired type, with

$$Q(t, z, \lambda) = \frac{1}{2\pi i} \oint_\gamma \frac{F(u, z, \lambda)\, du}{P_s(u, \lambda)(u-t)}$$

and
$$A_i(z, \lambda) = \frac{1}{2\pi i} \oint_\gamma \frac{P_{i-1}(u,\lambda) F(u, z, \lambda)}{P_s(u, \lambda)}\, du .$$

It is clear that these are analytic. Uniqueness is clear formally, keeping z and λ fixed.

The above argument actually proves a global result, not just a local one. Note, however, that though the corresponding local theorem in the real analytic case follows trivially, the global theorem here is false (see [6]): the complex singularities of F and zeros of P_s will then play a rôle.

2. The C^∞ case

It was observed by Thom about 1960 (see [2]) that if an analogous theorem were true for C^∞-maps, it would have important applications in singularity theory, giving as a start simple proofs of some canonical forms due to Whitney. Under pressure from Thom, such a theorem was proved by Malgrange in 1962 (see [3], [4], [5]). The rather technical proof involved a detailed study of the local geometric structure of analytic sets. A stronger result was obtained by a direct analytic argument by Mather about 1966 (see [6]) . Four further proofs (the sum total to date) appear below : in distinction to the earlier ones, they are inspired more by the complex variable proof above.

Here is how the actual results are related (as pointed out by Mather in a lecture expounding his paper below). The reader may amuse himself constructing a concordance between the different notations. Another trivial distinction is whether the functions take real or complex values. Of more substance is the question whether the variables are regarded as real or complex. The earlier proofs, and that of Lojasiewicz below, take λ as real: for the papers of Nirenberg, Mather and Glaeser below, they can be complex.

However, the real distinction in depth is between global and local forms of the theorem: the original proof of Malgrange, and that of Nirenberg below are only local. (But this does simplify the proof: Nirenberg's is the shortest known and the local version suffices for all known applications except the result that infinitesimal stability implies (globally) stability, for proper maps.) A concomitant difference is that in the global statements Q (and R) are chosen to depend on F by means of a continuous linear map. This is a non-trivial addition, since in the C^∞-case the division is not unique (as is easily seen). Watch this point in reading the papers below, and you will better understand the need for some of the devices employed.

In fact, some form of Whitney's extension theorem is used in each case. But for the full theorem, continuous linear maps do not exist. This point is discussed in the final section of Glaeser's paper.

3. Algebraic Formulation

As in the complex analytic case, the division theorem (local form) for C^∞-functions implies a preparation theorem of Weierstrass type, and hence a more general division theorem. This can be stated in a more algebraic way. Let $\mathcal{E}_{k+1}$ be the algebra of C^∞-functions of $(k+1)$ variables, G such a function which is regular of order s . Consider the quotient $\mathcal{E}_{k+1}/\langle G\rangle$ of $\mathcal{E}_{k+1}$ by the ideal generated by G. The theorem amounts to saying that this, as $\mathcal{E}_k$-module, is generated by the classes of the functions t^i ($0 \leq i < s$). We now give the algebraic form of the theorem, which I will quote below [10].

Malgrange preparation theorem Let $f : (N, x) \to (P, y)$ be a C^∞ map-germ; A a finitely generated $C_x^\infty(N)$ - module. If $\dim_{\mathbb{R}} (A/f^*\mathfrak{m}_y \, . \, A) < \infty$, then A is finitely generated as $C_y^\infty(P)$ - module.

Here, the module structure is induced by f^*. Malgrange's own statement refers to "differentiable algebras" which are quotients of the $C_x^\infty(N)$: for the above slightly more general version I am following Mather [7].

Proof of Malgrange's theorem Choose a local chart $\phi : (N, x) \to (\mathbb{R}^n, 0)$ and factorise f as the composite

$$(N, x) \xrightarrow{(f,\phi)} (P \times \mathbb{R}^n, (y, 0)) \xrightarrow{\pi_n} \dots \to (P \times \mathbb{R}, (y, 0)) \xrightarrow{\pi_1} (P, y) ,$$

where π_i is the projection defined by forgetting the last coordinate. Since (f, ϕ) is an embedding, $C^\infty_{(y,0)}(P \times \mathbb{R}^n)$ maps onto $C_x^\infty(N)$, so A is certainly finitely generated over it. The result will thus follow by induction if we can prove it with f replaced by π_i - or, changing notation, if we can prove it for $\pi_1 = \pi$.

Let $\alpha_1, \dots, \alpha_r \in A$ be a set of $C^\infty_{(y,0)}(P \times \mathbb{R})$ - generators, whose images in $A/\pi^*\mathfrak{m}_y \, . A$ span this $\mathbb{R}$ - vector space. Then any $x \in A$ can be written in the form

$$x = \sum_{i=1}^{r} (c_i + z_i) \, \alpha_i ,$$

where $c_i \in \mathbb{R}$, $z_i \in \pi^*\mathfrak{m}_y \, . \, C^\infty_{(y,0)}(P \times \mathbb{R})$. In particular if t is the function defined by projection on $\mathbb{R}$,

$$t\alpha_i = \sum_{j=1}^{r} (c_{ij} + z_{ij}) \, \alpha_j \qquad (1 \leq i \leq r) .$$

Let Δ be the determinant $|t\delta_{ij} - c_{ij} - z_{ij}|$. Since on $y \times \mathbb{R}$ we have $z_{ij} = 0$, so this is a monic polynomial in t, Δ is regular of some order $s \leq r$.

By the division theorem above, then, $C^{\infty}_{(y,0)}(P \times \mathbb{R}) / \langle\Delta\rangle$ is generated over $C^{\infty}_{y}(P)$ by the $t^i (0 \leq i < s)$.

But by Cramer's rule, $\Delta \alpha_i = 0$ for each i, so A is a module over $C^{\infty}_{(y,0)}(P \times \mathbb{R}) / \langle\Delta\rangle$, and it is generated over it by the α_i. Hence the $\alpha_i t^j (1 \leq i \leq r,\ 0 < j < s)$ generate A over $C^{\infty}_{y}(P)$.

REFERENCES

0. S. Bochner and W.T. Martin, Several complex variables. Princeton, 1948.
1. R.C. Gunning and H. Rossi, Analytic functions of several complex variables, Prentice-Hall, 1965.
2. H.I. Levine, Singularities of differentiable mappings. This volume, pp. 1 - 89.
3. B. Malgrange, Le théorème de préparation en géometrie différentiable. Sém. H. Cartan 15 (1962-63) exps 11, 12, 13, 22.
4. B. Malgrange, The preparation theorem for differentiable functions. In "Differential Analysis (papers presented at Bombay Colloquium, 1964)". Oxford, 1964, 203-208.
5. B. Malgrange, Ideals of differentiable functions. Oxford, 1966.
6. J.N. Mather, Stability of C^{∞} mappings I. The division theorem. Ann. of Math. 87 (1968) 89-104.
7. J.N. Mather, Stability of C^{∞} mappings III. Finitely determined map-germs. Publ.Math. I.H.E.S. 35 (1968) 127-156.
8. W. Rückert, Zum Eliminationsproblem der Potenzreihenideale. Math. Ann. 107 (1933) 259-281.
9. H. Späth, Der Weierstraßsche Vorbereitungssatz. Crelle's Jour. 161 (1929) 95-100.

10. C.T.C. Wall, Introduction to C^{∞} stability and classification. This volume, pp. 178 - 206.

11. K. Weierstrass, Vorbereitungssatz. Werke, vol. 2, 135- .

A PROOF OF THE MALGRANGE PREPARATION THEOREM

L. Nirenberg*

Malgrange's preparation theorem [3], [4] is an extension to real C^∞ functions of the Weirstrass preparation theorem. In this paper we give a short, rather direct proof of the theorem which applies also to complex functions:†

Theorem 1. *Let* $f(t, x)$ *be a* C^∞ *complex-valued function defined in a neighbourhood of the origin in* R^{n+1}, *here* $t \in R^1$, $x \in R^n$, *and let* $p > 0$ *be the first integer such that*

$$(1) \qquad (\frac{d}{dt})^p f(0, 0) \neq 0 .$$

Then in a neighbourhood of the origin one has the factorization

$$(2) \qquad f(t, x) = Q(t, x)P ,$$

where

$$(3) \qquad P = t^p + \sum_1^p \lambda_j(x) t^{p-j}$$

and Q *and* λ_j *are* C^∞ *complex-valued functions with* $Q(0, 0) \neq 0$.
If f *is real there is such a factorization with* Q *and* P *real* .

---

*This paper is a slightly expanded version of a letter written to B. Malgrange in November 1969. The work was supported by the U.S. Air Force Contract AF-49(638), 1719.

†Malgrange informed me that the complex case can be derived from the results in [4]. In addition, G.I. Eskin, using the results of [4], observed independently that for each positive integer k the complex function f admits the factorization (2) with Q and λ_j of class C^k. See also the papers in this volume by S. Lojasiewicz and G. Glaeser.

For $p = 1$ the theorem states that if $f(0, 0) = 0$, $\frac{\partial f}{\partial t}(0, 0) \neq 0$ then one has the C^∞ factorization

(4) $$f(t, x) = Q(t, x)(t + \lambda_1(x)) , \quad Q(0, 0) \neq 0 .$$

If f is real this follows easily from the implicit function theorem, and the factorization is then unique. However for complex f this simple looking result seems to be nontrivial. Furthermore, the factorization is not unique, for the function $f = t + i(x + e^{-1/x^2})$, here $n = 1$, admits the factorization $f = 1 \cdot f$ and $f = Q(t, x)(t + ix)$ - one easily sees that Q belongs to C^∞, and $Q(0, 0) = 1$.

The preparation theorem is based on the Malgrange division theorem which we state for complex functions :

Theorem 2.† Let $f(t, x)$ be a C^∞ complex-valued function in a neighbourhood of the origin in R^{n+1} and let

(5) $$P(t, \lambda) = t^p + \sum_1^p \lambda_j t^{p-j} , \quad \lambda = (\lambda_1, \dots, \lambda_p)$$

be a generic monic polynomial in t with constant complex coefficients λ_j. Then in a neighbourhood of the origin one has division by $P(t, \lambda)$:

(6) $$f(t, x) = q(t, x, \lambda) P(t, \lambda) + r(t, x, \lambda) ,$$

where q and r are complex-valued C^∞ functions in a neighbourhood of the origin in $R^{n+1} \times \mathbb{C}^p$, and r is a polynomial in t of degree less than p. Again if f and the λ_j are real one can obtain this with q and r real.

---

†The functions q and r are not unique. In the real case a stronger form of Theorem 2 was proved by J. Mather [5]. He showed that for every f one can find q and r in (6) in such a way as to depend linearly on f. In addition he established a global form of the theorem. Our proof does not yield q and r depending linearly on f. In a paper in this volume Mather shows how to modify our proof (in particular he presents an improved version of our Lemma 1) so as to obtain q and r depending linearly on f.

Proof of Theorem 1 : This follows [3] . Let f be as in Theorem 1 and consider the division (6) by the generic polynomial $P(t, \lambda)$ of (5) . The remainder r has the form

$$r = \sum_{1}^{p} r_j(x, \lambda)\, t^{p-j} .$$

We shall solve the p complex equations

$$(7) \qquad r_j(x, \lambda) = 0 , \quad j = 1,\dots, p$$

for $\lambda = \lambda(x)$, regarding this system as $2p$ real equations for $2p$ real unknown $\operatorname{Re} \lambda_j$, $\operatorname{Im} \lambda_j$, $j = 1 ,\dots, p$. We then obtain the desired factorization

$$f(t, x) = q(t, x, \lambda(x))\, P(t, \lambda(x)) .$$

From the definition of p we see that $r_j(0, 0) = 0,\ j = 1,\dots, p$ and $q(0,0,0) \neq 0$ hence to solve (7) it suffices to show that the Jacobian matrix $\dfrac{\partial(\operatorname{Re} r_j, \operatorname{Im} r_j)}{\partial(\operatorname{Re} \lambda_k, \operatorname{Im} \lambda_k)}$ is nonsingular at $x = 0$, $\lambda = 0$. In fact we shall show that the (complex) $p \times p$ matrix $\dfrac{\partial r_j}{\partial \lambda_k}(0, 0)$ is nonsingular while $\dfrac{\partial r_j}{\partial \bar{\lambda}_k}(0,0) = 0$, from which the desired result follows easily. Here $\dfrac{\partial}{\partial \lambda_k} = \dfrac{1}{2}\left(\dfrac{\partial}{\partial \mu_k} - i \dfrac{\partial}{\partial \nu_k}\right)$, $\dfrac{\partial}{\partial \bar{\lambda}_k} = \dfrac{1}{2}\left(\dfrac{\partial}{\partial \mu_k} + i \dfrac{\partial}{\partial \nu_k}\right)$, for $\lambda_k = \mu_k + i\nu_k$. Applying $\partial/\partial\bar{\lambda}_k$ to (6) we find

$$\sum_{j=1}^{p} \frac{\partial r_j}{\partial \bar{\lambda}_k}(0,0)\, t^{p-j} = - t^p \frac{\partial}{\partial \bar{\lambda}_k} q(t, 0, 0) , \quad k = 1,\dots, p ,$$

for all t , from which it follows that $\dfrac{\partial r_j}{\partial \bar{\lambda}_k}(0, 0) = 0$. Applying $\partial/\partial\lambda_k$ to (6) we find

$$q(t, 0, 0)t^{p-k} + t^p \frac{\partial}{\partial \lambda_k} q(t, 0, 0) + \sum_{j=1}^{p} \frac{\partial r_j}{\partial \lambda_k}(0,0)\, t^{p-j} = 0 ,$$

$$k = 1,\dots, p .$$

Suppose that for some complex vector $\xi = (\xi_1,\dots, \xi_p)$,

$$\sum_k \frac{\partial r_j}{\partial \lambda_k}(0,0)\,\xi_k = 0 \quad , \quad j = 1,\dots, p \;;$$

then we find

$$q(t,\,0,\,0) \sum_1^p t^{p-k}\ \xi_k = O(t^p) \ , \quad \text{as} \quad t \to 0 \ .$$

Since $q(0,0,0) \neq 0$ this implies that $\sum_1^p t^{p-k}\,\xi_k \equiv 0$ for all t which means $\xi \equiv 0$. Hence $\partial r_j/\partial\lambda_k$ $(0,\,0)$ is nonsingular. Q.e.d.

Our proof of Theorem 2 is an adaptation of one of the well known proofs of the Weierstrass Division Theorem (Theorem 2 in the analytic case), which we repeat here. If f is analytic we have, by Cauchy's integral theorem

(8)
$$f(t,\,x) = \frac{1}{2\pi i} \oint_{\partial D} \frac{f(z,x)}{z-t}\, dz$$

where D is a disc about the origin in the complex z plane containing t , and all the roots of $P(z,\,\lambda)$ for $|\lambda|$ small, in its interior. The function

$$R(t,\,z,\,\lambda) = \frac{P(z,\,\lambda) - P(t,\,\lambda)}{z - t}$$

is analytic, and a polynomial in t and z of degree less than p . Replacing $1/(z-t)$ in the proceding integral by

(9)
$$\frac{1}{z-t} = \frac{P(t,\,\lambda)}{(z-t)\,P(z,\lambda)} + \frac{R(t,\,z,\,\lambda)}{P(z,\lambda)} \quad ,$$

we find

$$f(t,x) = P(t,\lambda)\,\frac{1}{2\pi i} \int_{\partial D} \frac{f(z,x)}{(z-t)P(z,\lambda)}\, dz + \frac{1}{2\pi i} \int_{\partial D} \frac{f(z,\,x)}{P(z,\,\lambda)}\, R(t,\,z,\,\lambda)\, dz \ .$$

This is the desired form (6) with

$$q = \frac{1}{2\pi i} \int_{\partial D} \frac{f(z, x)}{(z - t) P(z, \lambda)} dz , \quad r = \frac{1}{2\pi i} \int_{\partial D} \frac{f(z, x)}{P(z, \lambda)} R(t, z, \lambda) dz .$$

This proof was also used by Mather in [5] .

We shall make use of the following extension lemma.

Lemma 1.† <u>Let</u> $f(t, x)$ <u>be a</u> C^∞ <u>complex-valued function in a neighbourhood of the origin in</u> R^{n+1}. <u>There exists a</u> C^∞ <u>complex function</u> $F(z, x, \lambda)$ <u>defined in a neighbourhood of the origin for</u> $z \in \mathbb{C}^1$, $x \in R^n$, $\lambda \in \mathbb{C}^p$, <u>satisfying</u>

(i) $F(t, x, \lambda) \equiv f(t, x)$ <u>for</u> t <u>real</u>

(ii) $F_{\bar z}$ <u>vanishes of infinite order on</u> $\text{Im } z = 0$ <u>and on</u> $P(z,\lambda) = 0$.

<u>Proof of Theorem 2 :</u> We wish to apply the analogue of (8) to our extension F . This is the well known generalization of the Cauchy integral theorem for a smooth , but non-holomorphic function of a complex variable z (see for instance Theorem 1.2.1. in [1]):

$$f(t, x) = F(t, x, \lambda) = \frac{1}{2\pi i} \oint_{\partial D} \frac{F(z,x,\lambda)}{z - t} dz + \frac{1}{2\pi i} \iint_D F_{\bar z} (z,x,\lambda) \frac{dz \wedge d\bar z}{z - t} ,$$

where the area of integration is a disc D about the origin containing in its interior the point t and all roots z of $P(z, \lambda)$ for $|\lambda|$ small. Substituting (9) for $1/(z-t)$ in these integrals we obtain (6) with :

$$q = \frac{1}{2\pi i} \int_{\partial D} \frac{F(z,x,\lambda)}{(z-t)P(z,\lambda)} dz + \frac{1}{2\pi i} \iint_D \frac{F_{\bar z}(z,x,\lambda)}{(z-t)P(z,\lambda)} dz \wedge d\bar z ,$$

$$r = \frac{1}{2\pi i} \int_{\partial D} F(z,x,\lambda) \frac{R(t,z,\lambda)}{P(z,\lambda)} dz + \frac{1}{2\pi i} \iint_D F_{\bar z} (z,x,\lambda) \frac{R(t,z,\lambda)}{P(z,\lambda)} dz \wedge d\bar z .$$

---

† In his paper in this issue Mather proves a global version of this and obtains F depending linearly on f ; the F constructed here depends on f in a nonlinear way.

The denominators in the double integrals are zero when $z - t$ or $P(z, \lambda)$ vanishes - but the numerators vanish of infinite order there, by Lemma 1. Hence those integrals converge absolutely. Furthermore one may differentiate under the integral signs with respect to t, x, and λ ; the resulting integrals are all absolutely convergent. Thus q and r are of class C^∞ and Theorem 2 is proved for the complex case. In case f is real we may use $\frac{1}{2}(F(z,x,\lambda) + F(\bar{z},x,\lambda))$ in place of F in the preceding; it is then readily verified that q and r are real for λ real.

We remark that Hörmander [2], in quite a different connection, has used the device of extending smooth functions of real variables to complex variables as functions satisfying the Cauchy-Riemann equations to infinite order on the real subvariety.

In proving Lemma 1 we shall make use of a special, simple case of the Whitney extension theorem for which we include a proof :

Lemma 2. Let $u(y)$, $v(y)$ be C^∞ complex-valued functions in a neighbourhood of the origin in R^N satisfying (here D^α represents any derivative of order $|\alpha|$)

$$D^\alpha(u - v) = 0 \text{ on } y_1 = \dots = y_k = 0 \text{ for all } \alpha .$$

Then there is a C^∞ function $F(y)$ satisfying

$$D^\alpha(F - u) = 0 \text{ on } y_1 = \dots = y_j = 0 \text{ for all } \alpha ,$$

$$D^\alpha(F - v) = 0 \text{ on } y_{j+1} = \dots = y_k = 0 \text{ for all } \alpha .$$

Proof : The function

$$F(y) = v(y) + \sum \frac{y^\alpha}{\alpha!} D^\alpha(u-v)(0,\dots,0,y_{j+1},\dots,y_N) \cdot \phi(t_{|\alpha|} \sum_1^j y_i^2) ,$$

where the sum is over $\alpha = (\alpha_1,\dots, \alpha_j, 0,\dots, 0)$, has the desired property, where $\phi(s)$ is a C^∞ function on $s \geq 0$ which is 1 on $0 \leq s \leq 1/2$ and zero for $s > 1$, and where $t_j \nearrow \infty$ is a rapidly increasing sequence tending to infinity. The choice of the t_j depends on the functions u and v , and the resulting F

therefore does not depend linearly on u, v.

Proof of Lemma 1 : Induction on p ; Lemma 1 is true for $p = 0$. Assuming it has been proved for $p-1$ we wish to prove it for p. First we make a change of variable in (z, λ) space to straighten out the variety $P(z, \lambda) = 0$. Leave z and $\lambda' = (\lambda_1, \ldots, \lambda_{p-1})$ alone and introduce a new variable μ to replace λ_p :

$$\mu = P(z, \lambda) .$$

Then in (z, λ', μ) space the variety $P = 0$ becomes $\mu = 0$, and the operator $\partial/\partial\bar{z}$ now becomes, in the new coordinates,

(10) $$L = \frac{\partial}{\partial\bar{z}} + \overline{P'(z, \lambda)} \frac{\partial}{\partial\bar{\mu}} ,$$

where P' represents the derivative of P with respect to z - note that P' is independent of λ_p.

We shall construct two C^∞ functions $v(z, x, \lambda')$ and $u(z, x, \lambda', \mu)$ such that

(11) $v(t, x, \lambda') = f(t, x)$ for t real ;

(12) $v_{\bar{z}}$ vanishes of infinite order on $\operatorname{Im} z = 0$;

and

(13) $u = v$ on $\mu = 0$

(14) Lu vanishes of infinite order on $\mu = 0$

(15) $D^\alpha(u - v) = 0$ on $\operatorname{Im} z = \operatorname{Re} \mu = \operatorname{Im} \mu = 0$ for all α.

Using Lemma 2 we can then put these functions together to obtain the desired function $F(z, x, \lambda)$.

The induction hypothesis is used to construct v, namely $v(z, x, \lambda')$ is a C^∞ function satisfying (11), (12) and also satisfying :

(16) $v_{\bar{z}}$ vanishes of infinite order on

$$P'(z,\lambda') = p\, z^{p-1} + (p-1)\lambda_1 z^{p-2} + \ldots + \lambda_{p-1} = 0 .$$

Then we set

$$u = \sum_{0}^{\infty} \left(-\frac{1}{\overline{P'}} \frac{\partial}{\partial \bar{z}}\right)^j v(z,\lambda') \frac{\bar{\mu}^j}{j!} \phi(|\mu|^2 t_j)$$

where $t_j \nearrow \infty$ is rapidly increasing (depending on the function v) .

Because $v_{\bar{z}}$ vanishes of infinite order on the set where $P' = 0$ one sees that for t_j rapidly increasing the function u belongs to C^∞ and satisfies (13).

Next

$$Lu = \overline{P'} \left(\frac{1}{\overline{P'}} \frac{\partial}{\partial \bar{z}} + \frac{\partial}{\partial \bar{\mu}}\right) u$$

$$= P' \sum_{0}^{\infty} \left(-\frac{1}{\overline{P'}} \frac{\partial}{\partial \bar{z}}\right)^{j+1} v(z, \lambda') \frac{\bar{\mu}^j}{j!} [\phi(|\mu|^2 t_{j+1}) - \phi(|\mu|^2 t_j)]$$

$$+ P' \sum_{0}^{\infty} \left(-\frac{1}{\overline{P'}} \frac{\partial}{\partial \bar{z}}\right)^j v(z, \lambda') \frac{\bar{\mu}^j}{j!} t_j \mu \phi'(|\mu|^2 t_j) ;$$

and one sees that Lu vanishes of infinite order on $\mu = 0$, i.e. (14) holds. Finally, to verify (15) we see from (12) that

$$u - v(z, \lambda') \phi(|\mu|^2 t_0)$$

vanishes of infinite order on $\operatorname{Im} z = 0$. Since $v(z,\lambda')(\phi(|\mu|^2 t_0 - 1)$ vanishes of infinite order on $\mu = 0$, we infer that (15) holds.

REFERENCES

[1] L. Hörmander, An Introduction to Complex Analysis in Several Variables. D. Van Nostrand Co., Inc., Princeton, 1966

[2] L. Hörmander, On the singularities of solutions of partial differenti-al equations. Conf. on Functional Analysis and Related Topics, Tokyo, 1969, 31-40.

[3] B. Malgrange, The preparation theorem for differentiable functions. Differential Analysis, Oxford Univ. Press, 1964, 203-208.

[4] B. Malgrange, Ideals of differentiable functions, Oxford Univ.Press, 1966.

[5] J.N. Mather, Stability of C^{∞} mappings : I. The division theorem, Annals of Math. 87, 1968, 89-104.

WHITNEY FIELDS AND THE MALGRANGE-MATHER PREPARATION THEOREM

S. Łojasiewicz

The purpose of this note is to give a proof of the Malgrange-Mather preparation theorem using some elementary properties of Whitney fields, especially symmetric ones.

1. Preliminaries

A Taylor field is a map $A : E \ni x \to A_x = \Sigma a_p(x) X^p \in \mathbb{C}[[X]], (X^p = X_1^{p_1} \dots X_n^{p_n})$, where $E \subset \mathbb{C}^n$, the a_p being called coefficients of the field. It is called a regular or Whitney field iff each of its derivatives $B = D^p A : x \to D^p A_x$ satisfies the following Whitney regularity condition :

$$r_B^k(x, x') = b_0(x') - B_x^{(k)}(x' - x) = o(|x' - x|^k)$$

as $x', x \to a$, for any $k \in \mathbb{N}$, $a \in E$ (one puts $(\Sigma c_p X^p)^{(k)} = \sum_{|p| \leq k} c_p X^p$). We denote the set of all Whitney fields on E by $J(E)$.

For any $\phi \in C^\infty(\Omega)$ = the space of complex valued C^∞ functions on Ω open in $\mathbb{R}^n \subset \mathbb{C}^n$, the field $[\phi] : x \to \Sigma \frac{1}{p!} D^p \phi(x) X^p$ is regular, and $C^\infty(\Omega) \ni \phi \to [\phi] \in J(\Omega)$ is an isomorphism of topological algebras with derivations. The same holds for $\phi \in \mathcal{H}(\Omega)$ = the space of holomorphic functions on Ω open in $\mathbb{C}^n$; observe that $[\phi]_{\mathbb{R}} = [\phi_{\mathbb{R}}]$.

Given a C^∞ map $\Psi : \Omega \to \mathbb{R}^n$, where Ω is open in $\mathbb{R}^m \subset \mathbb{C}^m$, and a field A on $E \subset \mathbb{C}^n$ we define $A \circ \Psi : \psi^{-1}(E) \ni y \to A_{\Psi(y)} \circ ([\Psi]_y - \Psi(y))$; then the properties : $(A \circ \Phi) \circ \Psi = A \circ (\Phi \circ \Psi)$, $[\phi] \circ \Psi = [\phi \circ \Psi]$ and A regular $\Rightarrow A \circ \psi$ regular trivially hold. The same is true for a holomorphic map; observe that $(A \circ \Psi)_{\mathbb{R}^m} = A \circ \Psi_{\mathbb{R}^m}$.

Given a field A on $E \cup F$, where E, F are closed and regularly separated i.e. $\rho(x,E) + \rho(x,F) \leq c\,\rho(x, E \cup F)^N$ in any compact with some $c, N > 0$, then the regularity of A_E, A_F implies that of A.

2. An extension operator

A real-situated subspace of $\mathbb{C}^n$ is the image of $\mathbb{R}^n \subset \mathbb{C}^n$ by a complex linear automorphism of $\mathbb{C}^n$. Among the real subspaces Π of (real) dimension n they are characterized by each of the following properties : Π does not contain any complex line; Π is not contained in any complex hyperplane; every real base of Π is a complex base of $\mathbb{C}^n$. Denote by $\theta = \theta(\mathbb{C}^n)$ the set of all finite unions of real-situated subspaces of $\mathbb{C}^n$, and observe that any two sets of θ are regularly separated.

<u>Proposition 1.</u> <u>For any two sets</u> $\Gamma \subset \tilde{\Gamma}$ <u>of</u> θ <u>there is a continuous linear map</u> $J(\Gamma) \ni H \to \tilde{H} \in J(\tilde{\Gamma})$ <u>such that</u> $H \subset \tilde{H}$. *

<u>Proof</u> Clearly, it is sufficient to know that for each real-situated Π there is a continuous linear $J(\Gamma) \ni H \to H^\Pi \in J(\Pi)$ such that $H^\Pi = H$ on $\Pi \cap \Gamma$. Proceed by induction on the number k of real-situated subspaces whose union is Γ.

Assuming $k > 1$ and the statement true for $k-1$, write $\Gamma = \Gamma_0 \cup \Pi_1 \cup \Pi_2$ with $\Gamma_0 \in \theta$, Π_i real-situated; put $\Lambda_i = \Pi_i \cap \Pi$, $\Lambda = \Lambda_1 \cap \Lambda_2$, and observe that it is sufficient to have a continuous linear $\sigma : J(\Pi)^2 \cap \{H_1 = H_2 \text{ on } \Lambda\} \to J(H)$ such that $\sigma(H_1, H_2) = H_i$ on $\Lambda_i \cup \{H_1 = H_2\}$, since then our statement holds with $H^\Pi = \sigma((H|(\Gamma_0 \cup \Pi_1))^\Pi, (H|(\Gamma_0 \cup \Pi_2))^\Pi)$. To construct σ, assume that $\Pi = \mathbb{R}^n$ and (using the isomorphism $C^\infty(\mathbb{R}^n) \to J(\mathbb{R}^n)$) take the (linear continuous) map

$$C^\infty(\mathbb{R}^n) \cap \{\phi_1 - \phi_2 \text{ flat on } \Lambda\} \ni (\phi_1, \phi_2) \to \phi = \begin{cases} \phi_2 & \text{on } \Lambda \\ \phi_2 + \alpha(\phi_1 - \phi_2) & \text{on } \mathbb{R}^n \setminus \Lambda \end{cases} \in C^\infty(\mathbb{R}^n)$$

* The Whitney extension theorem says that any Whitney field on a closed subset of $\mathbb{R}^n$ has an extension on $\mathbb{R}^n$, thus every Whitney field on Γ has an extension on $\tilde{\Gamma}$.

where α is C^∞ on $\mathbb{R}^n \setminus \Lambda$, equal to 1 resp. 0 on a neighbourhood of $\Lambda_1 \setminus \Lambda_2$ resp. $\Lambda_2 \setminus \Lambda$, and such that $(D^s\alpha)(x) = O(\rho(x,\Lambda)^{-|s|})$ for every s. (One can take $\alpha = \beta \circ p$ where p is the orthogonal projection on the orthogonal complement M of Λ, $\beta(v) = \gamma(v/|v|)$, and γ is C^∞ on $S = M \cap \{|x| = 1\}$ and equal to 1 resp. 0 on a neighbourhood of $\Lambda_1 \cap S$ resp. $\Lambda_2 \cap S$).

There remains the case $k = 1$ (in which Γ is a real-situated subspace). Proceed by induction on the codimension ℓ of $\Pi_0 = \Gamma \cap \Pi$ in Γ. Assuming $\ell > 1$ and the statement true for codimensions $< \ell$, we take $\Pi^* \supset \Pi_0$ of codimension 1 in Π and $b \in \Gamma \setminus (\Pi^* + i\Pi^*)$, then $\Pi' = \Pi^* + \mathbb{R}b$ is real-situated, $\Pi' \cap \Pi$ of codimension $< \ell$, and our statement holds with $H^\Pi = (H^{\Pi'})^\Pi$. The remaining case of $\ell = 1$ (that of $\ell = 0$ being trivial) is reduced to that in which $\Gamma \cap \mathbb{C}\Pi \setminus \Pi_0 \neq \emptyset$, (take $b \in \Gamma \setminus (\Pi_0 + i\Pi_0)$, $a \in \Pi \setminus (\Pi_0 + i\Pi_0)$, so $b = ica + v' + iv$ with $c \in \mathbb{C} \setminus \{0\}$, $v, v' \in \Pi_0$, and $\Pi' = \Pi_0 + \mathbb{R}(ca + v)$ real-situated, $b - v' \in \Gamma \cap \mathbb{C}\Pi' \setminus \Pi_0$, $ca \in \Pi' \cap \mathbb{C}\Pi \setminus \Pi_0$; the statement holds with $H^\Pi = (H^{\Pi'})^\Pi)$, hence we can assume

$$\Gamma = \mathbb{R}^n \quad \text{and} \quad \Pi = \mathbb{R}c \times \mathbb{R}^{n-1} \quad \text{with} \quad c \neq 0$$

(take $a \in \Gamma \cap \mathbb{C}\Pi \setminus \Pi_0$, $c \in \mathbb{C} \setminus \{0\}$ such that $ca \in \Pi$, and apply an automorphism L such that $L(\Pi_0) = 0 \times \mathbb{R}^{n-1}$ and $L(a) = (1,0)$).

Since the map $J(\mathbb{R}^n) \ni H \to [\phi]H \in J(\mathbb{R}^n)$, with $\phi \in C^\infty(\mathbb{R}^n)$ verifying $\phi(x, y) = 1$ resp. 0 for $|x| \leq 1$ resp. $|x| \geq 2$, is linear and continuous, it is sufficient to construct a map $J(\mathbb{R}^n) \cap \{H = 0 \text{ for } |x| \geq 2\} \ni H \to H^\Pi \in J(\Pi)$ linear, continuous and such that $H^\Pi = H$ on $0 \times \mathbb{R}^{n-1}$.* Let $\lambda_i \in \mathbb{C}$, $a_i \in \mathbb{R} \setminus \{0\}$ satisfy $\sum_1^\infty |\lambda_i| a_i^k < \infty$ for every $k \in \mathbb{N}$. Take the automorphisms $L_i : (z, w) \to (b_i z, w)$, where $b_i = a_i c^{-1}$. Then for any $H \in J(\mathbb{R}^n)$ vanishing on $\{|x| \geq 2\}$, we have $H \circ L_i \in J(\Pi)$,

$$H^\Pi = \sum_1^\infty \lambda_i H \circ L_i \in J(\Pi)$$

and the map $H \to H^\Pi$ is linear and continuous (because $\Sigma|\lambda_i|b_i^k < \infty$ for every

---

* We follow now an idea of Seeley.

$k \in \mathbb{N}^*$). At any $w \in 0 \times \mathbb{R}^{n-1}$ we have $H_w = \Sigma\, a_{kp} Z^k W^p$ with some a_{kp} , and

$$(H^\Pi)_w = \sum_{k,\,p} a_{kp} \, (\sum_{i=1}^{\infty} \lambda_i b_i^k) \, Z^k W^p \quad ;$$

therefore in order to have $H^\Pi = H$ on $0 \times \mathbb{R}^n$ it is sufficient to prove the following

<u>Lemma (Seeley)</u> $\sum_1^\infty \lambda_i a_i^k = c^k$ <u>and</u> $\sum_1^\infty |\lambda_i a_i^k| < \infty$ <u>for</u> $k \in \mathbb{N}$, <u>if</u> $a_i = 2^i$ <u>and</u> $\lambda_s = \prod_{i \neq s} \frac{c - a_i}{a_s - a_i}$.

To check this, observe that $\sum_1^N \lambda_{iN} a_i^k = c^k$, $k = 1, \dots, N-1$, with $\lambda_{sN} = \prod_{i \neq s}^{N} \frac{c - a_i}{a_s - a_i}$ (by Lagrange's interpolation formula) ; now $\lambda_{sN} \to \lambda_s$ as $N \to \infty$, and $|\lambda_{sN}| \leq K 2^{-\binom{s}{2}}$, since $\prod_{i \neq s}^{N} (1 - \frac{c}{a_i})$ is bounded (with respect to N, s) and converges as $N \to \infty$,

$$\prod_{s+1}^{N} (1 - \frac{a_s}{a_i}) = \prod_{1}^{N-s} (1 - \frac{1}{2^i}) > d > 0, \; | \sum_{1}^{s} (1 - \frac{a_s}{a_i}) | \geq \prod_{1}^{s} 2^{i-1} .$$

3. Some extension and division properties

Let $E \subset F \subset \mathbb{C}^n$; we say that E has property (e) in F if each field continuous on F (i.e. whose coefficients are continuous) and regular on E is regular on F . If Π is real-situated and $w \neq 0$ is a polynomial then $\Pi \cap \{w \neq 0\}$ has property (e) in Π . (Indeed, we can assume $\Pi = \mathbb{R}^n$ and $\mathbb{R}^n \cap \{w = 0\}$ not containing any line parallel to an axis, and observe that the

---

* Observe that $J(\Pi)$, isomorphic to $C^\infty(\mathbb{R}^n)$, is complete.

regularity of $x \to \Sigma \frac{1}{p!} h_p(x)X^p$ on an open set means simply

$\frac{\partial h_p}{\partial x_i} = h_{p+(0,.,1,.,0)}$ for every p).

We say that a field $P \in J(E)$ has property (d) if for any field G on E we have PG regular => G regular. Observe that the product of fields with property (d) has property (d), and if E, F are closed and regularly separated then the property (d) of P_E, P_F implies that of $P_{E \cup F}$ (for P defined on $E \cup F$).

Lemma 1. $[\ell]_\Gamma$ has property (d), where $\ell \neq 0$ is a complex linear form and $\Gamma \in \theta$.

Proof. We can assume that Γ is a real-situated subspace. Put $\Pi = \{\ell = 0\} \cap \Gamma$. It is sufficient to prove that $G[\ell^{k-1}]_\Gamma$ regular => G continuous (as $\Gamma \setminus \Pi$ has property (e) in Γ). Since

$$\frac{\partial}{\partial X_i}(G[\ell^k]_\Gamma) = \frac{\partial G}{\partial X_i}[\ell^k]_\Gamma + cG[\ell^{k-1}]_\Gamma ,$$

we have

$$G[\ell^{k-1}]_\Gamma \text{ regular} \Rightarrow \frac{\partial G}{\partial X_i}[\ell^k]_\Gamma \text{ regular}$$

and (by induction)

$$G[\ell]_\Gamma \text{ regular} \Rightarrow (D^pG)[\ell^{|p|+1}]_\Gamma \quad \text{regular.}$$

Thus, as the 0-th coefficient of D^pG is a p-th one of G multiplied by $p!$, it is sufficient to show that $G[\ell^k]_\Gamma$ regular implies g_0 continuous. Now, the continuity on $\Gamma \setminus \Pi$ being trivial, let $z_0 \in \Pi$; take $z \in \Gamma$ and let z' be the orthogonal projection of z on Π ; if $z \to z_0$, then $z' \to z_0$, and (since $(G[\ell^k]_{z'}^{(k)} = g_0(z')\ell(X)^k)$ we have *

$$r^k_{G[\ell^k]}(z', z) = g_0(z)\ell(z)^k - g_0(z')\ell(z)^k = O(|z' - z|^{k+1}) = O(|\ell(z)|^{k+1})$$

which gives (the case $\ell(z) = 0$ being trivial) $|g_0(z) - g_0(z')| \to 0$, but

---

* If A is regular on $E \ni a$ then $r^k_A(x,x') = O(|x'-x|^{k+1})$ as $x,x' \to a$, and the coefficients of A are locally Lipschitz.

$|g_0(z') - g_0(z_0)| = O(|z' - z_0|)$($g_0$ being equal on Π up to a constant factor - to a coefficient of $G[\ell^k]$), which implies $g_0(z) \to g_0(z_0)$.

4. On symmetric fields.

Lemma 2 (Glaeser type). Let $\phi : \Omega \to \mathbb{C}^m$ be holomorphic, Ω an open subset of $\mathbb{C}^m$; let $F \subset \phi(\Omega)$, $E = \phi^{-1}(F)$ be such that $\phi_E : E \to F$ is proper, $[D]_E$ has property (d) where $D = \text{jac}\,\phi$, and $F_0 = \phi(\{D \neq 0\}) \cap F$ has property (e) in F. Then any field G on F with $G \circ \phi$ regular is regular.

Proof. Since $\frac{\partial}{\partial Z_i}(G \circ \phi) = \sum_j \left(\frac{\partial G}{\partial X_j} \circ \phi\right)[\frac{\partial \phi_j}{\partial z_i}]_E$ we have

$\sum_i [X_{ij}]_E \frac{\partial}{\partial Z_i}(G \circ \phi) = \left(\frac{\partial G}{\partial X_j} \circ \phi\right)[D]_E$ with some X_{ij} holomorphic in Ω, which gives the regularity of $\frac{\partial G}{\partial X_j} \circ \phi$ and (by induction) of any $(D^p G) \circ \phi$; this implies that G is continuous (as $g_p \circ \phi$ and hence g_p are). Thus it is sufficient to have the regularity of G_{F_0}, but this follows from $G_{\phi(U)} = (G \circ \phi) \circ (\phi_U)^{-1}$ for a neighbourhood U of some c such that $\phi_U : U \to \phi(U)$ is a holomorphic isomorphism, $\phi(c)$ being any given point of F_0 .

Define $\pi_\alpha : \mathbb{C}^{n+r} \ni (z, t) \to (z_{\alpha_1}, \ldots, z_{\alpha_n}, t) \in \mathbb{C}^{n+r}$ for every permutation α, and $\sigma : \mathbb{C}^{n+r} \ni (z, t) \to (\sigma_1(z), \ldots, \sigma_n(z), t) \in \mathbb{C}^{n+r}$ where σ_i are the elementary symmetric functions. A field G is symmetric iff $G \circ \pi_\alpha = G$ for each α ; then the domain of G is symmetric. (This is, of course, the symmetry with respect to Z). Denote by $\mathbb{C}[[z, T]]^E_\Sigma$ the set of all symmetric fields on E , and by $J(E)_\Sigma$ that of regular symmetric ones. Let $E = \sigma^{-1}(F)$.

Proposition 2. $\mathbb{C}[[S,T]]^F \ni H \to H \circ \sigma \in \mathbb{C}[[Z,T]]^E_\Sigma$ is an isomorphism of algebras.

Proof. We need the following fact : given different $a_i \in \mathbb{C}^m$ and $A_i \in \mathbb{C}[[X]]$, $i = 1, \ldots, \ell$, there is a sequence $P_\nu \in \mathbb{C}[X]$ such that $P_\nu(X + a_i) \to A_i$ in the Krull topology. (To check it, observe that for $Q \in \mathbb{C}[X]$ such that $Q(a) = 0$, $Q(b) = 1$ we have

$$(1 - Q^\nu)^\nu(X + a) \to 1, \; (1 - Q^\nu)^\nu(X + b) \to 0 \;\; ;$$

thus taking $P_v^{ij} \in \mathbb{C}[X]$ such that

$$P_v^{ij}(X + a_i) \to 1, \; P_v^{ij}(X + a_j) \to 0$$

the statement holds with $P_v = \sum_i A_i^{(v)}(X - a_i) \prod_j P_v^{ij})$.

Now, let $G \in \mathbb{C}[[Z, T]]_\Sigma^E$; it is sufficient to find for any $w \in F$ some $H \in \mathbb{C}[[S,T]]$ such that $w = \sigma(z)$ implies $G_z = H \circ ([\sigma])_z - w)$. We have $P_v(X + z) \to G_z$ with symmetric $P_v = \frac{1}{n!} \sum_\alpha P'_v \circ \pi_\alpha$ where $P'_v \in \mathbb{C}[Z, T]$ satisfy $P'_v(X + \pi_\alpha(z)) \to G_{\pi_\alpha(z)}$. By the symmetric function theorem $P_v = Q_v \circ \sigma$ with some $Q_v \in \mathbb{C}[S,T]$. Hence $P_v(X + z) = Q_v(Y + w) \circ ([\sigma]_z - w) \to G_z$, which implies $Q_v(Y + w) \to H$ * and $H \circ ([\sigma]_z - w) = G_z$ with some $H \in \mathbb{C}[[S,T]]$. Now, if $\sigma(z') = w$, we have $z' = \pi_\alpha(z)$ and

$$G_{z'} = (G \circ \pi_\alpha^{-1})_{z'} = G_z \circ \pi_\alpha^{-1} = H \circ ([\sigma]_{z'} - w) .$$

Observing that σ is a proper surjection, $D(z) = \text{jac}\, \sigma(z) = \prod_{i<j} (z_i - z_j)$, and $\sigma(\{D \neq 0\}) = \{\Delta \neq 0\}$ where Δ is the discriminant of $\zeta^n - s_1 \zeta^{n-1} + \ldots$, we get by lemma 2 :

<u>Proposition 3.</u> <u>If</u> $[z_i - z_j]_E$ <u>have property</u> (d) <u>and</u> $\{\Delta \neq 0\} \cap F$ <u>has property</u> (e) <u>in</u> F, <u>then</u> $J(F) \ni H \to H \circ \sigma \in J(E)_\Sigma$ <u>is an isomorphism of algebras; it is bicontinuous if</u> J(E) <u>and</u> J(F) <u>are complete.</u>

In particular, if $F = \mathbb{R}^{n+r}$ and $E = \Gamma = \sigma^{-1}(\mathbb{R}^{n+r}) = \bigcup_{\substack{2p \leq n \\ \alpha \text{ a permutation}}} \Pi_{\alpha p}$ where $\Pi_{\alpha p}$ is the real-situated subspace defined by

---

* as $\mathbb{C}[[X]] \ni H \to H \circ B \in \mathbb{C}[[X]] \circ B$ is a bicontinuous 1 - 1 map provided that $D = \text{jac}\, B \neq 0$. Indeed, assume $H_v \circ B \to 0$; by derivation we get $\Sigma C_{ij} \frac{\partial}{\partial x_i} (H_v \circ B) = \frac{\partial H_v}{\partial X_j} \circ B)D$ which gives $\frac{\partial H_v}{\partial X_j} \circ B \to 0$ and (by induction) $D^p H_v \circ B \to 0$ for any p ; now it is sufficient to observe that the constant term of $D^p H_v \circ B$ is equal to the p-th coefficient of H_v multiplied by p!

$$\operatorname{Im}(z_{\alpha_\nu} + z_{\alpha_{\nu+p}}) = \Re e(z_{\alpha_\nu} - z_{\alpha_{\nu+p}}) = 0 , \quad \nu = 1, \dots, p ;$$

$$\operatorname{Im} z_{\alpha_\mu} = 0 , \quad \mu = 2p+1, \dots, n ; \quad \operatorname{Im} t_i = 0, \ i = 1, \dots, r ;$$

we have (by lemma 1 and completeness of $J(\Gamma)$) :

<u>Proposition 4.</u> $J(\mathbb{R}^{n+r}) \ni H \to H \circ \sigma \in J(\Gamma)_\Sigma$ <u>is an isomorphism of topological algebras.</u>

As a corollary we get the

<u>Glaeser-Newton symmetric function theorem.</u> <u>For every symmetric</u> $\phi \in C^\infty(\mathbb{R}^{n+r})$ <u>there is some</u> $\psi \in C^\infty(\mathbb{R}^{n+r})$ <u>such that</u> $\phi = \psi \circ \sigma_{\mathbb{R}}$; <u>the map</u> $\phi \to \psi$ <u>can be chosen linear and continuous</u>.

(Extend $[\phi]$ by $G \in J(\Gamma)_\Sigma$, taking $G = \frac{1}{n!} \sum_\alpha G' \circ \Pi_\alpha$ where $[\phi] \subset G' \in J(\Gamma)$; then $G = H \circ \sigma$ and $H = [\psi]$).

5. The Malgrange-Mather preparation theorem.

Consider $W : \mathbb{C}^m = \mathbb{C}^n \times \mathbb{C} \times \mathbb{C}^r \ni (\gamma, z, t) \to (z - \gamma_1) \dots (z - \gamma_n) \in \mathbb{C}$. Let $p : \mathbb{C}^m \ni (\gamma, z, t) \to (\gamma, t) \in \mathbb{C}^{m-1}$ and $r_j : \mathbb{C}^{m-1} \ni (\gamma, t) \to (\gamma, \gamma_j, t) \in \mathbb{C}^m$.

<u>Proposition 5.</u> <u>Let</u> $E \subset \mathbb{C}^m$ <u>be such that</u> $r_j(p(E)) \subset E$ <u>for each</u> j <u>and</u> $[z - \gamma_i]_E$ have property (d). Then, putting $E_0 = p(E)$,

$$J(E) \times J(E_0) \ni (Q, A) \to Q[W]_E + \sum_0^{n-1} (A_i \circ p)_E [z^i]_E \in J(E)$$

<u>is a linear isomorphism; it is bicontinuous if</u> $J(E), J(E_0)$ <u>are complete</u>.

<u>Proof.</u> <u>Injectivity</u> : Assuming the value at (Q, A) equal to 0 we get (by substituting r_j and since $E_0 \subset r_j^{-1}(E)$) $\sum_{i=0}^{n} A_i [\gamma_j]^i_{E_0} = 0$ which gives $A_i = 0$ and hence $Q = 0$.

<u>Surjectivity</u> : By induction, it is sufficient to find for any $G \in J(E)$ some $Q \in J(E)$ and $A \in J(E_0)$ such that $G = Q[z - \gamma_j]_E + (A \circ p)_E$, or (by property (d) and since $r_j(p(E)) \subset E$) some $Q \in \mathbb{C}[[C, Z, T]]^E$ satisfying

$Q[z-\gamma_j]_E = G-(G \circ r_j \circ p)_E$. The existence of Q in $E \cap \{z \neq \gamma_j\}$ being trivial, take $a = (\overset{\circ}{\gamma}, \overset{\circ}{\gamma}_j, \overset{\circ}{t}) \in E$, then

$$(G - G \circ r_j \circ p)_a = G_a - G_a \circ (C, C_j, T) = Q_a(Z - C_j) = Q_a[z-\gamma_j]_a$$

with some $Q_a \in \mathbb{C}[[C,Z,T]]$.

<u>Lemma 3.</u> <u>If</u> $\Gamma = \Gamma_* \times \mathbb{R}^{r+1}$ <u>with</u> $\Gamma_* \in \theta(\mathbb{C}^n)$ <u>symmetric, then there is a symmetric</u> $\tilde{\Gamma} \supset \Gamma$ <u>such that</u> $\tilde{\Gamma} \in \theta(\mathbb{C}^m)$, $p(\tilde{\Gamma}) \in \theta(\mathbb{C}^{m-1})$ <u>and</u> $r_j(p(\tilde{\Gamma})) \subset \tilde{\Gamma}$ <u>for each</u> j . (Take $\tilde{\Gamma} = \Gamma \cup \bigcup_{i=1}^{n} \tau_i^{-1}(\Gamma)$, where $\tau_i : (\gamma,z,t) \to (\gamma, z-\gamma_i, t)$ and observe that $r_j(\tau_j(p(\tilde{\Gamma}))) \subset \Gamma$ since $\tau_j \circ r_j \circ p = \tau_j \circ r_j \circ p \circ \tau_i^{-1} : (\gamma,z,t) \to (\gamma,0,t)$).

Consider

$$P : \mathbb{R}^m \ni (c_1 x, t) \to x^n - c_1 x^{n-1} + .. + (-1)^n c_n \in \mathbb{C}$$

and let $q : (c,x,t) \to (c,t)$.

<u>Malgrange-Mather preparation theorem.</u> <u>There is a continuous linear map</u>

$$C^\infty(\mathbb{R}^m) \ni Q \to (\chi, \alpha_0, \ldots, \alpha_{n-1}) \in \mathbb{C}^\infty(\mathbb{R}^m) \times \mathbb{C}^\infty(\mathbb{R}^{m-1})^n$$

<u>such that</u>

$$Q = \chi P + \sum_0^{n-1} (\alpha_i \circ q)\, x^i .$$

<u>Proof</u>. Let $\Gamma = \sigma^{-1}(\mathbb{R}^m)$ where $\sigma : (\gamma,z,t) \to (\sigma(\gamma),\ldots,\sigma_n(\gamma),z,t)$ and take $\tilde{\Gamma} \supset \Gamma$ as in lemma 3 ; there is a linear continuous $J(\Gamma)_\Sigma \ni G \to \tilde{G} \in J(\tilde{\Gamma})_\Sigma$ such that $G \subset \tilde{G}$ (proposition 1 and symmetrization). For any $\phi \in C^\infty(\mathbb{R}^m)$, take $G = [\phi] \circ \sigma \in J(\Gamma)$, its corresponding extension $\tilde{G} \in J(\tilde{\Gamma})_\Sigma$, write (Proposition 5):

$$\tilde{G} = \tilde{Q}\,[W]_{\tilde{\Gamma}} + \sum_0^{n-1} (\tilde{A}_i \circ p)_{\tilde{\Gamma}}\, [z^i]_{\tilde{\Gamma}}$$

with $\tilde{Q} \in J(\tilde{\Gamma})_\Sigma$ and $\tilde{A}_i \in J(\tilde{\Gamma}_0)_\Sigma$ with $\tilde{\Gamma}_0 = p(\tilde{\Gamma})$ (using the uniqueness in Proposition 5 to check the symmetry of $\tilde{Q}$ and $\tilde{A}_i$), take the restrictions $Q = \tilde{Q}_\Gamma$, $A_i = (\tilde{A}_i)_{\Gamma_0}$ where $\Gamma_0 = p(\Gamma)$, and finally (Proposition 3) write $Q = [\chi] \circ \sigma$ and $A_i = [\alpha_i] \circ \sigma_*$ (as $\Gamma_0 = \sigma_*^{-1}(\mathbb{R}^{m-1})$) where

$\sigma_* : (\gamma,t) \to (\sigma_1(\gamma), \ldots, \sigma_n(\gamma),t)$. Then the composed map $\phi \to (\chi, \alpha_0,\ldots,\alpha_{n-1})$ is linear and continuous. Since $[W]_\Gamma = [P] \circ \sigma$ and $[z^i]_\Gamma = [x^i] \circ \sigma$,

$$G = Q[W]_\Gamma + \sum_0^{n-1} (A_i \circ p)_\Gamma [z^i]_\Gamma$$

gives

$$[\phi] \circ \sigma = ([\chi][P] + \sum_0^{n-1} [\alpha_i \circ q][x^i]) \circ \sigma$$

and $\phi = \chi P + \sum_0^{n-1} (\alpha_i \circ q)x^i$.

REFERENCES

G. Glaeser, Fonctions composées différentiables, Annals of Math. 77 (1963), 193-209.

B. Malgrange, Ideals of differentiable functions, Oxford, 1966.

J. Mather, Stability of C^∞ mappings : I. The division theorem, Annals of Math. 87 (1968), 89-104.

R.T. Seeley, Extension of C^∞ functions defined in a half space, Proc.Amer.Math.Soc. 15 (1964), 625-626.

ON NIRENBERG'S PROOF OF MALGRANGE'S PREPARATION THEOREM

J. N. Mather *

Recently, Nirenberg has given a new proof of Malgrange's preparation theorem [1]. The difficult part of his proof is the proof of an extension lemma. In this paper we will give an alternative proof of Nirenberg's extension lemma by the method which we used in [2] to prove Malgrange's preparation theorem. This proof stands in the same relation to Nirenberg's proof as our proof in [2] stood in relation to Malgrange's original proof of his preparation theorem: our proof gives a linear operator which provides the extension (in Glaeser's terminology, an <u>extensor</u>), whereas Nirenberg's construction is non-linear.

To give the idea of the proof we first prove the following result.

<u>Lemma 0</u> <u>Let $f(t)$ be a C^∞ complex-valued function of a real variable t. There exists a C^∞ complex-valued function $F(z)$ on $\mathbb{C}$ such that</u>

(α) $F(t) = f(t)$, <u>for all</u> $t \in \mathbb{R}$

(β) $F_{\bar{z}}(z)$ <u>vanishes to infinite order on</u> $\mathbb{R}$.

<u>Proof</u>. We will consider only the case when f has compact support. The general case can be reduced to this case by means of a partition of unity.

Let $\hat{f}(\xi)$ denote the Fourier transform of f, normalized so that Fourier's integral formula takes the form

$$f(t) = \int_{-\infty}^{\infty} e^{i\xi t}\, \hat{f}(\xi)\, d\xi .$$

Let ρ be a C^∞ function on $\mathbb{R}$, with values in $[0,1]$, which has support $\subseteq [-1,1]$, and is identically 1 in a neighbourhood of 0. Let

---

* The author is grateful to Lojasiewicz and Glaeser for helpful discussions. In particular Glaeser pointed out the possibility of linearizing Nirenberg's lemma.

$$F(z) = \int_{-\infty}^{\infty} \rho(y\xi)\ e^{i\xi z}\ \hat{f}(\xi)d\xi$$

for all $z \in \mathbb{C}$, where y denotes the imaginary part of z.

We assert that F has the required properties. First, F is infinitely differentiable. This is proved by differentiating under the integral sign. For if we apply $\frac{\partial}{\partial z^\alpha} \frac{\partial}{\partial \bar{z}^\beta}$ to the quantity under the integral sign we get a function which has the following form:

$$R(y,\xi)e^{i\xi z}\ \hat{f}(\xi)$$

where $R(y,\xi)$ is a C^∞ function whose absolute value is bounded by a polynomial in ξ and $R(y,\xi) = 0$ for $|y\xi| > 1$. Since $|e^{i\xi z}| \leq e$ when $|y\xi| \leq 1$, it follows that $|R(y,\xi)e^{iz\xi}|$ is bounded by a polynomial in ξ. Since $\hat{f}(\xi)$ is in the Schwartz class it follows that the integral

$$\int_{-\infty}^{\infty} R(y,\xi)e^{i\xi z}\ \hat{f}(\xi)d\xi$$

is uniformly absolutely convergent. Hence we can differentiate under the integral sign and we get that F is C^∞.

Now (α) follows immediately from Fourier's integral formula and the fact that $\rho(0) = 1$. To prove (β), we first observe that (ii) is equivalent to

(ii') $$\frac{\partial^{\alpha+\beta}F}{\partial z^\alpha \partial \bar{z}^\beta}(z) = 0 \qquad \text{for } z \in \mathbb{R},$$

for all integers $\alpha \geq 0$ and $\beta \geq 1$. However (ii') follows immediately from differentiation under the integral sign.

Note that the construction of F we have made in this proof is linear: F depends linearly and continuously (in, e.g., the C^∞ topology) on f.

Now we state Nirenberg's lemma.

Let $P(z,\lambda) = z^p + \lambda_1 z^{p-1} + \dots + \lambda_p$ be the generic polynomial of order p.

<u>Lemma 1</u> (Nirenberg's extension lemma)

<u>Let $f(t,x)$ be a C^∞ complex valued function defined for $t \in \mathbb{R}$ and $x \in \mathbb{R}^n$. There exists a C^∞ complex valued function $F(z,x,\lambda)$ defined for $z \in \mathbb{C}$, $x \in \mathbb{R}^n$, and $\lambda \in \mathbb{C}^p$, satisfying</u>

(i) <u>$F(t,x,\lambda) = f(t,x)$, for t real,</u>

(ii) $F_{\bar{z}}$ <u>vanishes to infinite order on $\mathrm{Im}\, z = 0$ and on $P(z,\lambda) = 0$.</u>

(We have formulated this result globally rather than locally as Nirenberg did. The global form of the lemma trivially implies the local form).

Let $\delta(y,\lambda) = \mathrm{Inf}\ \{|y - \mathrm{Im}\, z| : P(z,\lambda) = 0\}$ for any $y \in \mathbb{R}$ and $\lambda \in \mathbb{R}^p$.

<u>Lemma 2</u>. <u>There exists a continuous function</u> $\rho(\xi,\lambda,y)$, <u>taking values in</u> $[0,1]$, <u>defined for</u> $\xi \in \mathbb{R}$, $\lambda \in \mathbb{C}^p$, <u>and</u> $y \in \mathbb{R}$, <u>such that</u>

(a) $\rho(\xi,\lambda,y) = 1$ <u>in a neighbourhood of</u> $y = 0$.

(b) $\rho(\xi,\lambda,y) = 0$ <u>when</u> $|\xi y| \geq 1$.

(c) $\rho_y(\xi,\lambda,y) = 0$ <u>in a neighbourhood of</u> $\delta(y,\lambda) = 0$

(d) <u>The function</u> $\rho(\xi,\lambda,y)$ <u>is infinitely differentiable with respect to</u> λ <u>and</u> y <u>and its derivatives are continuous with respect to all variables and satisfy</u>

$$\left|\frac{\partial}{\partial\lambda^\alpha}\ \frac{\partial}{\partial\bar{\lambda}^\beta}\ \frac{\partial}{\partial y^\gamma}\ \rho(\xi,\lambda,y)\right| \leq C\left(1 + |\xi|^{(2p+1)(1 + |\alpha| + |\beta| + \gamma)}\right)$$

<u>where</u> $C > 0$ <u>depends only on the multi-indices</u> α <u>and</u> β <u>and the non-negative integer</u> γ.

<u>Proof of Lemma 1</u> We will omit the variable x, since it plays no role except as a parameter. We will suppose f has compact support. The general form of lemma 1 can be deduced from the compact support case by means of a partition of unity argument.

We define F by

$$F(z,\lambda) = \int_{-\infty}^{\infty} \rho(\xi,\lambda,y)e^{i\xi z}\ \hat{f}(\xi)d\xi,$$

where y denotes the imaginary part of z.

This integral is uniformly absolutely convergent and we can differentiate under the integral sign, i.e., for any multi-indices α and β and any non-negative numbers γ and δ, the integral

$$\int_{-\infty}^{\infty} \frac{\partial}{\partial\lambda^\alpha}\ \frac{\partial}{\partial\bar{\lambda}^\beta}\ \frac{\partial}{\partial z^\gamma}\ \frac{\partial}{\partial\bar{z}^\delta}\ \left(\rho(\xi,\lambda,y)e^{i\xi z}\right)\ \hat{f}(\xi)d\xi$$

is uniformly absolutely convergent. For, by (b) and (d),

$$\left|\frac{\partial}{\partial\lambda^{\alpha}}\frac{\partial}{\partial\bar{\lambda}^{\beta}}\frac{\partial}{\partial z^{\gamma}}\frac{\partial}{\partial\bar{z}^{\delta}}\left(\rho(\xi,\lambda,y)e^{i\xi z}\right)\right|$$

is uniformly bounded by a polynomial in ξ, and $\hat{f}(\xi)$ is in the Schwartz class.

Hence $F(z,\lambda)$ is C^{∞}. Condition (i) follows from (a) and Fourier's integral formula. Condition (ii) follows from (a),(c) and differentiation under the integral sign. Q.E.D.

Remark. The F which we have constructed depends linearly and continuously on f, since it is given by an integral formula.

We will only outline the proof of lemma 2, referring to [2] for details, since it is very similar to an argument given there.

Proof of lemma 2 Let

$$\sigma(\eta,\lambda) = \tfrac{1}{2}\pi \int_{-\infty}^{\infty} \left|\frac{d}{dx}\log P(x+\eta i,\lambda)\right|^{2}dx$$

for $\eta \in \mathbb{R}$ and $\lambda \in \mathbb{C}^{p}$. One may verify that $\sigma(\eta,\lambda)$ satisfies the following estimates

(1) $1/2\delta(\eta,\lambda) \leq \sigma(\eta,\lambda) \leq p^{2}/2\delta(\eta,\lambda)$ (see [2], §6) and

(2) $\left|\frac{\partial}{\partial\lambda^{\alpha}}\frac{\partial}{\partial\bar{\lambda}^{\beta}}\frac{\partial}{\partial\eta^{\gamma}}\sigma(\eta,\lambda)\right| \leq C(1+\delta(\eta,\lambda)^{-2p(1+|\alpha|+|\beta|+\gamma)})$,

where $C > 0$ depends only on α,β, and γ (compare [2] §7).

Let ρ_0 be a C^{∞} function defined on the non-negative real numbers, satisfying

$$\begin{aligned} &\rho_0(x) = 1 && \text{for } 0 \leq x \leq 4p^3 \\ &0 \leq \rho_0(x) \leq 1 && \text{for } 4p^3 \leq x \leq 8p^3 \\ &\rho_0(x) = 0 && \text{for } x \geq 8p^3. \end{aligned}$$

and let ρ_1 be a second C^{∞} function defined on the non-negative real numbers, satisfying

$$\begin{aligned} &\rho_1(x) = 0 && \text{for } x \leq \epsilon \\ &0 \leq \rho_1(x) \leq 1 && \text{for } \epsilon \leq x \leq 1-\epsilon \\ &\rho_1(x) = 1 && \text{for } x \geq 1-\epsilon \end{aligned}$$

For $y \geq 0$, we define $\rho(\xi,\lambda,y)$ as follows. Let $I(\xi)$ be the interval

$[1/(1+|\xi|), 1/2(1+|\xi|)]$ and let

$$\begin{aligned} \rho(\xi,\lambda,y) &= 0 & y \geq 1/(1+|\xi|) \\ &= 1 & y \leq 1/2(1+|\xi|) \\ &= \rho_1 \frac{\int_y^{1/(1+|\xi|)} \rho_0(\sigma(\eta,y)/(1+|\xi|)d\eta}{1/4(1+|\xi|)}, & y \in I(\xi). \end{aligned}$$

We will verify $\rho(\xi,\lambda,y)$ so defined for $y \geq 0$ has all the desired properties. We can define $\rho(\xi,\lambda,y)$ similarly for $y \leq 0$ and verify the desired properties. This part of the proof will be omitted, however, since it has exactly the same as the case $y \geq 0$.

It is an easy consequence of (1) and the definition of ρ_0 that

$$\int_{I(\xi)} \rho_0(\eta,\lambda)/(1+|\xi|))d\eta \geq 1/4(1+|\xi|)$$

(compare [2], §8). Hence the partial derivatives of ρ of all orders with respect to the variables λ and y exist. Moreover, it follows from (1), (2) and the definition of ρ_0 that

$$\left|\frac{\partial}{\partial\lambda^\alpha} \frac{\partial}{\partial\bar{\lambda}^\beta} \frac{\partial}{\partial\eta^\gamma} \rho(\xi,\eta,\lambda)\right| \leq C\left(1+|\xi|^{(2p+1)(|\alpha|+|\beta|+\gamma+1)}\right).$$

This verifies condition (d). Conditions (a), (b), and (c) are easily verified.

Q.E.D.

References

[1] L. Nirenberg, A new proof of the preparation theorem. This volume, pp.97 - 105.

[2] J. N. Mather, Stability of C^∞ mappings: I. The division theorem. Annals of Math., 87 (1968) pp. 89-104.

SUR LE THÉORÈME DE PRÉPARATION DIFFÉRENTIABLE

par

G. Glaeser

I. Introduction.

Les fonctions $\mathcal{C}^\infty$ considérées ici dépendent de $1+n+r$ variables, au plus. La première (réelle ou complexe selon le contexte) sera désignée par x, z, X ou Z; les n suivantes, qui seront complexes, par $\underline{\gamma} = (\gamma_1, \gamma_2 \dots \gamma_n)$ ou par $\underline{\sigma} = (\sigma_1, \sigma_2 \dots \sigma_n)$. On décomposera parfois γ_k en partie réelle et imaginaire pure $\gamma_k = (\alpha_k + i\alpha_{n+k}) = \alpha_k + \beta_k$. On posera $n' = 2n$ et $\underline{\alpha} = (\alpha_1, \alpha_2 \dots, \alpha_{n'}) = (\alpha_1, \alpha_2 \dots \alpha_n, -i\beta_1, -i\beta_2 \dots -i\beta_n)$. Les fonctions dépendent, en outre, de certains paramètres $\underline{t} = (t_1, t_2, \dots t_r)$, utiles dans les applications, mais qui ne jouent qu'un rôle de figuration dans la démonstration qui suit : nous nous bornons à les mentionner dans l'enoncé suivant pour ne plus en parler ensuite.

Le théorème de division I est la clé du théorème de préparation de Malgrange-Mather.

Théorème 1. - Si $P(x,\underline{\sigma}) = x^n + \sum_{k=1}^{n} (-1)^k \sigma_k x^{n-k}$ est le polynôme générique à coefficients complexes, il existe des opérateurs linéaires continus $\mathcal{C}^\infty(\mathbb{R}^{1+r}; \mathbb{C}) \to \mathcal{C}^\infty(\mathbb{R}^{1+n+r}; \mathbb{C}) \times \{\mathcal{C}^\infty(\mathbb{R}^{n+r}; \mathbb{C})\}^n$ qui à $f \in \mathcal{C}^\infty(\mathbb{R}^{1+r}; \mathbb{C})$ associent $(Q, A_0, A_1 \dots A_{n-1})$ en sorte que

$$(1) \qquad f(x,\underline{t}) \equiv P(x,\underline{\sigma})\, Q(x,\underline{\sigma},\underline{t}) + \sum_{0 \le k \le n-1} x^k A_k(\underline{\sigma}, \underline{t}) .$$

J. Mather a démontré l'existence d'un tel "opérateur de préparation" (un préparateur*) dans la division par un polynôme à coefficients réels ([5]). L. Nirenberg a généralisé le théorème préparatoire [6] dans le cas des polynômes complexes; mais

---

* Nous utilisons la désinence -teur comme synonyme de opérateur linéaire continu de ... A côté de projecteur les néologismes prolongateur, interpolateur, régularisateur, convoluteur, etc... nous paraissent commodes.

pas plus que B. Malgrange [4] dans le cas réel, il n'a prouvé l'existence d'un préparateur.

La démonstration qui suit est étroitement inspirée de la démonstration cijointe dûe à S. Łojasiewicz dont elle conserve l'idée essentielle. J. Mather a modifié sa méthode de [5] de facon à obtenir la division par un polynôme à coefficients complexes. Les racines complexes du polynôme P sont à la source des difficultés. On utilisera un prolongement $f \mapsto \hat{f}$ qui a $f \in \mathcal{E}^{\infty}(\mathbb{R};\mathbb{C})$, considéré comme appartenant à $\mathcal{E}^{\infty}(\mathbb{R}\times\mathbb{C}^n;\mathbb{C})$ grâce à l'injection canonique $\mathbb{R} \simeq \mathbb{R}\times\{0\} \to \mathbb{R}\times\mathbb{C}^n$, associe une fonction $\hat{f} \in \mathcal{E}^{\infty}(\mathbb{C}^{1+n};\mathbb{C})$. Cela veut dire que $\hat{f}$ est indéfiniment dérivable par rapport aux $2+2n$ coordonnées réelles de $\mathbb{R}^{2+2n}$.

Mais en fait, on n'utilisera que le <u>champ taylorien</u> $\ddot{f}$ que $\hat{f}$ induit sur un certain sous-ensemble $\mathbb{E}$ de $\mathbb{C}^{1+n}$, qui sera décrit plus loin. Et le passage $f \mapsto \ddot{f}$ s'obtiendra à l'aide d'un <u>prolongateur</u> dont la construction constitue le noeud de la démonstration.

<u>Champs Tayloriens. Complexification.</u>

<u>Un champ de séries formelles</u> défini sur un fermé $\mathbb{F} \subset \mathbb{R}^p$ est une fonction, définie sur $\mathbb{F}$, à valeurs dans l'espace $\mathbb{C}[[X_1, X_2, \dots X_p]]$. Si ce champ satisfait aux inégalités classiques de Whitney (cf [4]), il s'identifie au champ des séries de Taylor d'une fonction $\in \mathcal{E}^{\infty}(\mathbb{R}^p;\mathbb{C})$: on dit alors que c'est un <u>champ Taylorien</u> (par abus de langage, certains parlent d'une "fonction" au sens de Whitney). L'ensemble des champs Tayloriens $\mathcal{E}^{\infty}$ défini sur $\mathbb{F}\subset\mathbb{R}^p$ se note $W^{\infty}(\mathbb{F}\subset\mathbb{R}^p;\mathbb{C})$. C'est un quotient de $\mathcal{E}^{\infty}(\mathbb{R}^p;\mathbb{C})$. Celui-ci sera muni de sa topologie de Fréchet habituelle (convergence uniforme d'ordre r, pour tout r et sur tout compact) et $W^{\infty}(\mathbb{F}\subset\mathbb{R}^p;\mathbb{C})$ de la topologie quotient. Si $\mathbb{F}$ est un fermé quelconque, cette topologie est plus fine que la convergence uniforme sur tout compact de chacun des coefficients des séries formelles du champ Taylorien : cependant si $\mathbb{F}$ est convexe (et plus généralement régulier au sens de Whitney) ces deux topologies sont les mêmes.

A toute $f \in \mathcal{E}^{\infty}(\mathbb{R}\times\mathbb{R}^{n'};\mathbb{C})$ correspond canoniquement son champ Taylorien trivial $f \in W^{\infty}(\mathbb{R}\times\mathbb{R}^{n'} \subset \mathbb{R}\times\mathbb{R}^{n'};\mathbb{C})$ qui a $(x, \underline{\alpha})$ associe la série formelle

$$\dot{f}_{(x,\underline{\alpha})} = \sum_{(k,\underline{\omega})} (k!\underline{\omega}!)^{-1} \frac{\partial^{k+\omega} f}{\partial x^k \partial \underline{\alpha}^{\underline{\omega}}} (x,\underline{\alpha})\, X^k \underline{A}^{\underline{\omega}}$$

(où $\underline{\omega} = (\omega_1, \omega_2, \ldots, \omega_{n'})$ et $\underline{A} = (A_1, A_2, \ldots, A_{n'})$.

On lui associe canoniquement son champ de séries formelles complexifiees qui a $(x,\underline{\alpha})$ associe la série formelle

$$\ddot{f}_{x,\underline{\alpha}} = \sum_{k,\underline{\omega}} (k!\underline{\omega}!)^{-1} \frac{\partial^{k+\omega} f}{\partial x^k \partial \underline{\alpha}^{\underline{\omega}}} (x,\underline{\alpha})\, (X + iY)^k \underline{A}^{\underline{\omega}} \tag{I.2}$$

Proposition 1. - Le champ complexifié d'un champ Taylorien défini sur $\mathbb{R} \times \mathbb{R}^{n'}$ est Taylorien et formellement holomorphe.

Il est Taylorien car les inégalités de Whitney pour $\ddot{f}$ sont à peu près les inégalités de Whitney pour f (certaines de ces inégalités auront cependant à être répétées plusieurs fois).

Le champ complexifié satisfait à l'égalite de Cauchy-Riemann

$$i \frac{\partial \ddot{f}}{\partial x} = \frac{\partial \ddot{f}}{\partial y}$$

($\frac{\partial}{\partial x}$ et $\frac{\partial}{\partial y}$ désigne ici des dérivations partielles dans l'anneau des séries formelles) : on dit que f est formellement holomorphe sur $\mathbb{R} \times \mathbb{R}^{n'}$.

Nous noterons $W^{\infty}_{\mathbb{C}}(\mathbb{F} \subset \mathbb{C} \times \mathbb{R}^{n'}; \mathbb{C})$ le sous-espace fermé de $W^{\infty}(\mathbb{F} \subset \mathbb{C} \times \mathbb{R}^{n'}; \mathbb{C})$ constitué par les champs Tayloriens formellement holomorphes en Z.

Proposition 2. - Si M est un hyperplan réel de $\mathbb{C}^{1+n}$, image de $\mathbb{R} \times \mathbb{C}^n$ par un automorphisme linéaire réel, qui est holomorphe par rapport à la première variable z, les algèbres topologiques (avec dérivation) $\mathcal{E}^{\infty}(M; \mathbb{C})$ et $W^{\infty}_{\mathbb{C}}(M \subset \mathbb{C}^{1+n}; \mathbb{C})$ sont isomorphes.

Dans le cas, où $M = \mathbb{R} \times \mathbb{C}^n$, nous venons de décrire cet isomorphisme $f \to \ddot{f}$. Mais comme l'équation de Cauchy-Riemann est invariante par toute transformation holomorphe en z, la proposition est démontrée.

L'ensemble $\mathbb{E}$. - Soit $M_o = \mathbb{R} \times \mathbb{C}^n$ et $M_k (k \leq n)$ l'hyperplan réel défini par $z - \gamma_k \in \mathbb{R}$.

L'ensemble $\mathbb{E}$ évoqué ci-dessus est la reunion des M_k (pour $0 \leq k \leq n$). Il contient l'ensemble de $\mathbb{C}^{1+n}$ défini par $(z - \gamma_1) \times (z - \gamma_2) \times \ldots \times (z - \gamma_n) = 0$.

II. Où l'on commence par la fin.

Nous démontrons plus loin le théorème suivant.

Théorème 2. Il existe une prolongateur Λ de $\mathcal{E}^\infty(M_0;\mathbb{C})$ dans $W^\infty_{\mathbb{C}}(\mathbb{E}\subset\mathbb{C}^{1+n};\mathbb{C})$.

Nous allons montrer d'abord comment le théorème 1 se déduit du théorème 2. Soit $f\in\mathcal{E}^\infty(\mathbb{R};\mathbb{C})$, considérée comme une fonction définie sur $M_0=\mathbb{R}\times\mathbb{C}^n$ (mais ne dépendant pas des $\underline{\gamma}$). Nous lui associons, grâce au théorème 2, un champ Taylorien $\ddot{f}\in W^\infty_{\mathbb{C}}(\mathbb{E}\subset\mathbb{C}^{1+n};\mathbb{C})$ (qui induit la fonction f sur M_0).

Division.

Proposition 3. - Il existe des champs Tayloriens $\ddot{q}\in W^\infty_{\mathbb{C}}(\mathbb{E}\subset\mathbb{C}^{1+n};\mathbb{C})$ et $\ddot{a}_k\in W^\infty(\mathbb{C}^n\subset\mathbb{C}^{1+n};\mathbb{C})$, déterminées d'une façon unique, tels que

$$\ddot{f}(z,\underline{\gamma})=(z-\gamma_1)(z-\gamma_2)\dots(z-\gamma_n)\ddot{q}(z,\underline{\gamma})+\sum_{k<n}z^k\ddot{a}_k(\underline{\gamma})$$

Montrons d'abord qu'il existe un champ Taylorien $\ddot{q}_1\in W^\infty(\mathbb{E}\subset\mathbb{C}^{1+n})$ tel que

$$\ddot{f}(z,\gamma)-\ddot{f}(\gamma_1,\underline{\gamma})=(z-\gamma_1)\ddot{q}_1(z,\underline{\gamma}).$$

C'est un cas particulier du theoreme de Łojasiewicz-Hörmander (cf [4]). En chaque point de $\mathbb{E}$ la série formelle $\ddot{f}(z,\underline{\gamma})-\ddot{f}(\gamma_1,\underline{\gamma})$ est divisible par $z-\gamma_1$: c'est évident, en tout point $(z,\underline{\gamma})$ où $z\neq\gamma_1$. Et si $z=\gamma_1$ c'est une conséquence de l'holomorphie formelle : le champ Taylorien de $z-\gamma_1$ s'y réduit a $X+iY=Z$ (d'apres (1,2)) et le champ Taylorien de $\ddot{f}(z,\underline{\gamma})-\ddot{f}(\gamma_1,\underline{\gamma})$ y contient Z en facteur. On notera que dans ces conditions la démonstration du théorème de Łojasiewicz-Hörmander, généralement énoncé pour les modules sur l'anneau $\mathcal{E}^\infty(\mathbb{R}^{1+n'};\mathbb{R})$ à valeurs dans un espace de dimension réelle finie $\mathbb{V}$, reste valable sans changement lorsqu'on remplace l'anneau de base par $\mathcal{E}^\infty(\mathbb{R}^{1+n'};\mathbb{C})$. On constate que le champ $\ddot{q}_1$ est formellement holomorphe sur $\mathbb{E}$: on peut le rediviser à son tour par $z-\gamma_2$, et la proposition s'obtient par récurrence finie. L'unicité résulte de la construction même.

Fonctions symetriques.

Comme f ne dépend pas des variables $\underline{\gamma}$, on peut l'envisager comme une fonction trivialement symétrique par rapport aux variables $\underline{\gamma}$. Il en sera de même de $\ddot{f}\in W^\infty_{\mathbb{C}}(\mathbb{E}\subset\mathbb{C}^{1+n};\mathbb{C})$ (car s'il n'en était pas ainsi, on pourrait faire suivre le prolongateur Λ d'un symétrisateur.) En vertu de l'unicité de la division précédente, les champs $\dot{a}_k(\underline{\gamma})$ sont également symétriques par rapport aux variables $(\gamma_1,\gamma_2 \dots \gamma_n)$. Comme les $\dot{a}_k$ sont definis sur $\mathbb{C}^n$ ils correspondent canoniquement à des fonctions $a_k \in \mathcal{E}^\infty(\mathbb{C}^n; \mathbb{C})$ symétriques.

Considérons l'application de Newton complexe $\bar{\theta} : (\gamma_1,\gamma_2 \dots \gamma_n) \mapsto (\sigma_1,\sigma_2 \dots \sigma_n)$ qui associe aux γ_k ses fonctions symétriques élémentaires.
Considérée comme une application définie sur $\mathbb{R}^{2n}$, elle ne doit être confondue avec l'application de Newton réelle θ. En particulier $\bar{\theta}$ est surjective (d'après le théoremè de d'Alembert-Gauss) alors que θ ne l'est pas.

Proposition 4. - A toute fonction $a \in \mathcal{E}^\infty(\mathbb{C}^n; \mathbb{C})$ symétrique par rapport aux variables $\underline{\gamma}$ correspond une fonction $A \in \mathcal{E}^\infty(\mathbb{C}^n; \mathbb{C})$ unique, telle que $a = A \circ \bar{\theta}$.

Ceci est un théorème de Newton différentiable complexe, plus aisé à démontrer que son analogue réel [1] : les propositions I, II, III, IV de [1] prouvent l'existence d'un champ de séries formelles A unique tel que $\dot{a} = \dot{A} \circ \bar{\theta}$, dont les coefficients varient continuement sur $\mathbb{C}^n$. Dans le cas du théorème de Newton differentiable réel, l'application θ n'était pas surjective et c'est le prolongement du champ défini sur $\theta(\mathbb{R}^n)$ qui constituait la principale difficulté du théorème.
Ici, au contraire, on peut stratifier $\mathbb{C}^n$ à l'aide du discriminant de $P(z,\underline{\sigma})$: sur chaque strate la multiplicité des racines de $P(z,\underline{\sigma})$ reste constante et l'on remarque que le champ de séries formelles A est Taylorien, en restriction à chaque strate. Un lemme élémentaire, mais utile, d'Hesténes (cf [2]) permet de conclure immédiatement que $\dot{A}$ est Taylorien et provient donc d'une façon canonique d'une fonction $A \in \mathcal{E}^\infty(\mathbb{C}^n; \mathbb{C})$.

La démonstration du théorème I s'acheve alors ainsi : la fonction $f(x) - \sum_{k<n} x^k A_k(\underline{\sigma})$ appartient à $\mathcal{E}^\infty(\mathbb{R} \times \mathbb{C}^n; \mathbb{C})$, et par construction elle appartient ponctuellement à l'idéal principal engendré par $P(x;\underline{\sigma})$. D'après la théorème de division de Łojasiewicz-Hörmander il existe une fonction $Q(x,\underline{\sigma})$ de classe $\mathcal{E}^\infty(\mathbb{R} \times \mathbb{C}^n; \mathbb{C})$ satisfaisant à (1,1).
Il ne reste plus qu'à démontrer le théorème 2.

III. Le prolongateur.

Définition. - Deux champs Tayloriens $\dot{f}_i = W^\infty(\mathbb{A}_i \subset \mathbb{R}^p; \mathbb{C})$ $(i = 1, 2)$ sont compatibles sur une partie $\mathbb{B} \subset \mathbb{A}_1 \cap \mathbb{A}_2$, si leurs restrictions à $\mathbb{B}$ sont egales.

Premier lemme de recollement.

Si $\mathbb{A}_1$ et $\mathbb{A}_2$ sont deux parties régulièrement séparées contenues dans $\mathbb{A} \subset \mathbb{R}^p$, il existe un recollateur C qui applique l'espace des couples $(\dot{f}_1, \dot{f}_2) \in W^\infty(\mathbb{A} \subset \mathbb{R}^p; \mathbb{C})^2$ compatibles sur $\mathbb{A}_1 \cap \mathbb{A}_2$, dans $W^\infty(\mathbb{A} \subset \mathbb{R}^p; \mathbb{C})$ en sorte que $C(\dot{f}_1, \dot{f}_2)$ soit compatible avec f_i sur $\mathbb{A}_i$ $(i = 1, 2)$.

Cet opérateur de recollement est de la forme $(\dot{f}_1, \dot{f}_2) \mapsto \dot{f}_1 + \dot{\alpha}(\dot{f}_2 - \dot{f}_1)$, ou $\dot{\alpha}$ est un champ Taylorien défini sur $\mathbb{B} - \{\mathbb{A}_1 \cap \mathbb{A}_2\}$, qui est un multiplicateur rugueux pour les champs tayloriens définis sur $\mathbb{A}$ et plat sur $\mathbb{A}_1 \cap \mathbb{A}_2$ (cf. [4] p.54). A priori, le produit $\check{\alpha}(\dot{f}_2 - \dot{f}_1)$ n'est pas défini sur $\mathbb{A}_1 \cap \mathbb{A}_2$; mais par définition, puisque $\dot{f}_2 - \dot{f}_1$ est plat sur $\mathbb{A}_1 \cap \mathbb{A}_2$, le produit se prolonge par continuité en un champ taylorien sur $\mathbb{A}$, plat sur $\mathbb{A}_1 \cap \mathbb{A}_2$.

En outre $\dot{\alpha}$ est égal a 0 (resp. 1) au voisinage de $\mathbb{A}_1 - \mathbb{A}_2$ (resp. $\mathbb{A}_2 - \mathbb{A}_1$).

Un tel multiplicateur rugueux (dont les dérivées successives doivent être majorees d'une facon convenable (cf. [4] p.54) peut s'obtenir grâce au théorème de Whitney et le complément 3.3 p.8 de [4].

Second lemme de recollement.

Si $\mathbb{A}_1$ et $\mathbb{A}_2$ sont régulièrement séparés dans $\mathbb{R}^p$, deux champs tayloriens $\dot{f}_i \in W^\infty(\mathbb{A}_i \subset \mathbb{R}^p; \mathbb{C})$ compatibles sur $\mathbb{A}_1 \cap \mathbb{A}_2$ définissent canoniquement un champ Taylorien sur $\mathbb{A}_1 \cup \mathbb{A}_2$.

Par un prolongement de Whitney (non nécessairement linéaire) on se ramène immédiatement au lemme précédent avec $\mathbb{A} = \mathbb{A}_1 \cup \mathbb{A}_2$. Mais le recollement final ne dépend évidemment pas du prolongement de Whitney choisi. Le multiplicateur α est alors l'unique champ Taylorien induit par 1 sur $\mathbb{A}_1 - \mathbb{A}_2$ et par 0 sur $\mathbb{A}_2 - \mathbb{A}_1$.

Remarque.

Ces deux lemmes (que l'on se gardera de confondre) se trouvent implicitement dans [4] p.12-16; mais il convient d'expliciter le caractère linéaire des opérations qui y sont décrites.

Proposition 5. - Il existe des prolongateurs

$$W^{\infty}_{\mathbb{C}}(M_o \subset \mathbb{C}^{1+n};\mathbb{C}) \hookrightarrow W^{\infty}_{\mathbb{C}}(M_o \cup M_k \subset \mathbb{C}^{1+n};\mathbb{C})$$

permettant de prolonger de M_o à $M_o \cup M_k$.

1ère démonstration (Mather)

On peut utiliser le lemme 0 de Mather (cf. article ci-joint) pour prolonger tout $\ddot{f}_{M_o} \in W^{\infty}_{\mathbb{C}}(M_o \subset \mathbb{C}^{1+n};\mathbb{C})$ en une fonction $\hat{f} \in \mathcal{E}^{\infty}(\mathbb{C}^{1+n};\mathbb{C})$. La fonction $\hat{f}$ induit un champ $\ddot{f}_{M_k}$ sur M_k, grâce à la proposition 2. Et l'on peut recoller $\ddot{f}_{M_o}$ et $\ddot{f}_{M_k}$ d'apres le second lemme de recollement.

2ème démonstration. (inspirée de Seeley [7] qui l'appliquait aux automorphismes $(x,\underline{y}) \mapsto (-2^p x,\underline{y})$ au lieu des automorphismes L_p suivants). (Łojasiewicz utilise une autre variante de cette méthode dans son article ci-joint).

Lemme 1. - Les automorphismes linéaire réels L_p, holomorphes en z, de $\mathbb{C}^{1+n}$ définis par $L_p : (z,\underline{\gamma}) \mapsto (z-\beta_k(1+2^p i),\underline{\gamma})$ appliquent M_k sur M_o et induisent l'identité sur $M_k \cap M_o$.

En effet, si $(z,\underline{\gamma}) \in M_k$, $z-\gamma_k$ est réel, et comme β_k est la partie imaginaire de γ_k, $z-\beta_k$ et $i\beta_k$ sont réels. De plus $\beta_k = 0$ lorsque $(z,\underline{\gamma}) \in M_k \cap M_o$.

Corollaire. Les automorphismes H_p réels, holomorphes en z définis par

$$(z,\gamma) \mapsto (z-(\beta_k-\beta_{k'})\ (1+2^p i),\ \underline{\gamma})$$

appliquent M_k sur $M_{k'}$ et induisent l'identité sur $M_k \cap M_{k'}$.

Pour démontrer la proposition 5 choisissons d'abord une fonction $\phi \in \mathcal{E}^{\infty}(\mathbb{R};\mathbb{C})$ à support compact égale à 1 au voisinage de l'origine : considérée comme une fonction definie sur $\mathbb{R} \times \mathbb{C}^n$ (mais indépendante de $\underline{\gamma}$) elle définie un champ taylorien complexifié $\ddot{\phi}$ sur M_o. L'application $\ddot{f} \mapsto \ddot{g} = \ddot{f} \times \ddot{\phi}$ est linéaire continue.

Nous allons définir le champ $\ddot{F} \in W^{\infty}_{\mathbb{C}}(M_k \subset \mathbb{C}^{1+n};\mathbb{C})$ grâce à la formule

III.1 $$\ddot{F} = \sum_{p \in \mathbb{N}} \lambda_p (\ddot{g} \circ L_p)$$

où les L_p sont les automorphismes du lemme 1, et où les λ_p sont des coefficients complexes qui devront être choisis convenablement. En dehors de $M_k \cap M_o$, grâce au facteur $\ddot{\varphi}$, l'on obtient un champ taylorien pour un choix quelconque des coefficients λ_p : la série (III.1) ne comporte alors qu'un nombre fini de termes non nuls.

Pour que $\ddot{F}$ soit partout défini et taylorien (même sur $M_k \cap M_o$) il suffit que les λ_p définissent une fonction entière $E(\zeta) = \sum_{p \in \mathbb{N}} \lambda_p \zeta^p$. En effet, dès que $\sum_{p \in \mathbb{N}} |\lambda_p| (2^k)^p < \infty$, pour tout k, $\ddot{F}$ définit bien un élément de $W^\infty_{\mathbb{C}}(M_k \subset \mathbb{C}^{1+n}; \mathbb{C})$ en vertu du calcul des dérivées successives de $\ddot{g} \circ L_p$ et du critère de Cauchy (rappelons que $W^\infty_{\mathbb{C}}(M_k \subset \mathbb{C}^{1+n}; \mathbb{C})$ isomorphe à $\mathcal{E}^\infty(M_k; \mathbb{C})$ (proposition 2), est complet).

<u>Lemme 2.</u> - <u>Pour que</u> $\ddot{F}$ <u>soit compatible avec</u> $\mathfrak{g}$ <u>(et par conséquent</u> $\ddot{f}$<u>) sur</u> $M_k \cap M_o$, <u>il faut et il suffit que</u> $E(2^p) = i^p$ <u>pour tout</u> $p \in$.

Pour exprimer cette compatibilité en un point $(\mathring{z}, \mathring{\gamma}) \in M_o \cap M_k$ (avec $\mathring{z}$ réel, $\mathring{\gamma}_k$ réel (i.e. $\mathring{\beta}_k = 0$)) considérons le développement de Taylor de $\ddot{f}$ en ce point. Il est de la forme

$$\sum_{\mu,\nu,\underline{\omega}} A_{\mu,\nu,\underline{\omega}} Z^\mu B^\nu \underline{A}^{\underline{\omega}}, \quad \text{ou } B = \beta_k - \mathring{\beta}_k \quad \text{et}$$

$\underline{A}$ représente le $2n - 1$ coordonnées de $\underline{\alpha} - \mathring{\underline{\alpha}}$, autre que β_k.
indice à $2n - 1$ termes. Alors le développement correspondant de $\ddot{F}$ est

$$\sum_{\mu,\nu,\underline{\omega}} A_{\mu,\nu,\underline{\omega}} \Sigma \lambda_p (Z - B(1 + 2^p i))^\mu B^\nu \underline{A}^{\underline{\omega}}$$

En écrivant $(Z-B(1 + 2^p i))^\mu B^\nu = \sum_{\mu' \leq \mu} \binom{\mu}{\mu'} Z^{\mu-\mu'} B^{\mu'+\nu} (1 + 2^p i)^{\mu'}$ la compatibilité signifie que le coefficient de $A_{\mu,\nu,\underline{\omega}} Z^{\mu-\mu'} B^{\mu'+\nu} \underline{A}^{\underline{\omega}}$ doit être égal à 1, si $\mu' = 0$ et à 0 si $\mu' \neq 0$. Autrement dit

$$\sum_{p \in \mathbb{N}} \lambda_p (1 + 2^p i)^{\mu'} = \begin{cases} 1 & \text{si } \mu' = 0 \\ 0 & \text{si } \mu' \neq 0 \end{cases}$$

Il est remarquable que ces conditions ne dépendent plus de μ, avec $\mu \geq \mu'$.
Utilisant la fonction entière $E(\zeta)$ cela s'écrit

$$\sum_{h \leq \mu'} \binom{\mu'}{h} i^h E(2^h) = \begin{cases} 1 & \text{si } \mu' = 0 \\ 0 & \text{si } \mu' \neq 0 \end{cases}$$

Ce système infini d'équations aux inconnues $E(2^p)$ admet manifestement comme unique solution $E(2^p) = i^p$ comme le montre le développement de $(1+i^2)^{\mu'}$. Le lemme 2 est démontré.

Une variante classique du théorème de Mittag-Leffler (cf. [3] p.192) affirme qu'il existe des fonctions entières prenant des valeurs arbitrairement choisies (ici i^p) en une suite de noeuds d'interpolation (ici 2^p) sous réserve que ceux-ci n'aient pas de points d'accumulation à distance finie (c'est le cas ici).

L'existence de coefficients λ_p définissant $\ddot{F} \in W^\infty_{\mathbb{C}}(M_k \subset \mathbb{C}^{1+n}; \mathbb{C})$ compatible avec $\ddot{f}$ sur $M_o \cap M_k$ est ainsi prouvée. Grâce au second lemme de recollement on obtient un prolongateur qui passe de M_o a $M_o \cup M_k$.

<u>Corollaire.</u> - <u>Il existe, de même, des prolongateurs passant de</u> M_k <u>a</u> $M_k \cap M_k$,

<u>Demonstration du Theoreme 2.</u> - Désignons par (P_j) la proposition suivante que nous démontrerons par récurrence finie sur j $(1 \leqslant j \leqslant n)$.

<u>Si</u> $\mathbb{E}_j$ <u>et</u> $\mathbb{E}_{j+1} = \mathbb{E}_j \cup M$ <u>sont des parties de</u> $\mathbb{C}^{1+n}$ <u>contenant</u> M_o, <u>réunions de</u> j (<u>resp.</u> $j+1$) <u>hyperplane pris parmi les</u> M_k, <u>il existe un prolongateur</u>

$$(P_j) \qquad W^\infty_{\mathbb{C}}(\mathbb{E}_j \subset \mathbb{C}^{1+n}; \mathbb{C}) \to W^\infty(\mathbb{E}_{j+1} \subset \mathbb{C}^{1+n}; \mathbb{C})$$

<u>passant de</u> $\mathbb{E}_j$ <u>a</u> $\mathbb{E}_{j+1}$.

Le cas $j = 1$ constitue la proposition 5.

Pour $j > 1$, soient $\mathbb{E}_j$ et $\mathbb{E}_{j+1} = \mathbb{E}_j \cup M$ des réunions d'hyperplans telles qu'elles viennent d'être précisées.

Alors $\mathbb{E}_j$ est une réunion $\mathbb{E}_j = \mathbb{E}' \cup \mathbb{E}''$ où $M_o \subset \mathbb{E}'$ et $M_o \subset \mathbb{E}''$ et que $\mathbb{E}'$ (resp. $\mathbb{E}''$) soit une réunion de moins de j hyperplans préleves parmi les M_k.

L'hypothèse de récurrence assure qu'il existe des prolongateurs Λ' et Λ'' passant de $\mathbb{E}'$ a $\mathbb{E}' \cup M$ (resp. de $\mathbb{E}''$ a $\mathbb{E}'' \cup M$). Soit $\ddot{f} \in W^\infty_{\mathbb{C}}(\mathbb{E}_j \subset \mathbb{C}^{1+n}; \mathbb{C})$. Après restriction à $\mathbb{E}'$ (resp. $\mathbb{E}''$) on le prolonge a $\mathbb{E}' \cup M$ (resp. $\mathbb{E}'' \cup M$) grâce à Λ' et Λ''. Ces champs Tayloriens $\Lambda'(\ddot{f})$ et $\Lambda''(\ddot{f})$ restreints à M sont compatibles sur $(M \cap \mathbb{E}') \cap (M \cap \mathbb{E}'') = M \cap \mathbb{E}' \cap \mathbb{E}'' \supset M \cap M_o$ car ils s'y réduisent a $\ddot{f}$.

Le premier lemme de recollement permet de les recoller linéairement sur M.

Soit $\tilde{\Lambda}$ un opérateur tel que $\tilde{\Lambda}(\overset{\bullet}{f})$ se réduise a $\Lambda'(\overset{\bullet}{f})$ (resp. $\Lambda''(\overset{\bullet}{f})$) sur $M \cap \mathbb{E}'$ (resp. $M \cap \mathbb{E}''$), c'est-à-dire a $\overset{\bullet\bullet}{f}$.

$\tilde{\Lambda}(f)$ se réduit à $\overset{\bullet\bullet}{f}$ sur $M \cap (\mathbb{E}' \cup \mathbb{E}'')$. Le second lemme de recollement permet alors de recoller $\overset{\bullet\bullet}{f}$ défini sur $\mathbb{E}_j = \mathbb{E}' \cup \mathbb{E}''$ et $\tilde{\Lambda}(\overset{\bullet}{f})$ défini sur M, compatibles sur $M \cap \mathbb{E}_j$.

Le théorème 2 est établi dès que j atteint n.

IV. Prolongement de Whitney et prolongateur.

Nous profitons de cette rédaction pour rappeler des faits connus (mais diffi - cilement accessibles) concernant le prolongement des fonctions différentiables.

Pour tout entier r, le théorème du prolongement de Whitney construit explicitement un prolongateur $W^r(\mathbb{F} \subset \mathbb{R}^p; \mathbb{C}) \to \mathcal{E}^r(\mathbb{R}^p; \mathbb{C})$ pour tout fermé $\mathbb{F} \subset \mathbb{R}^p$.

Mais pour $r = \infty$, il n'existe pas, en général, de prolongateur linéaire réalisant ce prolongement.

Le contre-exemple classique concerne le cas où $\mathbb{F}$ se réduit à un seul point: $W^\infty(\{0\} \subset \mathbb{R}^p; \mathbb{C})$ est l'anneau des séries formelles $\mathbb{C}[[X_1, X_2 \ldots X_p]]$ muni de la topologie de Frechet isomorphe à la topologie produit sur l'espace des coefficients.

Le théorème d'Emile Borel (1898), généralisé à p variables, affirme que toute série formelle est la serie de Taylor d'une fonction $\mathcal{E}^\infty$. Mais ce prolongement s'obtient par approximation succesives dépendants de la série formelle étudiée.

L'impossibilité d'un prolongement linéaire continu est une conséquence de la remarque suivante : Si $\mathbb{K}$ est une boule compacte au voisinage de l'origine $\mathbb{K} \subset \mathbb{R}^p$, $\mathcal{E}^\infty(\mathbb{K}; \mathbb{C})$ est un espace de Frechet dont la topologie est définie par une suite infinie de normes, alors que $\mathbb{C}[[X_1, X_2 \ldots X_p]]$ a une topologie définie par une suite de semi-normes (mais non de normes). Dans le second espace tout voisinage de 0 contient des droites, contrairement au premier.

Il en résulte que $\mathbb{C}[[X_1, X_2 \ldots X_p]]$, quotient de $\mathcal{E}^\infty(\mathbb{K}; \mathbb{C})$, ne saurait être isomorphe à un sous-espace fermé de $\mathcal{E}^\infty(\mathbb{K}; \mathbb{C})$. Autrement dit, l'idéal des fonctions $\mathcal{E}^\infty$ infiniment plates à l'origine, n'a pas de supplémentaire topologique dans $\mathcal{E}^{\infty}(\mathbb{K};\mathbb{C})$.

Cependant B. Mityagin a montré (cf. Approximative dimension and bases in nuclear spaces - Russian Math. Surveys 16, 1961) qu'il existait des prolongateurs de $\mathcal{E}^{\infty}(K;\mathbb{C})$ dans $\mathcal{E}^{\infty}(\mathbb{R}^p;\mathbb{C})$ et ce résultat a été retrouvé par Seeley [7].

Notons enfin que le relèvement d'une série formelle dans l'espace des fonctions $\mathcal{E}^{\infty}$ peut s'obtenir par un opérateur de classe $\mathcal{E}^{\infty}$ (non linéaire) défini sur l'espace de Fréchet $\mathbb{C}[[X_1,X_2 \dots X_p]]$ à valeurs dans $\mathcal{E}^{\infty}(\mathbb{R}^p;\mathbb{C})$. Ceci résulte d'une étude attentive du procédé de prolongement de Whitney dans le cas particulier d'E. Borel.

BIBLIOGRAPHIE

[1] G. Glaeser — Fonctions composées différentiables, Annals of Math. 77, 193-209 (1963)

[2] G. Glaeser — Racine carrée d'une fonction différentiable. Annales de l'Institut Fourier. XIII, 203-210 (1963)

[3] A. Guelfond — Calcul des différences finies. Dunod, Paris, (1963)

[4] B. Malgrange — Ideals of differentiable functions Oxford University Press, (1966)

[5] J. Mather — Stability of $\mathcal{C}^\infty$ mappings - I. The division theorem. Annals of Math. 87, 89-104 (1968)

[6] L. Nirenberg — Lettre à Malgrange (ci - jointe).

[7] R. Seeley — Extension of $\mathcal{C}^\infty$ functions defined in a half-space. Proc. American Math. Soc. 625-629 (1964)

Voir en outre les articles de Lojasiewicz et Mather ci-joints.

STRATIFIED SETS : A SURVEY

C.T.C. Wall

Introduction

The study of stratified sets originated with work of Whitney and Thom [17],[12] on singularities of analytic varieties. Its relevance to the theory of singularities of smooth maps arises from certain analytic subvarieties of jet spaces, and their preimages in a manifold: the important examples are the $\Sigma(i_1,i_2,\ldots,i_k)$ defined by Boardman [0] and the closures of Mather's contact classes [6] which are orbits of an algebraic group action, so their closures are algebraic).

As with simplicial and CW complexes, the basic idea of stratified set is of a partition of a topological space

$$X = \bigcup_n E_n$$

where the E_n are manifolds, called the strata. Among several different but equivalent formulations, I prefer to have E_n an n-manifold (not necessarily connected or non-empty) and, for each n, $\cup\{E_r : r \leq n\}$ a closed subset of X. Some analytic restriction is needed : X compact Hausdorff is unnecessarily narrow, but will suffice for most purposes. But the crux of any working definition is the manner in which E_n is related to its closure, for induction on strata is the basic technique for proving results. However, even the naïve fact that an analytic variety has a stratification in the above sense proved important in the original proof of the preparation theorem [5].

The Whitney regularity conditions

There are two approaches to the topic of stratified sets : the analytic and the synthetic. For the former, X must be contained in a manifold which, since our problems are local, may be supposed to be euclidean space $\mathbb{R}^N$. We need a condition satisfied by important examples which will enable us to prove theorems. Such properties were discovered by Whitney [19] and termed by him 'regularity'. They refer to two strata E_m, E_n $(m < n)$ and a point - which we may take as the origin - of $E_m \cap Cl(E_n)$: for a regular or Whitney stratification one requires them for all such points and all pairs of strata.

(a) For $P \in E_n$ near 0, T_pE_n nearly contains T_oE_m

(b) For $P \in E_n$ near 0, T_pE_n nearly contains the vector OP.

An alternative condition, equivalent [7] to the conjunction of these, is

(c) If $y_i \in E_m \to 0$ and $x_i \in E_n \to 0$ and $T_{x_i}E_n$

converges to an n-plane τ and the directions x_iy_i converge to a direction ℓ, then $\ell \subset \tau$.

This, too, is sometimes known as condition (b). One property, which follows easily from axiom (a), is that if a manifold V meets E_m transversely at 0, then it is transverse to E_n in a neighbourhood of 0 : thus (for V compact and axiom (a) everywhere satisfied) the set of maps $V \to \mathbb{R}^N$ transverse to each stratum of X is open (as well as dense) in $C^\infty(V, \mathbb{R}^N)$.

Whitney himself showed that complex analytic sets had stratifications satisfying these axioms [20] after writing earlier papers [17],[18] on the real analytic case. In the real case, however, the notion of semi-analytic is more useful. A subset X of $\mathbb{R}^N$ is semi-analytic if for any $P \in \mathbb{R}^N$ (not merely for $P \in X$) there is a neighbourhood U such that $X \cap U$ belongs to the Boolean algebra of subsets of U generated by the subsets $\{Q \in U : f(Q) > 0\}$ with $f : U \to \mathbb{R}$ analytic. It was shown by Łojasiewicz [4] that any semi-analytic set has a Whitney stratification.

Since the proof is not very accessible, I will describe the main idea. Let $\dim X = n$; define $E_n(X)$ to be the set of points of X at which X is locally a smooth submanifold of $\mathbb{R}^N$ of dimension n . Set $Y = X - E_n(X)$. If we took $E_{n-1}(X) = E_{n-1}(Y)$, there would be no guarantee that Whitney's axioms held. Let B be the set of points of $E_{n-1}(Y)$ at which Whitney's axioms fail for the pair of analytic manifolds $(E_n(X), E_{n-1}(Y))$. It is easy to show that B is a semi-analytic set. Once we know $\dim B < (n - 1)$ (or some equivalent property), we can define $E_{n-1}(X) = E_{n-1}(Y) - B$, and then iterate the construction.

The kernel of the argument is thus the proof that for any pair of (disjoint) analytic manifolds E, F in $\mathbb{R}^N$ (with dim E > dim F), the set of points B in F (and necessarily in Cl(E)) at which Whitney's axioms fail for the pair (E, F) is nowhere dense. Now if, for example, F is a point, (a) is trivial and (b) states that as $P \in E$ tends to F, T_pE tends to contain the vector FP. Were this false, then by the Bruhat-Cartan-Wallace curve selection lemma ([1],[16]); see also [8]) there would be an analytic curve C in E starting at F and containing a sequence of such P . But the direction of FP now tends to coincidence with that of $T_pC \subset T_pE$: a contradiction. In the general case one uses the same idea, and a generalisation of the curve selection lemma (the 'wing lemma') due to Whitney [20].

The usefulness of the regularity property, already mentioned in connection with transversality, is emphasised by the 'isotopy lemmas' of Thom [12], which will be further mentioned below.

Abstract stratified sets

The above theory is still, however, somewhat unsatisfactory since it is not clear which of the intrinsic properties of a stratified set are important and how much depends on the embedding in euclidean space. One would like to substitute more internal structure for the Whitney axioms. Such a definition was proposed by Thom in [14] , and a modified form is expounded in detail in his paper below [15] , so there is no need for me to describe it in full. The underlying idea is that each E_n is the interior of a compact manifold $\bar{E}_n$, so X can be formed by iterating the process of attaching $\bar{E}_n$ by a map of its boundary (or using a mapping cylinder), but giving canonical forms for such maps near lower strata. The idea in the following paper by Rayner [9] is essentially a special case of this, but may be easier to follow. It is an important case, as although one often uses a nice function on a manifold to obtain a cell structure by the method of [11], it was not previously known that in fact this gave a CW structure ; the standard cell decomposition of a Grassmannian can be regarded as a case in point.

Digression : Symmetric spaces.

I will now digress to describe (in the spirit of [21]) a nondegenerate function on the classical Lie groups and Riemannian symmetric spaces. Several cases of this already exist in the literature (especially the case $r = 1$ below), but I think it is of interest to give a single general formulation.

Let A be a simple associative algebra of finite dimension over $\mathbb{R}$, I an involution of A and J an anti-involution commuting with I; write J' for the anti-involution IJ. Denote the positive and negative eigenspaces of I by A_o, A_1; of J by V_+, V_- and of J' by V'_+, V'_-. Let G be the Lie group

$$G = \{x \in A : Jx . x = 1\} ;$$

G_o the subgroup $G \cap A_o$. The usual map $x \mapsto Ix . x^{-1}$ induces an embedding of the symmetric space G/G_o as a union of components of $S = G \cap V'_+$. The interesting case, G compact, corresponds to having J a positive anti-involution. Finally let $t : A \to \mathbb{R}$ be the trace function. Since for $x \in A$, $t(x) = t(Ix) = t(Jx)$, this defines an inner product $\langle x, y\rangle = t(Jx.y)$ on A which is nondegenerate (positive if J is); then V_+, V_- are orthogonal complements and so are V'_+, V'_-.

For any $r \in A$, we may define a function $f : G \to \mathbb{R}$ by $f(x) = t(rx)$. It will be convenient to suppose $r \in V_+$. Then x is a critical point of f iff $t(ry)$ is constant for y on the tangent space of x, which is parallel to xV_-. This is equivalent to having $r = J(r)$ orthogonal to xV_-, i.e. in xV_+, and thus to $x^{-1}r = J(x^{-1}r) = rx$, or $r^2 = (rx)^2$.

If $r = 1$, we see that the critical points are the involutions of G. If, however, r^2 has distinct eigenvalues so that in representing A as a matrix ring we can take $r = \mathrm{diag}(r_1, \ldots, r_n)$, it follows that $rx = \mathrm{diag}(\pm r_1, \ldots, \pm r_n)$ and so that $x = \mathrm{diag}(\pm 1, \ldots, \pm 1)$: we get all 2^n involutions in the centraliser of r. (Indeed it is easy to see more generality that if the centralisers of r and r^2 coincide, X belongs exactly to the set of involutions which centralise r.) The cases which occur here are A the ring of $(n \times n)$ matrices over $\mathbb{R}$, $\mathbb{C}$ or $\mathbb{H}$ (I is not used yet), and f has 2^n critical points on O_n, U_n or Sp_n accordingly : all are nondegenerate.

We now consider the restriction of f to $S \subset G$. It is convenient here to suppose further that $r \in V'_+$ (i.e. $r \in A_0$). I claim that $x \in S$ is then critical for $f|S$ if and only if it is critical for $f|G$. Note first that (although G and V'_+ do not meet transversely) $T_x S = T_x G \cap V'_+ = xV_- \cap V'_+$: this follows since G acts on the left (by $g \cdot a = I(g)\, a\, g^{-1}$) on A, transitively on each component of S, and leaving V'_+ invariant. Now $T_x G = xV_-$ splits as the direct sum of this and of $xV_- \cap V'_-$ (since J' leaves xV_- invariant), and $r \in V'_+$ is orthogonal to this last summand anyway. Thus r is orthogonal to $T_x S$ iff it is to $T_x G$, which proves my claim.

It remains again to consider cases, and in fact these are determined by I, since the positive involution J commuting with I is unique up to automorphisms of A: indeed we will always take Jx to be the conjugate transpose of the matrix x.

If I is the inner automorphism by a, so a^2 is a scalar matrix λI with λ in the ground field ($\mathbb{R}$ or $\mathbb{C}$), and if (in the real case) $\lambda > 0$, one finds that S is a union of grassmannians :

$$S = \cup \{ O_n / O_p \times O_{n-p} : 0 \leq p \leq n \}$$

(indeed for one choice of I, S consists exactly of the involutions in G); similarly for U_n and Sp_n. Taking r to have distinct eigenvalues - the traditional choice is $r = \text{diag}(1,2,\dots, n)$ - we get a function on S with 2^n critical points. This gives (one can easily check) the usual cell structure on Grassmannians. Hence this does come from a C.W. complex. This was previously proved by Paul Schweizer.

There are, up to isomorphism, four more cases. In each case we obtain a function f on S with 2^n critical points, all diagonal matrices with ± 1 on the diagonal, and (as above) it can be verified without trouble that all these critical points are nondegenerate; in fact it is not hard to find their indices.

If $A = M_n(\mathbb{H})$, and I is conjugation by $i\,I$, we choose r as above. $S = Sp_n / U_n$.

If $A = M_n(\mathbb{C})$, and I is complex conjugation, again take r as above. Here $S = U_n / O_n$.

If $A = M_{2n}(\mathbb{R})$, and I is conjugation by the direct sum of n copies of $\begin{pmatrix} 0 & 1 \\ -1 & 0 \end{pmatrix}$, then $S = O_{2n}/U_n$. Here we must take $r = \mathrm{diag}\,(1, 1, 2, 2, \ldots, n, n)$ since it is to be invariant by I.

Finally if $A = M_{2n}(\mathbb{C})$, and I is conjugation by the same matrix followed by complex conjugation, we have $S = U_{2n}/Sp_n$, and can take r as above.

Abstract stratified sets (resumed)

A more formal treatment of abstract stratified sets is given by Mather in his notes [7]: these stay more consistently in the differential category. Working with the same definition as in Thom's paper below, Mather gives a formal proof that a Whitney stratified set in $\mathbb{R}^n$ admits this structure : this turns on no more than a very carefully stated version of the tubular neighbourhood theorem.

The definition gives for two strata E, F with $\dim E < \dim F$ a neighbourhood $L_E(F)$ of E in F and maps $g_E : L_E(F) \to \mathbb{R}$, $k_{EF} : L_E(F) \to E$ such that (g_E, k_{EF}) is (at least near E) a submersion onto $E \times \mathbb{R}$ and satisfying certain compatibility relations for triples of incident strata. Mather calls these control data, and then defines a controlled vector field as giving a vector field η_E on each stratum E, satisfying $\eta_F(g_E) = 0$ and $T k_{EF}(\eta_F) = \eta_E$. He shows that integrating this stratumwise gives a 1-parameter group of homeomorphisms of X.

Mather shows how, given a controlled submersion $f : X \to P$ (i.e. each $f|E_n$ is a submersion, and f is compatible with the local projections k_{EF}) to lift a vector field on P up to a controlled vector field on X, and deduces Thom's first isotopy theorem; namely, that in this case f is a locally trivial fibration. It follows, for example, that given two points in the same connected component of the same stratum of X, then X has a homeomorphism taking one to the other. It also follows that any point P of X has a neighbourhood homeomorphic to a cone (one can refine the stratification to make that point P a stratum, and then show g_P is a fibration).

Thom's second isotopy theorem, which gives an analogous (but more technical) result for morphisms $X \to Y$ of stratified sets (it is stated in [15] below) is the basis for Mather's proof that C^0-stable maps are dense in $C^\infty(N,P)$ for N, P smooth manifolds with N compact.

A rather different use of vector fields is made by Sullivan in his paper [10] below. If each non-empty stratum has odd dimension, he constructs inductively a nowhere zero controlled vector field. It follows that X has a fixed point free homeomorphism homotopic to the identity, and so zero Euler characteristic.

It would be interesting to see a formal proof that a stratified set had a presentation in Thom's sense [15]. Such a proof would go a long way towards showing that stratified sets are triangulable. There seems to me little doubt that they are; Łojasiewicz has already [3] triangulated semi-analytic sets.

REFERENCES

0. J.M. Boardman, Singularities of differentiable maps, Publ. Math. I.H.E.S. 33(1967) 21-57.

1. F. Bruhat and H. Cartan, Sur la structure des sous-ensembles analytiques réels, C.R. Acad. Sci. Paris, 244 (1957) 988-990.

2. T.C. Kuo, The ratio test for analytic Whitney stratifications. This volume, pp.141-149.

3. S. Łojasiewicz, Triangulation of semi-analytic sets, Ann. Scu. Norm. Sup. Pisa, Sc. Fis. Mat. Ser. 3, 18 (1964) 449-474.

4. S. Łojasiewicz, Ensembles semi-analytiques. Lecture notes, I.H.E.S. (Bures-sur-Yvette), 1965.

5. B. Malgrange, Ideals of differentiable functions, Oxford University Press, 1966.

6. J.N. Mather, Stability of C^∞ mappings, III : finitely determined map-germs, Publ. Math. I.H.E.S. 35 (1968) 127-156.

7. J.N. Mather, Notes on topological stability. Lecture notes, Harvard University, 1970.

8. J. Milnor, Singular Points of Complex Hypersurfaces, Ann. of Math. Study 61, Princeton University Press, 1968.

9. C.B. Rayner, Thom's cell decomposition as a stratified set. This volume, pp.150-152.

10. D. Sullivan, Combinatorial invariants of analytic spaces. This volume, pp.165-168.

11. R. Thom, Sur une partition en cellules associée à une fonction sur une variété, C.R. Acad. Sci. Paris, 228 (1949) 973-975.

12. R. Thom, La stabilité topologique des applications polynomiales, l'Enseignement Math. 8 (1962) 24 - 33.

13. R. Thom, Local topological properties of differentiable mappings, In Differential Analysis, Oxford University Press, 1964 191-202.

14. R. Thom, Ensembles et morphismes stratifiés, Bull. Amer. Math. Soc. 75 (1969) 240-284.

15. R. Thom, Stratified sets and morphisms : local models. This volume, pp.153-164.

16. A.H. Wallace, Algebraic approximation of curves, Canad. J. Math. 10 (1958) 242-278.

17. H. Whitney, Elementary structure of real algebraic varieties, Ann. of Math. 66 (1957) 545-556.

18. H. Whitney and F. Bruhat, Quelques propriétés fondamentales des ensembles analytiques-réels, Comm. Math. Helv. 33 (1959) 132-160.

19. H. Whitney, Local properties of analytic varieties, In Differential and Combinatorial Topology (ed.S.S.Cairns), Princeton University Press, 1965, 205-244.

20. H. Whitney, Tangents to an analytic variety, Ann. of Math. 81 (1965) 496-549.

21. C.T.C. Wall, Graded algebras, anti-involutions, simple groups and symmetric spaces. Bull. Amer. Math. Soc. 74 (1968) 198-202.

THE RATIO TEST FOR ANALYTIC WHITNEY STRATIFICATIONS

T.- C. Kuo

Let V be a real (or complex) analytic variety in R^n (or C^n) and let

$$V = M_1 \cup \ldots \cup M_s \tag{1}$$

be a given stratification. A problem of fundamental importance is to decide whether or not (1) is regular in the sense of Whitney ([6], §19, p.540). (The regularity conditions will be recalled in §2).

In this paper, we shall give a sufficient condition, called the ratio test, for regularity (Theorem 1, §1). In certain cases, this condition is also necessary (Theorem 2, §1). Some examples are given in §4.

The ratio test fails for non-analytic stratifications. An example is the "slow spiral" in §4.

The author wishes to thank S. Lojasiewicz, Y.C. Lu, F. Pham and R. Thom for many valuable conversations.

§1. The Results.

Since complex varieties in C^n can be considered as real varieties in R^{2n}, we shall only consider the real case. The regularity conditions (see §2) are of local nature, we can therefore restrict ourselves to the following situation in R^n : X is a semi-analytic set (see §2) and is a manifold ; Y is an analytic manifold containing the origin 0 ; and $0 \in \bar{X} - X$.

We can perform an analytic transformation of R^n near 0 , if necessary, so that Y coincides with its tangent space $T(Y, 0)$ at 0 .

<u>Theorem 1</u> (<u>Ratio Test</u>). <u>For</u> X, Y <u>as above</u>, <u>if for every</u> $\tau \in T(Y, 0)$,

$$(R) \qquad \lim_{x \to 0} \frac{|\pi_{N(X,x)}(\tau)| \; |x|}{|x - \pi_Y(x)|} = 0 \qquad (x \in X) ,$$

where $N(X, x)$ denotes the normal space of X at x, π_L, for any L, the projection of R^n onto L, then X is regular over Y at 0.

Of course, we may restrict τ to the members of a basis of $T(Y, 0)$.

The converse of Theorem 1 is probably false in general. However, we do have a partial converse.

Theorem 2 If Y is 1-dimensional, then (R) is also necessary for X to be regular over Y at 0.

Geometric Meaning of (R). The distance from (the end point of) τ to the tangent space $T(X, x)$ is

$$|\pi_{N(X,x)}(\tau)| \quad ; \quad \text{while} \quad |x|^{-1}|x - \pi_Y(x)|$$

is the distance from the unit vector $|x|^{-1}x$ to Y. Thus (R) asserts that for any $\tau \in T(Y, 0)$, the ratio of the above two quantities tends to 0 as $x \to 0$.

For linear subspaces A, B let us define a distance from A to B by

$$\delta(A, B) = \sup |a - \pi_B(a)| \ , \ a \in A \ , \ |a| = 1 \ .$$

In general, $\delta(A, B) \neq \delta(B, A)$. (If $\dim A = \dim B$ then $\delta(A, B) = \delta(B, A)$, see [6], §2, p. 498). One can show that if $\dim A \leqq \dim B$ then $\delta(A, B) \leqq \delta(B, A)$. Then (R) asserts that

$$\lim \frac{\delta(T(Y, 0) \ , \ T(X, x))}{\delta(L(x) \ , \ T(Y, 0))} = 0 \ ,$$

where $L(x)$ denotes the line spanned by x.

§2. The Regularity Conditions and Semi-Analytic Sets.

Let X, Y be two disjoint sub-manifolds of R^n. Let $T(Y, y)$ denote the tangent space of Y at y, $N(X, x)$ the normal space of X at x. Let $y \in Y \cap \bar{X}$ be a given point. To simplify notations, let us translate the origin 0 to y. We then say X is regular over Y at 0 if for any $\tau \in T(Y, 0)$ the following two conditions are satisfied :

$$\text{(a)} \quad \lim_{x \to 0} \pi_{N(X, x)}(\tau) = 0 \ ,$$

(b') $$\lim_{x \to 0} \pi_{N(X,x)}\left(\frac{x - \pi_{T(Y,0)}(x)}{|x - \pi_{T(Y,0)}(x)|} \right) = 0 .$$

Condition (a) means that $T(Y, 0)$ is nearly perpendicular to $N(X,x)$; condition (b') means that $N(X,x)$ is nearly perpendicular to the direction of the vector $x - \pi_{T(Y,0)}(x)$.

If X is regular over Y at every point, then we say X is regular over Y.

More about the regularity conditions can be found in [5] and in the preliminary notes [3] of a book by Mather (to be published by Springer).

A subset Z of R^n is a semi-analytic set if every $x \in R^n$ has a neighbourhood U for which $Z \cap U$ is defined by a finite set of analytic equalities and inequalities: namely, is a finite union of sets of the form

$$f_1 = 0 , \dots , f_p = 0 , \quad f_{p+1} > 0 , \dots , f_{p+q} > 0 .$$

Attention should be paid to points of $\bar{Z} - Z$. For instance, the set

$$y = \sin \frac{1}{x} \qquad x \neq 0 ,$$

in R^2 is not semi-analytic. Points $(0, y)$, $|y| \leq 1$, do not have the required property; but all other points do have the property.

In §3, the following lemma will be of vital importance.

<u>The Curve Selection Lemma.</u> (Bruhat-Cartan-Wallace, [2], p,103; [4] §3, p.25.)

<u>If</u> A <u>is a semi-analytic set in</u> R^n <u>and</u> $p \in \bar{A} - A$, <u>then we can select an analytic arc</u>

$$a(s) = (x_1(s), \dots , x_n(s)) , \quad 0 \leq s \leq \delta$$

(i.e. <u>each</u> $x_i(s)$ <u>is a convergent power series</u>) <u>such that</u> $a(0) = p$, <u>and</u> $a(s) \in A$ <u>for</u> $s > 0$.

§3 Proofs.

Consider the set $\tilde{X}$ in $R^n \times R^n$ consisting of all $(x, \nu(x))$ where $x \in X$ and $\nu(x)$ is a unit normal vector of X at x ; $\tilde{X}$ is a semi-analytic set (see [6], §16) .

Proof of Theorem 1.

Since

$$\frac{|x|}{|x - \pi_Y(x)|} \geq 1 ,$$

we have, by (R),

$$\lim_{x \to 0} \pi_{N(X,\, x)}(\tau) = 0 . \tag{3.1.}$$

Hence condition (a) is verified.

Now suppose (b') were not satisfied. Then there would exist $\epsilon > 0$ and points $(x, \nu(x)) \in \tilde{X}$ where x is arbitrarily close to 0, satisfying

$$\nu(x) \cdot \frac{x - \pi_Y(x)}{|x - \pi_Y(x)|} \geq \epsilon > 0 . \tag{3.2}$$

We shall derive a contradiction.

Clearing the denominators of (3.2), one finds that (3.2) defines a semi-analytic subset of $\tilde{X}$. Hence, by the Curve Selection Lemma (§2) we can select an analytic arc $a(s)$, $0 \leq s \leq \delta$, $a(s) \in X$, $a(0) = 0$, and unit normal vectors $\nu(s) \in N(X, a(s))$ for which (3.2) is satisfied. Here $\nu(s)$ is analytic; but we shall not use this fact.

We introduce a notation to simplify formulae :

For a vector $v \neq 0$, write $\mu(v) = |v|^{-1} v$.

Now let $\phi(s)$ denote the (orthogonal) projection of the vector $\nu(s)$ to the 2-dimensional subspace spanned by the unit vectors

$$v_1(s) = \mu(\tfrac{da}{ds}) , \quad v_2(s) = \mu(\pi_y(\tfrac{da}{ds})) .$$

(By replacing δ by a smaller number, if necessary, we may assume $\frac{da}{ds} \notin T(Y, 0)$ for all s. Then $v_1(s)$, $v_2(s)$ are linearly independent.) We shall write $v_i \equiv v_i(s)$ for simplicity. Then we have

$$\phi(s) = \frac{\nu(s) \cdot v_2}{|v_2 - \pi_{v_1}(v_2)|} \mu(v_2 - \pi_{v_1}(v_2)) \tag{3.3}$$

where π_v denotes the projection onto the line spanned by v. Note that

$\nu(s) \cdot v_1 = 0$, since $\nu(s)$ is a normal vector, One can verify (3.3) directly, or find a more general formula in [1], §3, (3.8).

Since v_1, v_2 are unit vectors ,

$$|v_1 - \pi_{v_2}(v_1)| = |v_2 - \pi_{v_1}(v_2)| . \tag{3.4}$$

Hence

$$|\phi(s)| = \frac{|\nu(s) \cdot v_2|}{|v_1 - \pi_{v_2}(v_1)|} . \tag{3.5}$$

Now write $a(s) = (p(s), y(s))$, where $y(s)$ is the component in $T(Y, 0)$, $p(s)$ that in the orthogonal complement of Y .

Then

$$|v_1 - \pi_{v_2}(v_1)| = \left|\frac{da}{ds}\right|^{-1} \left(\frac{dp}{ds} , 0\right) \tag{3.6}$$

By analyticity,

$$\left|s \frac{da}{ds}\right| \sim |a(s)| , \quad \left|s \frac{dp}{ds}\right| \sim |p(s)| \tag{3.7}$$

where $A \sim B$ means that $\frac{A}{B}$ lies between two positive constants.

On the other hand,

$$|\nu(s) \cdot v_2| \leq |\pi_{N(X,a(s))} (v_2)| .$$

Hence, by (3.5), (3.6), (3.7) ,

$$|\phi(s)| \leq \frac{|\pi_{N(X,a(s))} (v_2)| \, |\frac{da}{ds}|}{|\frac{dp}{ds}|}$$

$$\leq C \frac{|\pi_{N(X,a(s))}(v_2)| \, |a(s)|}{|p(s)|}$$

where C is a constant.

Hence, by (R) ,

$$\lim_{s \to 0} |\phi(s)| = 0 . \tag{3.8}$$

Since $\nu(s) - \phi(s)$ is perpendicular to both v_1 and v_2 ,

$$(\nu(s) - \phi(s)) \cdot \mu(v_1 - \pi_{v_2}(v_1)) = 0 \quad ,$$

or

$$(\nu(s) - \phi(s)) \cdot (\mu(\tfrac{dp}{ds}) \ , \ 0) = 0 \ . \tag{3.9}$$

By (3.8) ,

$$\lim \phi(s) \cdot (\mu(\tfrac{dp}{ds}) \ , \ 0) = 0 \ .$$

Hence, by (3.9) ,

$$\lim \nu(s) \cdot (\mu(\tfrac{dp}{ds}), 0) = 0 \ . \tag{3.10}$$

But by analyticity,

$$\lim \mu(\tfrac{dp}{ds}) = \lim \mu \ (p(s)) \ .$$

Hence

$$\lim \nu(s) \cdot (\mu(p(s)), 0) = 0 \ .$$

This contradicts (3.2) . The proof of Theorem 1 is complete .

Proof of Theorem 2

The proof is essentially the reverse of the last proof. Choose $\tau \in T(Y,0)$, $|\tau| = 1$. Since we assumed $\dim Y = 1$, τ generates $T(Y, 0)$ over R .

Assuming regularity, but that (R) fails, we shall derive a contradiction. By the Curve Selection Lemma (§2), there is an analytic arc $a(s)$, $a(0) = 0$, $a(s) \in X$, and unit vectors $\tau(s) \in T(Y, 0)$, $\nu(s) \in N(X, a(s))$ such that

$$\frac{|\nu(s) \cdot \tau(s)| \ |a(s)|}{|p(s)|} \geqslant \epsilon > 0 \ , \tag{3.11}$$

where ϵ is a constant.

As before, let $\phi(s)$ denote the projection of $\nu(s)$ to the space spanned by the vectors $v_1(s) = \mu(\frac{da}{ds})$ and $v_2 = \tau$.

Then, by (3.3), (3.4) ,

$$\phi(s) = \frac{\nu(s) \cdot \tau}{|v_1 - \pi_\tau(v_1)|} \mu(\tau - \pi_{v_1}(\tau)) \tag{3.12}$$

$$(\nu(s) - \phi(s)) \cdot (\mu(\tfrac{dp}{ds}), \ 0) = 0 .$$

But

$$\lim \nu(s) \cdot (\mu(\tfrac{dp}{ds}), 0)$$

$$= \lim \ \nu(s) \cdot (\mu(p(s)), 0) \qquad \text{by analyticity}$$

$$= 0 \qquad \text{by condition (b')} .$$

Hence

$$\lim \phi(s) \cdot (\mu(\tfrac{dp}{ds}), 0) = 0 ,$$

$$\lim \phi(s) \cdot \mu(v_1 - \pi_\tau(v_1)) = 0. \qquad (3.13) .$$

Now let us show that

$$\lim v_1(s) \in T(Y, 0). \qquad (3.14) .$$

Suppose (3.14) were false, then

$$\lim \frac{|p(s)|}{|a(s)|} \neq 0 \quad , \qquad \lim \frac{|a(s)|}{|p(s)|} < \infty .$$

By condition (a), §2 ,

$$\lim \ \nu(s) \cdot \tau(s) = 0 .$$

Hence

$$\lim \ \frac{|\nu(s) \cdot \tau(s)| \ |a(s)|}{|p(s)|} = 0 ,$$

contradicting (3.11) .

Now we may assume $\lim v_1(s) = \tau$, since $\dim Y = 1$. By a simple calculation, we have

$$\lim_{s \to 0} \ [\mu(v_1 - \pi_\tau(v_1)) + \mu(\tau - \pi_{v_1}(\tau))] = 0 . \qquad (3.15) .$$

By (3.13), (3.15) ,

$$\lim \phi(s) \cdot \mu(\tau - \pi_{v_1}(\tau)) = 0 .$$

Hence, by (3.12), (3.6) ,

$$\lim |\phi(s)| = \lim \frac{|\nu(s) \cdot \tau| \ |\frac{da}{ds}|}{|\frac{dp}{ds}|} = 0 . \qquad (3.16)$$

Thus, by (3.7),

$$\lim \frac{|\nu(s)\cdot\tau|\;|a(s)|}{|p(s)|} = 0 .$$

This contradicts (3.11). The proof of Theorem 2 is complete.

§4. Examples

(1) For the real variety V defined by

$$f(x, y, t) \equiv y^2 - t^2x^3 - x^5 = 0$$

in R^3, let Y be the t-axis, X the complement of Y (in V). Then X is regular over Y. The regularity is obvious at any point other than 0. We shall now show regularity at 0.

At $(x, y, t) \in V$, $(x, y) \neq (0, 0)$, the normal space is spanned by

$$\text{Grad } f = (-3t^2x^2 - 5x^4,\ 2y,\ -2tx^3) .$$

Since $(x, y, t) \in V$, $y^2 = t^2x^3 + x^5$ and hence

$$|\text{Grad } f| \geq \sqrt{4y^2} = 2x^{3/2}(t^2 + x^2)^{\frac{1}{2}} .$$

Since $T(Y, 0)$ is 1-dimensional, we may assume $\tau = (0, 0, 1)$. Then

$$\lim \frac{|\nu\cdot\tau|\sqrt{(x^2+y^2+t^2)}}{\sqrt{(x^2+y^2)}} \leq \lim |\nu\cdot\tau| + \lim \frac{|\nu\cdot\tau|\;|t|}{\sqrt{(x^2+y^2)}} .$$

But $|\nu\cdot\tau| \leq \dfrac{|tx^3|}{x^{3/2}(x^2+t^2)^{\frac{1}{2}}} \to 0$ as $(x, t) \to (0, 0)$.

Since $\sqrt{(x^2+y^2)} = \sqrt{(x^2+t^2x^3+x^5)} \geq \frac{1}{2}|x|$ for $|x|$ small, we have

$$\lim \frac{|\nu\cdot\tau|\;|t|}{\sqrt{(x^2+y^2)}} \leq \lim \frac{2|tx^3|\;|t|}{|x|^{5/2}(t^2+x^2)^{\frac{1}{2}}} = 0 .$$

Hence, by Theorem 1, X is regular over Y at 0.

S. Łojasiewicz has also established the regularity by a different method (unpublished).

Let V^* be the complex variety in C^3 defined by the same equation $f = 0$. Let Y^* be the complex t-axis, X^* its complement. Then the regularity condition

(a) is not satisfied at 0 . Carrying out a similar calculation to that above, one finds that

$$\lim \nu \cdot \tau \neq 0$$

along the complex path $t = i\,x$ on V^* .

(2) For the real variety

$$V(k, \ell, m) : \; y^{2k} - t^2 x^{2(k+\ell)+1} - x^{2(k+\ell+m)+1} = 0$$

in R^3 , where $k > 0$, $\ell \geqslant 0$, $m = 0$ are integers, let Y be the t-axis, X its complement. Again, X is regular over Y everywhere. The calculation is similar.

(3) The "slow spiral" in R^2 . Let $Y = \{0\}$, and X be defined by $r = e^{-\theta}$ in polar coordinates. Then condition (a) is satisfied, but condition (b') fails. Since $T(Y, 0) = 0$, (R) holds. This example shows that the ratio test fails for non-analytic stratifications.

(4) See [1] , Theorem 1, Part III .

REFERENCES

[1] T.C. Kuo, Criteria for v-sufficiency of Jets (to appear).

[2] S. Łojasiewicz, Ensembles Semi-Analytiques. Lecture notes, I.H.E.S., 1965.

[3] J.N. Mather, Notes on Topological Stability. Lecture notes, Harvard University, 1970.

[4] J. Milnor, Singular Points of Complex Hypersurfaces. Ann. of Math. Studies No.61. Princton University Press, 1968.

[5] R. Thom, Ensembles et Morphismes Stratifiés. Bull. Amer. Math. Soc. 75,(1969),240-284.

[6] H. Whitney, Tangents to an Analytic Variety. Ann. of Math, 81, (1965) 496-549.

THOM'S CELL DECOMPOSITION AS A STRATIFIED SET

C.B. Rayner †

Let M be a C^∞ compact n-manifold, $\partial M = \phi$, let X be a vector field on M and define

$$\Gamma = \{\beta \in M : X|_\beta = 0\} \subset M .$$

I shall assume that

(I) For each orbit $t \to \phi_t(x)$ of X,

$$\underset{t \to \pm\infty}{\mathrm{Lim}} \ \phi_t(x) \subset \Gamma$$

(II) At each $\beta \in \Gamma$ the eigenvalues of the matrix $(\partial X^i/\partial x^j)$ have non-zero real part.

(III) For all $\alpha, \beta \in \Gamma$, $W_\alpha \pitchfork W_\beta^*$, where W_α, W_β^* denote the unstable and stable manifolds of α, β respectively.

Notation :

(a) $f(\beta) = \dim W_\beta^* = $ index of β

(b) $\alpha < \beta \iff W_\alpha \cap W_\beta^* \neq \phi$

(c) $W_{\alpha\beta} = W_\alpha \cap W_\beta^*/\sim$

where $x \sim y \iff e^{tX}x = y$ for some $t \in \mathbb{R}$.

Thom [4] observed that I, II imply that the unstable manifolds partition M into cells ;

$$M = \bigcup_{\beta \in \Gamma} W_\beta .$$

† I am sad to report that Dr. Rayner died in a swimming accident in August 1970.

Smale [2] showed that "most" vector fields satisfying I, II have property III and moreover

(a) $\alpha < \beta \iff W_\beta \subset \operatorname{Fr} W_\alpha$

(b) $\operatorname{Fr} W_\alpha = \bigcup_{\beta > \alpha} W_\beta$.

He also showed [3] that there exists a "nice" function $f \in C^\infty(M)$ such that

(c) $df(X)|_x > 0 \iff X|_x \neq 0$ for all $x \in M$

(d) $f(\alpha) = \dim W^*_\alpha$ for all $\alpha \in \Gamma$.

Let II' denote condition II strengthened as follows.

(II') Near each $\beta \in \Gamma$ there is a chart $(x_1, \ldots, x_n)$ relative to which X has the form

$$X = \sum_{i=1} A_{ij} x_i \frac{\partial}{\partial x_j}$$

where (A_{ij}) is a constant matrix whose eigenvalues have non-zero real part. From Hartman [1] we have that, up to homeomorphisms of M, II $\iff$ II' . I have obtained a proof of the following theorem, the full details of which I hope to publish shortly in the Journal of Differential Geometry.

Theorem

(i) I, II, III imply that $\{M, W_\alpha\}$ can be given the structure of a CW complex in which the W_α are the open cells.

(ii) I, II', III imply that $\{M, W_\alpha\}$ can be given the structure of a stratified set in the sense of Thom [5] .

(It seems probable that condition II' can be sharpened to II). The idea of the proof is to construct a C^∞ manifold with corners, M', (which I call a gradient complex) and a surjection $\psi : M' \to M$ such that, for all $\delta \in \Gamma$,

$$\psi^{-1}(W_\epsilon) \cong \bigcup_{\alpha < \beta < \gamma < \ldots < \delta < \epsilon} W_{\alpha\beta} \times W_{\beta\gamma} \times \ldots \times W_{\delta\epsilon} \times W_\epsilon ,$$

where the union is disjoint, over all sequences for which $f(\alpha) = 0$. One builds up M' by attaching faces and by extending ψ' over each new face.

REFERENCES

[1] Hartman, P., On the local linearization of differential equations, Proc. Amer. Math. Soc. 14, 1963, 568-573.

[2] Smale, S., Morse inequalities for a dynamical system, Bull. Amer. Math. Soc. 66, 1960, 43-49.

[3] ________, On gradient dynamical systems, Ann. of Math. 74, 1961, 199-206.

[4] Thom, R., Sur une partition en cellules associée à une fonction sur une variété. C.R. Acad. Sci. 228, 1949, 973-975.

[5] ________, Ensembles et morphismes stratifiés, Bull. Amer. Math. Soc. 75, 1969, 240-284.

STRATIFIED SETS AND MORPHISMS : LOCAL MODELS

R. Thom

§1 The models

Definition 1. Graded linear map. (denoted G.L.M.)

Let E^i denote a real vector space of dimension i. Let n be a positive integer, $n = i_1 + i_2 + .. + i_k$ an ordered partition of n into k positive integers, symbolized by the symbol $\omega = i_1, i_2, \ldots, i_k$. A graded linear map F_ω, of symbol ω, is a sequence of surjective linear mappings :

(1) $E_0^n \overset{f_1}{\to} E_1^{n-i_1} \overset{f_2}{\to} E_2^{n-i_1-i_2} \to \ldots \to E_{k-1}^{i_k} \overset{f_k}{\to} 0$

Write $F_m = \ker f_m$, $G_m = \ker (f_m \circ \ldots \circ f_1)$, $(1 \leq m \leq k)$.

The sequence of linear subspaces

$G_1 \subset G_2 \subset \ldots \subset G_m \subset \ldots \subset G_k = E^n$ of respective dimensions

$i_1, i_1 + i_2, \ldots, i_1 + .. i_m, \ldots, n$ will be called the kernel flag of the graded linear map F.

Definition 2. G.L.M.s subordinate to a given G.L.M.

Let A be any subset of the set $\underset{\sim}{k} = \{1, 2, \ldots k\}$ of the first k integers. Suppose that, in the sequence (1) defining F, we delete all E_m whose index m belongs to A. Between two consecutive E_j whose indices do not belong to A, define the corresponding (surjective) linear map $E_j \to E_j$ by composing the arrows in (1) when going over a deleted E_α, $\alpha \in A$. For instance, if we have a three term G.L.M. :

$E^3 \overset{f_1}{\to} E^2 \overset{f_2}{\to} E^1 \overset{f_3}{\to} 0$, the subordinate G.L.M. obtained after deleting E^2 is : $E^3 \overset{f_2 \circ f_1}{\to} E^1 \overset{f_3}{\to} 0$.

Let (F_ω^A) be the subordinate G.L.M. obtained from F_ω by deletion of the subset E_α, with α in the subset A of $\underset{\sim}{k} = \{1, 2, ..k\}$. Clearly, the kernel flag of

(F_ω) is obtained by deleting in the kernel flag of F those spaces G_j whose index belongs to A.

Recall that the number of subordinate G.L.M.s to a graded linear map of length k (composed of k mappings) is the number of subsets in $\underset{\sim}{k} = \{1, 2, \dots k\}$, hence 2^k.

Tubular neighbourhood of a flag

We consider the unit cube I^k in Euclidean k-space $\mathbb{R}^k$, defined by $0 \leq t_i \leq 1$. Let J be the canonical flag defined by the sequence $0 \subset \mathbb{R}^2 \subset \dots \subset \mathbb{R}^j \subset \dots \mathbb{R}^k$, where $\mathbb{R}^j$ is defined by $t_{j+1} = t_{j+2} = \dots t_k = 0$. This flag J is associated to the following G.L.M. :

$$\mathbb{R}^k \overset{\hat{t}_k}{\to} \mathbb{R}^{k-1} \overset{\hat{t}_{k-1}}{\to} \mathbb{R}^{k-2} \to \dots \to \mathbb{R}^1 \overset{\hat{t}_1}{\to} 0$$

where the map $\hat{t}_j$ is the projection obtained by forgetting the coordinate t_j. The linear space $\mathbb{R}^j\,(t_{j+1} = t_{j+2} = \dots = t_k = 0)$ is the image of a splitting of the linear map

$$\mathbb{R}^{j+1} \overset{\hat{t}_{j+1}}{\to} \mathbb{R}^j$$

and indeed also of a splitting of $\hat{t}_{j+1} \circ \dots \circ \hat{t}_k$.

Put $X_j = \mathbb{R}^j - \mathbb{R}^{j-1}$ as subsets of $\mathbb{R}^k$. We want to reconstruct the total space $(\mathbb{R}^k)^+$ (defined by $t_j \geq 0$) from the disjoint union $\cup_i X_i$ matched together via the attaching maps $\hat{t}_j$. As explained in [EMS], the topology of the quotient at a point $x_t \in X_i$ is defined as follows. Let V^i_λ be a system of fundamental neighbourhoods of x in X_i, then in X_j, $j>1$, a system of fundamental neighbourhoods for x is given by the sets :

$$(\hat{t}_{i+1} \circ \hat{t}_{i+2} \dots \circ \hat{t}_{j-1} \circ \hat{t}_j)^{-1}\,(V^i_\lambda) \cap [r_j < \epsilon]\ ,$$

where r_j is some "carpeting function" for X_j.
Here we take for carpeting function a piecewise differentiable function with corners of the type $r_j = \inf \rho_i$, where

$$\rho_k = t_k,\ \rho_{k-1} = \sqrt{t_k^2 + t_{k-1}^2}\ , \dots,\ \rho_m = \Big(\sum_{k-m \leq j \leq k} (t_j)^2\Big)^{\frac{1}{2}}, \dots,\ \rho_1 = (\Sigma(t_i)^2)^{\frac{1}{2}}$$

and to obtain transverse intersections we take

$\rho_1 = \epsilon,\ \rho_2 = \epsilon^2, \ldots, \rho_m = \epsilon^m, \ldots, \rho_k = \epsilon^k$ for small ϵ.

Locally, this amounts to $t_1 = \epsilon,\ t_2 = \epsilon^2, \ldots\ t_k = \epsilon^k$.

As ϵ tends to zero, we see that the tubes around X_j have to be "flattened" in the neighbourhood of X_{j-i}.

For instance, for $k = 2$, we get a system of fundamental neighbourhoods of $R'(t_2 = 0)$ by considering the triangle bounded by the parabola $t_1 = \epsilon,\ t_2 = \epsilon^2$, by $t_2 = 0$, and by $t_1 = 1$. But one may define a C^0 map s of the unit square

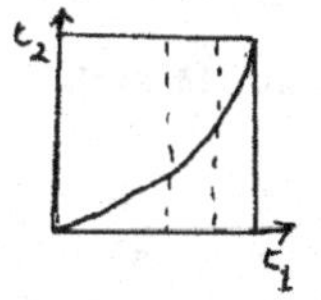

$0 < t_1 < 1,\ 0 < t_2 < 1$ onto this triangle, [or, equivalently, onto the triangle spanned by $(0,0), (1,0)(1,1)$]. The point $(1,1)$ may be taken as fixed points of s_2, and s_2 preserves the levels of the map t_1. Generalizing to k variables, we define a neighbourhood of 0 in $\mathbb{R}^k$ along the J-flag as the standard simplex Δ^k spanned by

$a_0(0,0,\ldots 0),\ a_1(1,0,0 \ldots 0),\ \ldots,\ a_j(1,1,\ldots 1,\ 0 \ldots 0),\ \ldots,\ a_k(1,1,1,\ldots 1)$.

j

The collapsing map s_k of the unit cube I^k $(0 \leq T_j \leq 1)$ onto the standard simplex Δ^k is given by the equations

$$t_1 = T_1,\ t_2 = T_1T_2,\ t_3 = T_1T_2T_3,\ \ldots,\ t_k = T_1T_2 \ldots T_k .$$

The restriction of this map to the subset where all $T_j \neq 0$ is a homeomorphism onto the complement of the face $t_k = 0$. Note that s_k admits the vertices a_i of Δ^k as fixed points. It may be extended to the complement of I^k to give a map (C^0 not C^1) which is the final map of a pseudoisotopy constant outside a neighbourhood of I^k.

<u>Definition 3.</u> <u>Canonical normal k-cross.</u>†

The canonical normal k-cross in k-space R^k is the figure formed by k hyperplanes in general position in k-space, for instance the hyperplanes $T_i = 1$ in (T)-space.

The k hyperplanes of a normal k-cross divide the space into 2^k connected components (as is immediately seen by induction on k). Let A be any subset of the

† <u>Note</u> I am indebted to Robert Williams for the consideration of "normal crosses". R. Williams introduced them in his theory of "branched manifolds".

set $\underset{\sim}{k} = (1, 2, .. k)$, $\bar{A}$ its complement. We define the region $M(A)$ associated to A as the set of points in T space for which $T_i \leq 1$ for $i \in A$, $T_j \geq 1$ for $j \in \bar{A}$.

It is clear that at any point of a normal k-cross there exists a neighbourhood homeomorphic to the product of a normal (k-m) cross by a factor E^m which is a Euclidean m-space. The adherent regions of the complement are at this point manifolds with corners; the integer $k-m$ is said to be the <u>codimension</u> of the corner. With this terminology, a smooth boundary is a "corner" of codimension one at each of its points. The region $M(A)$ associated to a set A of indices is a manifold with corners of codimension k.

Let B be a proper subset of A, $A - B$ its complement in A. The two regions $M(A)$, $M(A-B)$ intersect along a submanifold $N(A, B)$ defined by : $T_i \leq 1$ for $i \in A - B$, $T_j = 1$, for $j \in B$, $T_m \geq 1$ for $m \in \bar{A}$. $N(A, B)$ has zero codimension in a linear submanifold of codimension the cardinal of B. As pointed out in [EMS], if B, C are disjoint subsets of A, $N(A, B \cup C)$ is the transversal intersection of $N(A, B)$ by $N(A, C)$. In particular, $N(\underset{\sim}{k}, A)$ is defined by $T_i \leq 1$ for $i \notin A$ and $T_i = 1$ for $i \in A$; its intersection with I^k is mapped by the collapsing map s_k on a subsimplex of the standard simplex Δ^k, namely the convex hull of the vertices a_j whose indices belong to the complement $\bar{A} = \underset{\sim}{k} - A$.

<u>Definition 4.</u> <u>Normal cross associated to a graded linear map.</u>

Let F be a G.L.M. as defined in (1). Denote by r_m the Euclidean norm on the kernel F_m of f_m, extended to E_{m-1} by the orthogonal projection $E_{m-1} \to F_m$. Denote also by r_m the function induced on E_j (for any $j < m$) by the map $f_{i-1} \circ f_{i-2} \circ \dots \circ f_{j+1}$.

It is clear that the hypersurfaces $r_j = 1$ meet transversally in E_0^n. Any point e of this intersection admits a neighbourhood U which can be described as follows. The set of functions r_j defines a mapping $r : U \to R^k$, which is of maximal rank at e. In the target space $\mathbb{R}^k$, we consider the canonical normal k-cross (H) defined by $r_j = 1$; then U is diffeomorphic to the product of H by a euclidean fibre of dimension $n - k$.

To any subordinate G.L.M. F^A we may associate in U the counterimage by the map r of the region $M(A)$ in the canonical k-cross H. This region meets $M(\underset{\sim}{k})$ (the "interior" of the flag neighbourhood) along a subvariety $N(k;\bar{A})$ whose codimension is the cardinal of $\bar{A}$. We shall call this intersection $N(k;\bar{A})$ the <u>trace</u> of the subordinated G.L.M. F^A on the original map F_ω.

<u>Definition 5.</u> <u>Sector</u>

We consider a sequence of surjective maps between topological spaces :

(2) $$X_1 \overset{p_1}{\to} X_2 \quad \dots \to X_{k-1} \quad \to \quad \dots \overset{p_{k-1}}{\to} \dots \quad X_k$$

We recall that the join of the spaces X_i, denoted $C(X_i)$, is the quotient space of the product $P = X_1 \times X_2 \times \dots X_k$ of the spaces X_j and the standard (k-1)-simplex Δ_{k-1} by a collapsing map $h : P \to C$ defined as follows. Use barycentric coordinates $(t_1,\dots, t_k)$ on the simplex : $0 \leqslant t_i \leqslant 1$, $\Sigma\, t_i = 1$. Then two points $(x_1, \dots, x_k, t_1, \dots, t_k)$, $(x_1', \dots, x_k', t_1', \dots, t_k')$ of P are identified by h if $t_i = t_i'$ for all i, and $x_i = x_i'$ for all i with $t_i \neq 0$. Define in P the closed subset F formed by all systems of points : $x_1 \in X_1$, $x_2 \in X_2, \dots, x_k \in X_k$ such that $x_2 = f_1(x_1)$, ..., $x_k = f_{k-1}(x_{k-1})$. The image $S = h(F \times \Delta_{k-1})$ is, by definition, the <u>sector</u> associated to the sequence (2) of maps .

We then come back to our graded linear map F, defined by (1). The equations $r_j = 1$ define in each space E_m an imbedded manifold W_m, which is a product of spheres; f_{m+1} restricted to W_m is a fibering, with base space W_{m+1} and fiber the sphere $S^{i_m - 1}$ of dimension $i_m - 1$.

Consider the sequence of fiberings :

(3) $$W_0 \to W_1 \to \dots \to W_m \to \dots \to W_{k-1} .$$

Given any point w_0 in W_0, we can form its successive images $w_1 = f_1(w_0)$, $w_2 = f_2(w_1)$ .. $w_{k-1} = f_{k-1}(w_{k-2})$; if we use splittings g_j of the linear maps f_j, we may consider these points as imbedded in E_0^n, where they form independent vectors. In the vector space R_w^k spanned by these vectors w, we may define a collapsing map s_w. Then the image by s_w of the variety $M(\underset{\sim}{k})$ in R_w^k is a k-simplex (Δ_w). The union of all these simplices for variable w_0 in W_0 may be identified with the sector S associated to the sequence of maps (3).

To any subordinate G.L.M. F^A there is an associated subsequence of (3), where the W_j with $j \in A$ are deleted. Then the union of all the traces of all varieties $M_w(A)$ on $M_w(\underset{\sim}{k})$, w varying in W_1, has for image by the collapsing maps s_w the subsector $S(\underset{\sim}{k}, A)$ canonically imbedded in $S(\underset{\sim}{k})$.

§2. Stratified spaces

To define a stratified space A , we need first to have its incidence scheme. This is a finite oriented graph G , with a positive integral valued function d on its vertices, the dimension. If an oriented edge goes from Y to X , notation $Y \hookrightarrow X$, then $d(Y) > d(X)$. Moreover, if there are edges $Z \hookrightarrow Y$, $Y \hookrightarrow X$, there exists also $Z \hookrightarrow X$.

To any vertex X is associated a paracompact manifold (of class C^∞) diffeomorphic to the interior of a compact connected manifold with (possibly empty) boundary and dimension $d(X)$. This is a stratum of E also denoted X . The set of X such that $Y \hookrightarrow X$ is called the boundary of Y ; the set of all Z such that $Z \hookrightarrow Y$ is the star of Y .

Incidence strip, tubular functions and attaching maps.

Let $X \hookleftarrow Y$ be a pair of strata (X incident to Y). To this pair is associated an open subset $L_X(Y)$ of Y , the "incidence strip", and a C^∞ positive real valued function $g_X : L_X(Y) \to R^+$ with everywhere nonvanishing differential, and $g_X^{-1}(0)$ "at infinity" in Y (i.e. g_X is a proper map). Moreover, there is given a map $k_{XY} : L_X(Y) \to X$, with the property that the restriction of k_{XY} to each level hypersurfaces $g_X^{-1}(a)$, for a sufficiently small, is onto X , and of maximal rank. More precisely: to any compact $K \subset X$, there exists a positive b, such that for all $a < b$, the restriction $k_{XY} \mid g^{-1}(a) \cap k_{XY}^{-1}(K)$ is onto K and of maximal rank. These objects have to satisfy the following axioms :

i) Disjointness

If $X \hookleftarrow Y_1$, $X \hookleftarrow Y_2$ and $Y_1 \hookleftarrow U$, $Y_2 \hookleftarrow U$, and no stratum Z exists such that $X \hookleftarrow Z \hookleftarrow Y_1$ and $X \hookleftarrow Z \hookleftarrow Y_2$, then

$$k^{-1}_{Y_1 U} L_X(Y_1) \cap k^{-1}_{Y_2 U} L_X(Y_2) = \emptyset .$$

ii) Transitivity

If $X \hookleftarrow Y \hookleftarrow Z$, then $L_X(Z) \supset k^{-1}_{YZ} (L_X(Y) \cap L_Y(Z))$ and, on the right hand side :

$$k_{XZ} = k_{YZ} \circ k_{XY} .$$

iii) Tubular functions

If $X \hookleftarrow Y \hookleftarrow Z$, then the restriction of g_X to $k^{-1}_{YZ}(L_X(Y)) \cap L_Y(Z)$ is equal to $(g_X \mid Y) \circ k_{YZ}$.

iv) Compacity

For any stratum Y , the union of all $L_X(Y)$ where X describes the boundary of $Y(X \quad Y)$ is a neighbourhood of infinity in the open manifold X .

Remark This system of axioms differs from the presentation in [EMS] ; it replaces the "carpeting functions" of strata by "tubular functions", a dual notion. As pointed out to the author by H. Levine and J. Mather, this change may have some technical advantages.

As a consequence of Axiom iii), to any chain of strata $Y_1 \hookleftarrow Y_2 \hookleftarrow \quad \hookleftarrow Y_k$ there is associated a family of tubular functions g_Y such that the tubes $g_Y = \epsilon$ meet transversally for all sufficiently small ϵ . Hence any point of such an intersection belongs to a normal k-cross in Y_k . Moreover, the totality of all the maps $k_{Y_i Y_j}$ defines locally a graded linear map onto Y_1 .

Presentation of a stratified space.

To any vertex $\underline{a}$ of the incidence scheme G there is associated a compact connected manifold with corners $M(a)$. Let c be a chain of strata in the boundary of $\underline{a}$, for instance : $h \hookleftarrow g \hookleftarrow b \hookleftarrow a$. Let k be the length of c , that is the number of arrows included in the chain. To any such chain c is associated in the boundary $\partial M(a)$ a corner of codimension k, whose normal cross defines a manifold $N_c(a)$ of codimension k. The boundary $\partial M(a)$ is the disjoint union of all $N_c(a)$, where c describes the set of chains ending at a of length ≥ 1. Any subchain c' of c ending at a gives rise to a corner $N_{c'}(a)$, and its normal cross $N_{c'}(a)$ is locally subordinate to the normal cross $N_c(a)$, as explained in §1, Definition 4.

To any couple of strata x, y such that $x \hookleftarrow y$ is associated an attaching map k_{xy} having the following property. Let $A x B y$ be any chain of the incidence scheme G ending at y and containing x (here A and B stand for blocks of arrows); then k_{xy} maps $N_{AxB}(y)$ onto $N_A(x)$ by a surjective map, which looks locally like the normal projection of the normal cross $N_{AxB}(y)$ onto the subordinate normal cross $N_A(x)$, defined by forgetting the B, y coordinates.

These maps k_{xy} being given, the space E obtained from the disjoint union $\bigcup_a M(a)$ by identifying along each boundary $\partial M(a)$ each point $m \in \partial M(a)$ with its images by any map k_{ay} in $M(y)$ is the stratified space we started with. This way of constructing a stratified space E as a union of manifolds with corners, with identifying maps k_{xy} on their boundaries, will be called a "presentation" of E.

The presentation of a stratified space E plays - in stratification theory - an analogous role to the first derived barycentric subdivision in simplicial complex theory.

We may replace the brutal identification through the maps k_{xy} by adding to the boundary $M(x)$ all possible mapping cylinders of the maps k_{xy}. This leads precisely to the "sector" construction. Given a stratum $\{a\}$, we consider its star, that is, the set of all chains c starting at a : $a \hookleftarrow b \hookleftarrow \quad \hookleftarrow g \hookleftarrow h$. To any such chain we associate the sequence of corresponding maps k defined on the normal crosses :

$$N_{ab\ldots g}(h) \xrightarrow{k_{gh}} N_{ab\ldots f}(g) \xrightarrow{k_{fg}} \ldots \rightarrow N_a(b) \xrightarrow{k_{ab}} M(a) ,$$

and construct the corresponding "sector" S_c. We then form a tubular neighbourhood $T(a)$ of the stratum $\{a\}$ by taking the union of all sectors S_c (the trivial sector

$M(a)$ included) and identifying two sectors S_c, $S_{c'}$ along their traces $N_{c \cap c'}$. Then E can be defined as the space obtained from the union of all these $T(a)$ by identifications defined on the sectors by the subordination rule : $S_c \hookleftarrow S_c$, if c' is a subchain of c. An intelligent observer inhabiting a given stratum $\{a\}$ may foresee the local topological structure of the locally attached sectors at a point m of the boundary $\partial M(a)$ by applying the collapsing map s_k to the normal cross defined at m.

§3. Stratified mappings.

Weak definition. Let E, E' be two stratified spaces.

A map $f : E \to E'$ is said to be weakly stratified if

1°) f, continuous, is C^∞ on each stratum X of E

2°) f maps each stratum of E onto a stratum of E' by a map of maximal rank.

3°) for any two strata $X, Y, X \hookleftarrow Y$, let X', Y' be their images in E'. If $X' = Y'$, we require $f_{|X} \circ k_{XY} = f_{|Y}$ on $L_Y(X)$. If $X' \hookleftarrow Y'$, we demand that $f \quad L_X(Y) \subset L_{X'}(Y')$ and

$$f_X \circ k_{XY} = k_{X'Y'} \circ f_Y$$

This definition satisfies the transitivity condition, allowing us to form the weakly stratified mappings into a category. Nevertheless, as no hypotheses are made with respect to the way f behaves with respect to tubular functions, there is no way of relating the "presentations" of E, E' by normal crosses in the strata. Especially, a good deal of pathology in the maps cannot be avoided, and there is no hope of characterizing the topological type of the map by its effect on the incidence schemes of E, E'.

Nevertheless, the weak definition is strong enough to ensure that in a weakly stratified map $f : E \to E'$, the counter-image $f^{-1}(U')$ of any stratum in E' is a fiber space over U', f being the fibration. From this we may deduce the first isotopy theorem of [EMS].

A very natural assumption, to strengthen the definition, is that the local foliation defined by f in a stratum X of E behaves in a uniform way with resp-

-ect to the normal crosses of a "presentation" of E , when we stay inside the same stratum of E . A particular - and very important - case of this situation is given by the following definition.

<u>Definition 6.</u> <u>Gentle mappings</u> (applications douces, sans éclatement in [EMS]).

A weakly stratified mapping $f : E \to E'$ is <u>gentle</u> if for any pair $X \quad Y$ of strata in E with images X', Y' in E' , the map induced by k_{XY} on the tangent plane $(T\ f)^{Y}_{-}(y)$ to the level manifold $f^{-1}\ (f(y))$, $y \in Y$, $x = k_{XY}(y)$, satisfies:

$$(4) \qquad k_{XY} : (T\ f)^{Y}_{-}(y) \to (T\ f)^{X}(x)) \text{ is surjective .}$$

If a mapping $f : E \to E'$ is weakly stratified, surjective and gentle, it is possible to adapt the presentation of E to the mapping f in such a way that the presentation of E' in normal crosses appears as a quotient of the presentation of E . Roughly speaking, we replace the tubular functions g_X by functions $\tilde{g}_X = \text{Sup}\ (g_X\ ,\ g_{X'} \circ f)$. Hence the incidence strips, with these new tubular functions are manifolds with corners at those points where $g_X = g_{X'} \circ f$. Let us consider first the corresponding algebraic situation.

$$\text{Let} \quad X_1 \to X_2 \to \dots X_i \to \dots X_\ell \to 0$$

$$Y_1 \to Y_2 \to \dots Y_j \to \dots Y_k \to 0$$

be two graded linear maps. One may form the direct sum of these two maps by taking the sums $X_i \oplus Y_j$ and the corresponding surjective sum-maps. One gets that way a bigraded linear map. Now one may go from the first element $X_1 \oplus Y_1$ to the last $(0 \oplus 0)$ by a path involving in an arbitrary order horizontal (for Y) and vertical (for X) arrows .

This defines a G.L.M., say

$$E_1 \overset{g_1}{\to} E_2 \overset{g_2}{\to} E_3 \to \dots \to E_s \to 0 ,$$

which projects onto

$$Y_1 \to Y_2 \to \dots \to Y_i \to \dots \to Y_k \to 0$$

by surjective mappings, f_s . (Note that if a map f is a vertical arrow, source and target of this map are mapped on the same Y .) If we denote by N_i the kernel of f on E we observe that the induced map $g : N_i \to N_{i+1}$ is surjective, the condition (4) for gentle mappings. Conversely, if a G.L.M. (E) like

$E_1 \overset{g_1}{\to} E_2 \overset{g_2}{\to} E_s \overset{g_s}{\to} 0$ projects by maps f onto a G.L.M.

$Y_1 \to Y_2 \to Y_i \to 0$ (onto each Y_i), and if the maps $g_s : N_s \to N_{s+1}$ are surjective, where $N_s = \ker (f|E_s)$, then it is possible to generate E as a path in a bigraded linear map of the form $(\to Y \to) \oplus (\to X \to)$.

As a result we may define for E a normal cross presentation which is extracted from the product of the normal crosses associated to Y and to X. This is precisely the construction we obtain after adapting the tubular functions of E to the gentle mapping f . The gentle mappings have the following property: In the diagram : $E \overset{p}{\to} E' \overset{t}{\to} T$ let p be gentle, t weakly stratified. The counter images of a stratum U of T : $t^{-1}(U)$, $p^{-1}\, t^{-1}(U)$, are fibre spaces over U ; then any local trivialization of $t^{-1}(U)$ lifts up to a local trivialization of $p^{-1}\, t^{-1}(U)$. (Second isotopy theorem).

The gentle mappings may - probably - be characterized among all weakly stratified mappings by the following property : if $E \overset{p}{\to} E'$ is gentle, $X \hookleftarrow Y$ a pair of strata in E , $X' \hookleftarrow Y'$ their images in E' ; given a map h of the standard q-simplex Δ^q in X , any deformation of the projected map $p \circ h$ in $X' \cup Y'$ can be lifted up in $X \cup Y$. It is possible that lifting of arcs is already sufficient to characterize gentleness. The importance of gentle maps comes from the fact that any "generic" differentiable map is gentle.

§4 Strongly stratified mappings.

If a weakly stratified map $f : E \to E'$ is not gentle, then there exists a pair of strata $X \hookleftarrow Y$ in E such that the mapping

(4) $\qquad k_{XY}\, (T\ \ f)^{Y}(y) \to (T\ \ f)^{X}(x)\ ,\ \ y = k_{XY}\,(x)$

is not surjective. This shows that the map f collapses the small stratum X more than the big stratum Y (at least locally). This situation occurs, for instance, when projecting the surface of equation $y^2 - x^2 z = 0$ onto the xy - plane : X is the z-axis, Y its complement on the surface. For such "harsh" mappings, the theory is still very unsatisfactory.

A very natural assumption to make is that, on a stratum such as Y , the mapping k_{XY} has constant rank on $(T\ f)^Y\ (y)$; this would make us consider its image in $(T\ f)^X\ (x)$, if the rank is smaller than $\dim f^{-1}\ (x)$. Also we would have to introduce for the mapping k_{XY} the linear sub-bundle defined by the kernel, and take its image by f in $T_{Y'}\ (Y')$, which defines a factorization of $k_{X'Y'}$ depending on $y \in f^{-1}(y')$. This shows that, in such a general situation, the mapping $k_{X'Y'}$ has itself a graded structure (even if $X' \hookleftarrow Y'$ has no intermediary stratum). This may give a way of defining a "strongly stratified mapping". The maps k_{XY} have themselves a graded structure - in fact, they may have several finite dimensional continuous families of such - the factors of these graded structures are mapped surjectively (by surjective linear maps) onto themselves by the k_{XY} , and surjectively onto factors of the $k_{X'Y'}$ by f . Such a definition would allow transitivity. And there is good hope that such a "stratification" may be found for any analytic (or semi-analytic) proper morphism. With this definition, a stratified set which is a projection of a compact semi-analytic set would carry in its own definition of stratified set the trace, the memory, of all the collapsings which were needed to construct it, under the form of families of graded factors in the attaching maps k_{XY}.

But even with this "strong" definition, there is no hope that the incidence scheme map may characterise locally the topological type of the map, as this type may depend on continuous parameters [ST].

REFERENCES

[EMS] R. Thom, Ensembles et morphismes stratifiés. Bull.Amer. Math. Soc. 75 (1969) 240-284.

[ST] R. Thom, La stabilité topologique des applications polynomiales. l'Enseignement Math. 8 (1962), 24-33.

COMBINATORIAL INVARIANTS OF ANALYTIC SPACES

D. Sullivan

We will discuss some corollaries of the following observation :

Let V be a compact space which is stratified by odd dimensional manifolds (in the sense of Thom). Then the Euler characteristic of V is zero.

The proof proceeds by induction over the number of strata. At the nth stage one is attaching the boundary of an odd dimensional manifold to the previously constructed space to obtain a new stratum X_n. The attaching map is a union of fibrations over various spaces which have zero Euler characteristic by induction. It follows that ∂X_n and thus X_n each has zero Euler characteristic.

The proof actually shows that one can build up (starting from the lower strata) a fixed point free flow on V. †

Corollary 1 If X is a stratified space with only even dimensional strata then X is locally homeomorphic to the cone over a space with zero Euler characteristic.

Proof. It is easy to see that locally X is the cone over a stratified space with only odd dimensional strata.

Example i) Suppose that X can be defined locally in $\mathbb{C}^n$ by complex analytic equations. Then according to Thom and Whitney X may be stratified by complex analytic manifolds. These strata are of course even dimensional so that X satisfies the local topological condition of the corollary.

† This point owes much to remarks of A. Borel and the work of J. Mather.

Example ii) Let X be defined locally in R^n by real analytic equations. Consider the complex points in $\mathbb{C}^n$ defined by these equations $X_{\mathbb{C}}$ and the natural involution τ on these induced by complex conjugation of the co-ordinates in $\mathbb{C}^n$. Using the Lojasiewicz theorem we can triangulate $X_{\mathbb{C}}$ so that X is a subcomplex and τ is simplicial. It follows that locally

$$X \sim \text{cone } V$$

$$X_{\mathbb{C}} \sim \text{cone } V_{\mathbb{C}}$$

where τ is an involution of $V_{\mathbb{C}}$ with fixed subcomplex V. Since the number of simplices in $V_{\mathbb{C}}$ is clearly congruent modulo two to the number of simplices in V we have

Corollary 2 A real analytic space is locally homeomorphic to the cone over a polyhedron with even Euler characteristic.

Example a) For a one dimensional space to be real analytic there must be an even number of branches at each singular point.

Example b) The natural singularity type

$$(\text{cone over } \mathbb{C}P^2) \times (\text{euclidean space})$$

is not an analytic singularity.

Stiefel Whitney homology classes of analytic spaces.

One can now make a purely combinatorial discussion in the class of triangulated spaces V satisfying the local Euler characteristic condition of corollary 2. Following Whitney (1940) consider the mod 2 chains s_i defined as the sum of all the i-simplices in a first barycentric subdivision of V, $i = 0, 1, \ldots, \dim V$.

It is a pleasant combinatorial exercise to verify that these chains are cycles mod 2. In fact if $\sigma = (\tau_1 < \tau_2 < \ldots < \tau_i)$ is an $(i-1)$ - simplex of the subdivision, then the coefficient of σ in ∂s_i is just the mod 2 Euler characteristic of a deleted conical neighbourhood of the barycentre of τ_i, so the statement is equivalent to the local Euler characteristic condition. We obtain the Stiefel homology classes

$$s_i \in H_i\,(V\,;\,Z/2) \qquad\qquad i = 0, 1, \ldots, \dim V\,,$$

Now suppose that $f : V \to W$ is a semi-triangulable map, i.e. that the mapping cylinder of f is triangulable.* If the Euler characteristic of each fibre $f^{-1}(pt)$ is odd, then

$$f_*(s_i V) = s_i W\,.$$

For if M_f denotes the mapping cylinder of f with $W \times I$ added to the end, the condition on $f^{-1}(pt)$ implies that M_f has even local Euler characteristics in its its interior. The boundary formula above now implies

$$\partial\, s_{i+1}(M_f) = s_i(V) + s_i(W)\,.$$

<u>Example</u> In case V is a manifold, the Stiefel classes $s_i(V)$ are dual to the (Stiefel) Whitney cohomology classes $w_{n-i}(V)$ (Whitney 1940, Cheeger 1968).

In case $f : V \to W$ is obtained by blowing up a submanifold of odd codimension, then

$$f_*(s_i V) = s_i W\,.$$

<u>Historical Note</u> J. Cheeger and J. Simons were seeking (1967-68) a combinatorial formula for the rational Pontryagin classes of a manifold (solid angle, incidence, etc.) This problem is still unsolved, difficult, and extremely provocative. What resulted however was Cheeger's rediscovery of combinatorial Stiefel-Whitney classes for manifolds (the remarks of Whitney on the subject in 1940 are little known - there are no proofs) and the Chern-Simons work on $4\ell - 1$ forms in the principal frame bundle of a Riemannian manifold.

Cheeger proceeded by constructing specific k-fields with singularities on a triangulated manifold. He then identified the obstruction co-chain for removing the singularities to the dual of s_{k-1}. The identification was a formidable calculation involving the evaluation of a complicated integral and certain solid angle formulae.

---

* Proper algebraic maps are semi-triangulable using Lojasiewicz theory. So are stratified maps and it seems reasonable that proper analytic maps should be.

We wondered if there was a direct argument that these simply defined Stiefel chains were mod 2 cycles. The "even local Euler characteristic" condition was worked out with E.Akin at Berkeley.

Then it seemed natural to ask whether other classes of spaces besides manifolds satisfied this Euler condition and had natural Stiefel homology classes.

It was fairly clear that complex varieties of complex dimension one and two had vanishing local Euler characteristic - essentially Corollary 1 for a small number of strata. I asked P. Deligne if he could give an example of a complex algebraic variety not satisfying corollary 1. To my surprise he almost immediately replied with a convincing argument that no such example existed using Hironaka's local resolution of singularities. He has since outlined a general conjectural theory of Chern classes for singular varieties based on ideas of Grothendieck and this resolution idea. Progress on the theory now depends on the following conjecture.

If $V \xrightarrow{f} W$ is an algebraic map between complete non-singular complex algebraic varieties (V and W not necessarily connected), so that the Euler characteristic of each point inverse $f^{-1}(p)$ is one, then

$$f_* \text{(Chern class of } V) = \text{Chern class of } W .$$

Deligne's verification for complex varieties inspired me to finish the naïve (but complicated) geometrical discussion required for Corollary 1.

Finally one should point out that the generality of corollary 1 shows the result about complex varieties follows from "dimensional considerations". The mod 2 result about real varieties is however <u>geometrically</u> surprising. It depends on the existence of the associated complex variety. This was also true in the "mod 2 fundamental cycle" result of Borel and Haefliger. The sequence of cycles $s_0, s_1, \ldots, s_n$ $n = \dim V$ (or the local Euler condition) provides a generalization of their result - s_n being the fundamental class. One now wonders at the significance of the lower Stiefel homology classes of these analytic spaces - of course s_o is just the mod 2 Euler characteristic of V .

THE JET SPACE $J^r(n, p)$

T-C. Kuo

§1. The space $J^r(n, p)$

Given a local C^s-mapping $f = (f_1, \ldots, f_p) : R^n \to R^p$ with $f(0) = 0$, we may expand each f_i in a Taylor expansion about the origin. If we omit all the terms of degree $\geq r + 1$ $(r < s)$, what remains is a p-tuple of polynomials of degree r, which approximates f. Such a p-tuple is called an r-jet.

Definition An r-jet $z : R^n \to R^p$ is an ordered p-tuple of polynomials of degree r (with real coefficients) in n variables.

The above definition depends on the choice of a coordinate system.

Intrinsic Definition Two germs of C^s-mappings f and g, where $s \geq r + 1$, are called equivalent if their partial derivatives or order $\leq r$ at 0 coincide. An r-jet is an equivalence class.

The set of all r-jets is denoted by $J^r(n, p)$. (Strictly speaking, this space also depends on s.) For $f : R^n \to R^p$, let $j^{(r)}(f)$ denote the Taylor expansion of f up to and including the terms of degree r. Call f a realization of the jet $j^{(r)}(f)$.

§2. The r-extension of a mapping ([8] [14]).

For a local mapping $f : R^n \to R^p$, defined on an open set U, the r-extension of f

$$J^r(f) : U \to J^r(n, p)$$

is defined as follows. For $a \in U$, translate the origins to $\underline{a}$ and $f(a)$, then $J^r(f)(a)$ is the Taylor expansion of f at $\underline{a}$ up to degree r.

Example (2.1) $f(x) = x^3 : R^1 \to R^1$.

Near $\underline{a}$, $f(x) = a^3 + 3a^2(x - a) + 3a(x - a)^2 + (x - a)^3$.

Hence $J^3(x^3)(a) = 3a^2x + 3ax^2 + x^3 \in J^3(1, 1)$.

$J^3(1,1)$ can be identified with R^3 under the correspondence

$$\alpha x + \beta x^2 + \gamma x^3 \quad <-> \quad (\alpha, \beta, \gamma) \quad .$$

Thus $J^3(x^3)$ maps R^1 onto the parabola $\alpha = 3a^2$, $\beta = 3a$, $\gamma = 1$, in R^3 .

Exercise (2.2) For manifolds N^n, P^p, define the r-jet bundle $J^r(N, P)$ with base $N \times P$, fibre $J^r(n, p)$. Then define the r-extension $J^r(f)$ of a mapping $f : N \to P$. (See [8] [14].)

§3 Equivalence in $J^r(n, p)$ defined by coordinate transformations.

If h, h' are local C^s-diffeomorphisms of R^n and R^p respectively, then we say the jet $z \in J^r(n, p)$ is equivalent to $j^{(r)}(h \circ z \circ h')$ under coordinate transformations.

How many equivalence classes are there? In general, infinitely many. In $J^r(1, 1)$, however, there are only $r + 1$ classes, and they are represented by $0, x, \ldots, x^r$. The proof is obvious.

Thus in $J^3(1, 1)$ the four classes are :

C_0, the origin alone ;

C_1, the γ-axis with 0 deleted ;

C_2, the $\beta\gamma$-plane; with the γ-axis deleted ;

C_3, R^3 with the $\beta\gamma$-plane deleted .

(In $J^r(n, p)$, each equivalence class is always a submanifold. See [8], p.3, Proposition 1.)

In Example (2.1), the image of $J^3(x^3)$ meets the γ-axis; this is not a transversal intersection. By Thom's transversality theorem ([14], p.180), for almost all $g(x)$ near x^3 ,

$$\text{Im}(J^3(g)) \cap \gamma\text{-axis} \;=\; \phi \; .$$

Two typical "stable" $g(x)$ near x^3 are

$$g_1(x) \;=\; x^3 + ax \quad \text{and} \quad g_2(x) \;=\; x^3 - ax \; ,$$

where $a > 0$ is small.

§4. Thom's reduction lemma for elementary catastrophes ([16] Chapter 5).

Consider the linear space S of all $z \in J^r(n, 1)$ with $j^{(1)}(z) = 0$.

For $z \in S$, write

$$z = Q(x_1, \ldots, x_n) + (\text{terms of degree } \geq 3),$$

where Q is a quadratic form.

By a suitable linear transformation of the coordinate system (see [1], p.271, Theorem 16), we can write

$$Q = \sum_{i=k+1}^{n} \pm x_i^2 ,$$

where k = corank of Q, $n - k$ = rank of Q. In this new coordinate system,

$$z = \sum_{i=k+1}^{n} \pm x_i^2 + z_1$$

where $j^{(2)}(z_1) = 0$.

<u>Exercise</u> (4.1) Under a further coordinate transformation, we can write

$$Z = \sum_{i=k+1}^{n} \pm x_i^2 + \rho(x_1, \ldots, x_k)$$

where $j^{(2)}(\rho) = 0$. (This says that all terms in z_1 involving x_i, $i \geq k+1$, can be absorbed into the quadratic part.)

We call ρ the residual singularity ([16], Chapter 5, §5.2.D.)

Now for Q to have corank k, k given, the coefficients of $x_i x_j$, where $1 \leq i \leq j \leq k$, must satisfy certain conditions. This imposes $\frac{k(k+1)}{2}$ independent conditions. Let L_k denote the submanifold of S consisting of all Z with corank $Q = k$, then

$$\operatorname{codim} L_k = \frac{k(k+1)}{2} .$$

If $k \geq 3$, then codim $L_k \geq 6$; if $k \leq 2$, codim $L_k \leq 3$. Hence, by the transversality theorem, we have the

<u>Reduction Lemma</u>. (Thom [16], Chapter 5, §5.2.C. last corollary.)

<u>For a generic differentiable mapping</u>

$$\sigma : R^4 \to J^r(n, 1)$$

<u>with</u> $\operatorname{Im}(\sigma) \subset S$, <u>we can write</u>

$$\sigma(t) = \sum_{i=3}^{n} \pm x_i^2 + \rho(x_1, x_2; t)$$

in the neighbourhood of any given point of R^4.

The fact that the residual singularity ρ depends only on two variables x_1, x_2 is crucial to the proof (in [16], Chapter 5) that there are only seven elementary catastrophes.

§5. Sufficiency of jets.

For an r-jet $Z = (Z_1, \ldots, Z_p)$, the complete inverse image $Z^{-1}(0)$ of $0 \in R^p$ is the real algebraic variety defined by the polynomial equations

$$Z_1 = 0 , \ldots, Z_p = 0 .$$

Definition ([7], [15]) A jet $Z \in J^r(n, p)$ is called v-sufficient if for any C^{r+1}-function f with $j^{(r)}(f) = Z$, the germs of $Z^{-1}(0)$ and $f^{-1}(0)$ are homeomorphic. (Roughly speaking, this means that adding terms of degree $\geq r + 1$ to Z does not change the local topological picture of the variety near 0.)

Call Z C^o-sufficient if there exist local homeomorphisms h, h' such that

$$\begin{array}{ccc} R^n & \xrightarrow{f} & R^p \\ h\downarrow & & \downarrow h' \\ R^n & \xrightarrow{Z} & R^p \end{array}$$

is locally commutative.

Of course, C^o-sufficiency implies v-sufficiency .

(We say Z is C^q-sufficient if we can require h, h' to be local C^q-diffeomorphisms. However, Mather in [13] has constructed a jet $Z \in J^7(15, 14)$ having the unpleasant property that for any $Z' \in J^r(15, 14)$ with $j^{(7)}(Z') = Z$, Z' is not C^1-sufficient.)

The remaining sections are devoted to the problem of determining C^o - and v - sufficiency of jets . For C^∞-sufficient of jets, see Mather [12] ; for C^ω-sufficiency see Bochnak [2].

§6. Sufficiency in $J^r(n, 1)$.

Thom's Principle : For a germ of analytic function $f(x_1, \ldots, x_n) : R^n \to R^1$, having 0 as a topologically isolated singularity (i.e. Grad $f(x) = 0$ only for $x = 0$), the variety $f^{-1}(0)$ determines the function f.

An immediate consequence of this principle is the following

Conjecture (6.1) In $J^r(n, 1)$, v-sufficiency and C^0-sufficiency are equivalent notions.

Toward the end of this symposium, Bochnak and Lojasiewicz have given a proof of this conjecture; see [3]. Thom has also described a different proof.

Theorem (6.2) If for a germ of C^∞-function $f : R^n \to R^1$, there is an integer r and $\epsilon > 0$, $\delta > 0$ such that

$$|\text{Grad } f(x)| \geq \epsilon |x|^{r-\delta}$$

near $x = 0$, then $j^{(r)}(f)$ is C^0-sufficient.

This theorem, for $\delta = 1$, was first discovered by Kuiper in [4]. The proof given in [5] is valid for general δ (see [6], Theorem 0); moreover, the technique of the proof in [5] can be used for the following

Exercise (6.3) (Tougeron. See [16], Chapter 5, §5.2D) If a differentiable function $V(x_1, \ldots, x_n)$ is 2-flat (i.e. $j^{(2)}(V) = 0$) and if $g(x_1, \ldots, x_n) \in I^2$, where I denotes the ideal generated by $\frac{\partial V}{\partial x_1}, \ldots, \frac{\partial V}{\partial x_n}$, then $V + g$ is C^1-equivalent to V (i.e. there exists a C^1-diffeomorphism h such that $(V + g) \circ h = V$).

Theorem (6.4) (A special case of Corollary (8.2). See also [3].) If a jet $Z \in J^{r-1}(n, 1)$ is given, then for "almost all" r-forms $H_r(x_1, \ldots, x_n)$, we have

$$|\text{Grad}(Z + H_r)| \geq \epsilon |x|^{r-1} ,$$

where $\epsilon > 0$ is a constant depending on H_r ; hence $Z + H_r$ is a C^0-sufficient r-jet.

The set of all r-forms constitutes a Euclidean space R^N. By "almost all H_r" we mean all H_r except possibly those in a proper algebraic subvariety of R^N.

Exercise (6.5) We say 0 is an algebraically isolated singularity of

$f(x_1,\dots, x_n)$ if, for some k, every k-form $H_k(x_1,\dots, x_n)$ belongs to the ideal generated by the partial derivatives $\frac{\partial f}{\partial x_1},\dots,\frac{\partial f}{\partial x_n}$; and we say 0 is a topologically isolated singularity if for some r, $\epsilon > 0$,

$$|\operatorname{Grad} f(x)| \geqslant \epsilon |x|^r$$

for x near 0. Show that if 0 is an algebraically isolated singularity then it is also topologically isolated.

The converse (in the real case) is false. Take, for instance, $f(x,y) = (x^2+y^2)^2$.

One can derive from the Transversality Theorem, Theorem (6.4), and Exercise (6.5) the following

Proposition (6.6) ([16], Chapter 5, §5.2.B, last corollary).

Every mapping $F : M \times N \to R$ (M, N compact) can be approximated arbitrarily closely in the C^s-topology by a mapping $G(m, x)$, $m \in M$, $x \in N$, having the property that for each fixed m, $G(m, x) : N \to R$ has only topologically isolated singularities.

§7. Sufficiency in $J^r(2, 1)$

In $J^r(2, 1)$, C^0-sufficiency is equivalent to v-sufficiency. This is an immediate consequence of Theorem 1 in [6]. See also [3]. Following Lu ([10]), we say a polynomial $f(x, y)$ has degree of (C^0- or v-) sufficiency r if r is the smallest integer such that $j^{(r)}(f)$ is sufficient. We write $D_s(f) = r$. If no such r exists, write $D_s(f) = \infty$.

For a given polynomial $f(x, y)$, $D_s(f)$ can be calculated ([6][10]). The best step by step method is (probably) the following :

Step 1. By Theorems 1, 2 and 4 of Lu in [10], we may assume $f(x, y)$ has the form

$$f(x, y) = x^m + H_{m+1}(x, y) + \dots + H_n(x, y),$$

where the homogeneous forms H_i do not have terms involving any power x^k for $k \geqslant m - 1$.

Step 2. Apply the method in [6] to calculate $D_s(f)$ for the above f. The details will not be repeated here.

The degree of sufficiency is in general quite different from the degree of the polynomial.

Example (7.1) For $f(x, y) = x^3 - 3xy^k$, $D_s(f)$ is the smallest integer $> \frac{k}{2} + k - 1$ (see [6], §3).

When is $D_s(f) = \infty$? The answer is the following theorem.

Theorem (7.2) *A polynomial $f(x, y)$ has degree of sufficiency $D_s(f) = \infty$ if and only if $f(x, y)$ is divisible by $h(x, y)^2$, where $h(x, y)$ is a polynomial having zeros (in R^2) arbitrarily close to 0.*

The last condition can not be omitted.

Example (7.3) For $f(x, y) = (x^2 + y^2)^2$, $D_s(f) = 4$ by Theorem (6.2). But $h(x, y) = x^2 + y^2$ has no zero other than 0.

Proof of Theorem (7.2) The "if" part is obvious. For "only if", consider the intersection of the (real) curves $\frac{\partial f}{\partial x} = 0$, $\frac{\partial f}{\partial y} = 0$. By Bézout's Theorem ([17], Theorem 5.4, p.111), either 0 is an isolated intersection point (in R^2) or $\frac{\partial f}{\partial x}$, $\frac{\partial f}{\partial y}$ have a common (real) factor $h(x, y)$ which vanishes arbitrarily close to 0. If 0 were an isolated intersection point, we would have a Łojasiewicz inequality ([9], Theorem 2, p.85 ; [11], Theorem 4.1, p.59)

$$\left(\frac{\partial f}{\partial x}\right)^2 + \left(\frac{\partial f}{\partial y}\right)^2 \geq \epsilon(x^2 + y^2)^{k/2} .$$

Then, by Theorem (6.2), $j^{(r)}(f)$ would be sufficient for any $r > \frac{k}{2}$, $D_s(f) \neq \infty$. Now, we may assume h irreducible. By ([6], §3, Proposition 3) f is divisible by h^2.

§8. v-Sufficiency in $J^r(n, p)$

The main criterion has been described in an earlier talk ([7a]. Notice that in condition ($\mathcal{L}$), ∇f_i stood for Grad f_i.) We shall only add some corollaries here; the details will appear in [7] .

Corollary (8.1) *Suppose 0 is a topologically isolated singularity of (the germ of) an analytic variety V defined by $f_1 = 0, \dots, f_p = 0$ (i.e. $f_1(x) = \dots = f_p(x) = \mathrm{Vol}(\mathrm{Grad}\, f_1(x), \dots, \mathrm{Grad}\, f_p(x)) = 0$ only when $x = 0$). Then for some r, the hypothesis of the Corollary in [7a] is satisfied and so $j^{(r)}(f)$ is v-sufficient.*

Corollary (8.2) ([15], Theorem 3) Given an (r-1)-jet $z = (z_1, \ldots, z_p) \in J^{r-1}(n, p)$, then for almost all choices of p-tuples of r-forms $(H_r^{(1)}, \ldots, H_r^{(p)})$, the r-jet $(Z_1 + H_r^{(1)}, \ldots, Z_p + H_r^{(p)})$ satisfies the hypothesis of the Corollary in [7a] and is therefore v-sufficient.

(The p-tuples form a Euclidean space, "almost all" means all except possibly those in a proper algebraic subvariety.)

Now we need another notion to state one more corollary. For an r-form $H_r = \Sigma a_\omega x^\omega$, write $|H_r| = \Sigma |a_\omega|$. We say that $f = (f_1, \ldots, f_p) : R^n \to R^p$ satisfies the T-Condition at degree r near 0 if there is an $\epsilon > 0$ such that for any set of p r-forms $H_r^{(i)}$, $1 \leq i \leq p$, with $|H_r^{(i)}| < \epsilon$, the p surfaces (in R^n)

$$f_1 + H_r^{(1)} = 0, \ldots, f_p + H_r^{(p)} = 0$$

intersect transversally everywhere near 0 (except possibly at 0) in the sense that for x near 0, $x \neq 0$, if $f_i(x) + H_r^{(i)}(x) = 0$ for $1 \leq i \leq p$, then the vectors $\mathrm{Grad}(f_i(x) + H_r^{(i)}(x))$, $1 \leq i \leq p$, are linearly independent.

The T-condition is stronger than saying that the family of surfaces $f_1 = 0, \ldots, f_p = 0$ intersect transversally near 0 ; it also implies that no surface $f_i = 0$ can have any singularity (other than 0) lying on all the other surfaces.

Corollary (8.3) If a local analytic mapping $f = (f_1, \ldots, f_p) : R^n \to R^p$ satisfies the T-condition at degree r near 0, then $j^{(r)}(f)$ is v-sufficient.

Some ideas in connexion with Corollary (8.3) were suggested by R. Thom.

REFERENCES

[1] G. Birkhoff and S. MacLane. A survey of modern algebra, MacMillan, 1953.

[2] J. Bochnak and S. Łojasiewicz. Remarks on finitely determined analytic germs, This volume, pp.263-270

[3] ______________, A converse of the Kuiper-Kuo Theorem. This volume, pp. 254-261.

[4] N.H. Kuiper, C^1-equivalence of functions near isolated critical points. Symposium in Infinite Dimensional Topology. (Baton Rouge, 1967.)

[5] T.C. Kuo, On C^0-sufficiency of jets of potential functions. Topology 8, 1969, 167-171.

[6] T.C. Kuo, A complete determination of C^0-sufficiency in $J^r(2, 1)$. Inventiones Math. 8, 1969, 226-235.

[7] T.C. Kuo, Criteria for v-sufficiency of jets. (to appear)

[7a] T.C. Kuo, A criterion of v-sufficiency of jets, Liverpool Symposium 1969-70.

[8] H.I. Levine, Singularities of differentiable mappings I, This volume, pp.1-89.

[9] S. Łojasiewicz, Ensembles semi-analytiques, Lecture notes, I.H.E.S,, 1965.

[10] Y.C. Lu, Sufficiency of jets in $J^r(2, 1)$ via decomposition, Inventiones Math. 10, 1970, 119-127.

[11] B. Malgrange, Ideals of differentiable functions. Oxford University Press, 1966.

[12] J.N. Mather, Stability of C^∞ mappings III. Publ. Math. I.H.E.S. 35, 1968, 127-156.

[13] J.N. Mather, Stability of C^∞ mappings IV, Publ. Math. I.H.E.S. 37, 1969, 223-248.

[14] V.A. Poenaru, On the geometry of differentiable manifolds. Studies in Modern Topology, Edited by P.J. Hilton, MAA Studies in Mathematics, Vol.5.

[15] R. Thom, Local topological properties of differentiable mappings. Differential Analysis. (Bombay Colloquium 1964) Oxford University Press, 1965.

[16] R. Thom, Stabilité structurelle et morphogenèse. Benjamin. (to be published.) Translated by D.H. Fowler, University of Warwick.

[17] R. Walker, Algebraic curves. Princeton Univ. Press, 1950.

LECTURES ON C^{∞}-STABILITY AND CLASSIFICATION

C.T.C. Wall

Acknowledgement

Although I have used many other sources, this course of lectures is basically an exposition of the work of John Mather, as contained in his 6 papers on stability of C^{∞}-mappings.

Lecture 1 : Stability problems

I want to begin by discussing notions of stability and classification in rather general terms, and show which problems will be of interest in particular cases, and how to answer them in the easiest cases.

In general, suppose given a set S on which we have a topology $\mathcal{T}$ and an equivalence relation e . For most of our applications, S will be a function space.

An element $x \in S$ is stable if the e-equivalence class of x contains a neighbourhood of x. Thus the classes of stable elements contain disjoint open sets. If $\mathcal{T}$ is second countable, there will only be a countable number of them. Hence it is not unreasonable to ask in general for a classification of equivalence classes of stable elements by numerical or algebraic invariants. For other elements of S , this is much less likely, and a better idea there is to try again after changing e (or $\mathcal{T}$). Usually $(S, \mathcal{T})$ will be a Baire space : the intersection of a countable family of dense open sets is dense. Write $\mathcal{G}_S$ for the set of such intersections. A property P of elements of S will be called generic (or said to hold generically) if

$$P(S) = \{x \in S : x \text{ satisfies } P\}$$

contains an element of $\mathcal{G}_S$ as subset. A key problem on $(S, \mathcal{T}, e)$ is whether stability is generic.

To illustrate these concepts, I now given an example. Let S be a (finite-dimensional) manifold, $\mathcal{T}$ its natural topology. Let G be a differentiable transformation group acting (on the right) on S , and let e be the equivalence relation

$$x \ e \ y \iff \exists\, g \in G,\ x.g = y\ .$$

In case this is still too abstract, here are two very explicit examples. Let S be the vector space of complex $n \times n$ matrices, $G = GL_n(\mathbb{C})$ the group of invertible $n \times n$ matrices. Then G acts on S by matrix multiplication. In this case, $x\, e\, y$ if and only if the columns of y span the same space as those of x . Thus x is stable iff it is invertible; stability is generic. Second, G also acts on S by

$$x^g = g^{-1} x g\ .$$

The Jordan canonical form of x represents its equivalence class in this case. Since a small change in x will (in general) change the eigenvalues, no $x \in S$ is stable.

These assertions are clear from linear algebra, but can be recovered from the the differentiable viewpoint. The equivalence class of $x \in S$ is the orbit xG, the image of G under the map $r_x : G \to S$ defined by $r_x(g) = xg$. It is a sub-manifold of S; its tangent space at x is the image of that of G at 1 by the differential $T_1 r_x$. In particular, the dimension of this linear space is that of xG, and xG is open in S (i.e. x is stable) iff $T_1 r_x$ is surjective. For finite dimensional manifolds, it is now easy to compute $T_1 r_x$ and (theoretically but not necessarily computationally) easy to check for surjectivity.

If x is not stable, the standard way to study the action near x is to take a <u>slice</u> : this is any submanifold D of S, passing through x, whose tangent space at x is a complement of that of xG .

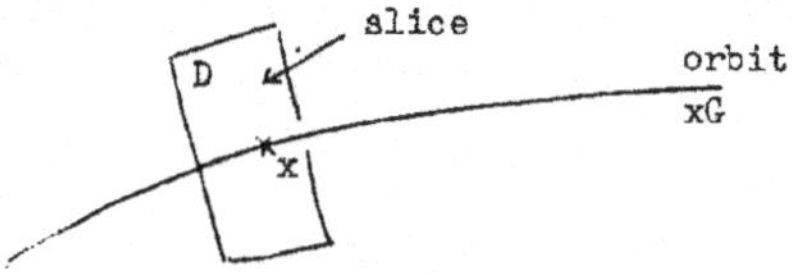

Then DG does contain a neighbourhood of x in S : one pictures the slice sweeping it out under the action of G. Suppose the orbit has dimension a , the slice dimension b : then S has dimension a + b . Although x is not stable, the slice is (locally) stable as a submanifold or, as a b-parameter family of elements of S . This is not hard to show using the local triviality of everything which comes from the differential structure. In this situation we say that x is of <u>codimension b</u> ; in particular, of finite codimension.

It is now time for our main example. Let N^n and P^p be manifolds; consider the set $S = C^\infty(N, P)$ of all smooth (i.e. C^∞-) maps $f : N \to P$. We give S the C^∞- topology, which I will define in lecture 2; I suppose N compact since although more general results can be obtained, I will not have time for the extra technical details. Write Diff N for the group of diffeomorphisms of N (smooth bijections g with g^{-1} smooth), and $G = \text{Diff } N \times \text{Diff } P$. Then G acts on S by $(f)(g,h) = h^{-1}fg$, and we take the equivalence relation given by the action. Thus $f, f' : N \to P$ are equivalent iff there are diffeomorphisms g of N, h of P such that the following commutes :

$$\begin{array}{ccc} N & \xrightarrow{f'} & P \\ \downarrow g & & \downarrow h \\ N & \xrightarrow{f} & P \end{array}$$

The idea of attacking this problem is to take the finite dimensional case above as a model. One must first think of S as a manifold and take the tangent space $\theta(f)$ at f ; we can make this vague idea explicit as follows .

A tangent vector is pictured as an infinitesimal deformation of a point. An infinitesimal deformation of a map $f : N \to P$ should specify for each $x \in N$ how $f(x)$ is to be deformed, i.e. give a tangent vector to P at $f(x)$. Hence we define $\theta(f)$ to be the space of C^∞-maps $N \to TP$ over f, i.e. such that the following diagram commutes

$$\begin{array}{ccc} & & TP \\ & \nearrow & \downarrow \\ N & \xrightarrow{f} & P \ . \end{array}$$

If we write N for the identity map of N, $\theta(N)$ is the tangent space at N to Diff N, so the r_x above should give an $\mathbb{R}$-linear map

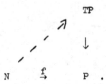

$$\theta(N) \oplus \theta(P) \rightarrow \theta(f)$$

and we expect f to be stable iff this is surjective and of (finite) codimension q iff the image is. More precisely, define $tf : \theta(N) \rightarrow \theta(f)$ and $wf : \theta(P) \rightarrow \theta(f)$ to be given by composition (on the left)

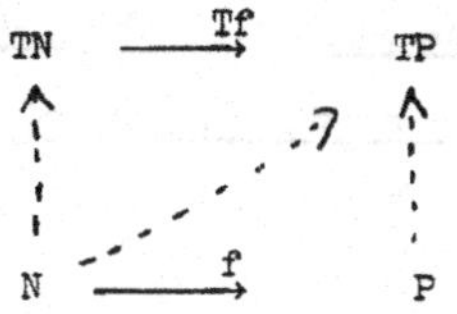

with Tf resp. composition (on the right) with f. This expectation is justified, but the proof is long and difficult, because the topological vector spaces involved are not Banach, and finite dimensional results do not just carry over. The crucial result is in fact the inverse (or implicit) function theorem, and some version of this is the key analytic step in the argument.

For explicit examples, it is simpler just to look at them locally and - since the local problem is much the hardest part of the global - this is not too misleading. First, examples with $n = p = 1$.

$y = x$ no singularity : stable

$y = x^2$ also stable

$y = x^3$ codimension 1.

A slice here is the family $y = x^3 + tx$. Note how there are two stable singularities for $t < 0$ which coalesce and disappear at $t = 0$. This can also be regarded as giving a map $\mathbb{R}^2 \rightarrow \mathbb{R}^2$:

$$y_1 = x_1^3 + x_1x_2 , \qquad y_2 = x_2 .$$

This has Jacobian $3x_1^2 + x_2$, so its only singularities are on the parabola given parametrically as $x_1 = t$, $x_2 = -3t^2$. The image of this in the y-plane is the semi-cubical parabola $y_1 = -2t^3$, $y_2 = -3t^2$, with a cusp at the origin. This is stable (locally) as a map $\mathbb{R}^2 \rightarrow \mathbb{R}^2$. Thom calls it the <u>universal</u> <u>unfolding</u> of the map $y = x^3$. By contrast, the map

$$y_1 = x_1^3 , \qquad y_2 = x_2$$

still has codimension 1 , and

$y_1 = 0$, $y_2 = x_2$

has "infinite codimension".

For dimension 2, there is a complete result, due to Whitney.

Theorem

The mapping $f : M^2 \to N^2$ is stable at the point x_0 if and only if it is equivalent in some neighbourhood of x_0 to one of three mappings

WI. $y_1 = x_1$, $y_2 = x_2$ (regular point),

WII. $y_1 = x_1$, $y_2 = x_2^2$ (fold) ,

WIII. $Y_1 = x_1$, $y_2 = x_1 x_2 - \frac{1}{3} x_2^3$ (cusp)

of a neighbourhood of 0 in the (x_1, x_2)-plane into a neighbourhood of 0 in the (y_1, y_2)-plane.

The stable mappings $f : M^2 \to R^2$ of a compact surface into the plane form an everywhere dense set in the space of all smooth mappings.

The smooth mapping $f : M^2 \to R^2$ is stable if and only if the following two conditions are satisfied :

(a) The mapping is stable at every point in M^2 .

(b) The image of folds intersect only pair-wise and at non-zero angles, whereas images of folds and cusps do not intersect.

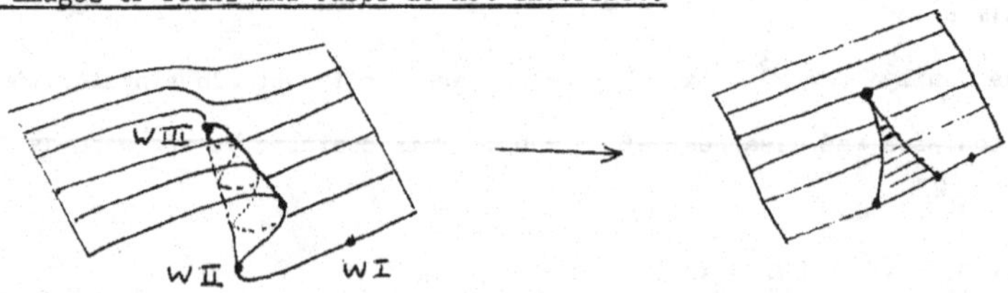

However, there is no such neat result in general; in fact stable maps are only dense in $C^\infty(N, P)$ if (approximately) $p < 7$ or $p \geq \frac{7}{6}(n-1)$. * For classification in other dimensions it is better to look at equivalence up to homeomorphism of range and domain, which is outside the scope of this course.

*Exceptions : include $(n, p) = (8,7), (8,8), (14,15), (20,22), (26,29)$ and exclude $(8,6)$. The proof is in the paper (Mather VI) immediately following this.

Lecture 2 : Jets and local algebras

For any spaces X, Y and point $x \in X$, two maps $f, g : X \to Y$ are said to define the smae germ at $x \in X$ if they agree in some neighbourhood of x. It is not even necessary that f, g should be defined everywhere on X. The point x is called the source of the map-germ, f(x) is the target.

In the differentiable case, we can localise even further. Let N^n, P^p be smooth manifolds; $f, g : N \to P$ smooth maps, $x \in N$ a base point and $r \geqslant 0$ an integer. We write $f \sim_r g$ if $f(x) = g(x)$ and all partial derivatives of order $\leqslant r$ of f and of g, calculated with respect to some choice of local coordinates at $x \in N$ and at $f(x) \in P$, agree. In other words, the Taylor series expansions of f and of g coincide up to (and including) the rth order terms. It is easy to see that this condition does not depend on the local charts. An equivalence class for $\sim_r$ is called an r-jet at x of maps $N \to P$, with target f(x). The r-jets with fixed source and target are parametrised by the

$$p\left\{ n + \binom{n+1}{2} + \binom{n+2}{3} + \dots + \binom{n+r-1}{r} \right\}$$

partial derivatives in question; that these can vary independently is clear by considering polynomial maps(given by the Taylor series up to order r). For example, the 1-jets are determined by the pn entries in the Jacobian matrix of f, so a 1-jet consists simply of source $x \in N$, target $y \in P$ and a linear map $T_xN \to T_yP$.

If we use local coordinates at source and target, together with the partial derivatives listed above, to give charts, we find that the space of all r-jets $N^n \to P^p$ can be considered as a smooth manifold. We denote it by $J^r(N, P)$. Forgetting the rth order partial derivatives defines a map (in fact, the projection of a vector bundle)

$$J^r(N, P) \to J^{r-1}(N, P) ;$$

the composite projection to $J^0(N, P) = N \times P$ just picks out source and target. Write $\pi_r : J^r(N, P) \to N$ for the projection determined by taking the source.

For any smooth map $f : N \to P$ and any $x \in N$, the equivalence class of f at x is an r-jet with source x, say j^rf_x. Taking these for all x defines a smooth section $j^r f$ of π_r, called the r-jet of f.* Of course, not all smooth sections of π_r are r-jets of smooth maps (except when $r = 0$) : there is an integrability condition to be satisfied.

It is convenient to give the definition of function space topologies at this point. The (fine) C^r-topology on the space of C^r-maps $N \to P$ has as base of open sets the sets

$$\mathcal{F}(U) = \{f : N \to P \text{ such that } j^r f(N) \subset U\}$$

for all open sets $U \subset J^r(N, P)$. The ordinary C^r-topology (which we will not use) on $C^r(N, P)$ is similarly induced from the compact-open topology on $C^0(N, J^r(N, P))$. For N compact, both topologies coincide, and convergence is equivalent to uniform convergence of f with all derivatives of order $\leq r$. The (fine) C^∞-topology is defined by taking the above for all r as a base of open sets. With either of these topologies, the space of smooth maps $N \to U$ is a Baire space: countable intersections of dense open sets are dense. Other desirable properties are given in Mather II. Consider in particular the case $P = \mathbb{R}$. Then the germs of smooth maps $(N, x) \to \mathbb{R}$ form an algebra $C_x^\infty(N)$ under pointwise sum and product. The germs with zero target form an ideal $\mathfrak{m}_x$ with quotient $\mathbb{R}$: thus $\mathfrak{m}_x$ is a maximal ideal.

<u>Lemma.</u> <u>Let</u> f, g <u>be germs at</u> x <u>of smooth maps</u> $N \to \mathbb{R}$. <u>Then</u>

$$f \sim_r g \iff f - g \in \mathfrak{m}_x^{r+1} .$$

<u>A set</u> $\{f_\alpha\}$ <u>of elements of</u> $\mathfrak{m}_x$ <u>generate it iff the differentials</u> df_α <u>span the cotangent space at</u> x .

<u>Proof</u> For the second statement, if $\{f_\alpha\}$ generate $\mathfrak{m}_x$ then for local coordinates x_i, we can write $x_i = \Sigma\lambda_\alpha f_\alpha$, and then $dx_i = \Sigma\lambda_\alpha(x)\,df_\alpha$, so the df_α do span the space. Conversely if they do, we can select a subset $f_1, \dots, f_n$ with the df_n a basis. By the inverse function theorem, we can take $(f_1, \dots, f_n)$ as local coordinates at x . So it is enough to show that for $0 \in \mathbb{R}^n$, $\mathfrak{m}_0$ is generated by $x_1, \dots, x_n$.

---

* In the original terminology of Ehresmann, this was called the flow induced by f .

Now for any $f \in C_0^\infty(\mathbb{R}^n)$, we have

$$f(x) - f(0) = \int_0^1 \frac{\partial}{\partial t} f(tx)\, dt$$

$$= \int_0^1 \sum_{i=1}^n x_i\, d_i f(tx)\, dt = \sum_{i=1}^n x_i g_i(x) ,$$

where $$g_i(x) = \int_0^1 d_i f(tx)\, dt ,$$

and so $g_i \in C_0^\infty(\mathbb{R}^n)$. Thus if $f \in \mathfrak{m}_0$ (i.e. $f(0) = 0$), f is in the ideal generated by the x_i.

To prove the first statement, again take local coordinates and write f in place of $f - g$. If $r = 0$, the result holds by definition. Otherwise we proceed by induction.

If $f \in \mathfrak{m}_0^{r+1}$, we have $f = \Sigma x_i f_i$ with $f_i \in \mathfrak{m}_0^r$. By induction, $f_i \sim_{r-1} 0$. Using the rule for differentiating a product, we find $x_i f_i \sim_r 0$, and so $f \sim_r 0$.

If $f \sim_r 0$, and g_i is defined as above, I claim $g_i \sim_{r-1} 0$. Then by induction $g_i \in \mathfrak{m}_0^r$, and $f = \Sigma x_i g_i \in \mathfrak{m}_0^{r+1}$. As to the claim, it follows really since g_i is constructed from the first derivatives of f; explicitly, the Taylor expansion of f determines that of g_j: each term $\lambda \Pi x_i^{a_i}$ for f contributes exactly $\lambda a_j \Pi x_i^{a_i} / x_j \Sigma(a_i)$ to the Taylor expansion for g_j.

This simple result marks the introduction of algebraic tools which play a decisive role in some deeper parts of the theory. Once the seed of algebra is sown it grows fast. For example, if $f : (N, x) \to (P, y)$ is a map-germ, composition with f defines an algebra homomorphism $f^* : C_y^\infty(P) \to C_x^\infty(N)$.

Then if g is another such,

$$f \sim_r g \iff (f^* - g^*)(C_y^\infty(P)) \subset \mathfrak{m}_x^{r+1} .$$

So the induced map

$$f^* : C_y^\infty(P)/\mathfrak{m}_y^{r+1} \to C_x^\infty(P)/\mathfrak{m}_x^{r+1}$$

depends only on $j^r f_x$.

We can now define the local algebra

$$Q_r(f) = C_x^\infty(N)/f^* \mathfrak{m}_y C_x^\infty(N) + \mathfrak{m}_x^{r+1} ,$$

which also depends only on $j^r f_x$. This may look complicated, but is fairly simple in practice. $C_x^\infty(N)/\mathfrak{m}_x^{r+1}$ is the quotient of a polynomial algebra in local coordinates $x_1, \ldots, x_n$ at $x \in N$ by the ideal generated by homogeneous polynomials of degree $(r + 1)$. Now adjoin further relations $y_i = 0$, one for each local coordinate at $f(x) \in P$, where the y's are expressed in terms of the x's by f (but we only need the terms of degree $\leq r$ of the Taylor series).

If we omit the term $\mathfrak{m}_x^{r+1}$, we obtain $Q(f)$, which may be considered the ring of functions on the preimage $f^{-1}(y)$ with induced structure. More useful however, because more easily computed, is the inverse limit $\hat{Q}(f)$ of the $Q_r(f)$. This is a quotient of the ring

$$\lim_{\leftarrow} \quad C_x^\infty(N)/\mathfrak{m}_x^{r+1} ,$$

which we can identify (using a chart) with the ring of formal power series

$$\mathbb{R}[[x_1, \ldots, x_n]] .$$

Note that any $f \in C_0^\infty(\mathbb{R}^n)$ has an (infinite) Taylor expansion, which is a formal power series. There is no need for this to converge or, if it does, for it to converge to f. In fact, we have the

<u>Lemma</u> (E. Borel). (For proof see Malgrange's book)

<u>The map</u> $C_0^\infty(\mathbb{R}^n) \to \mathbb{R}[[x_1, \ldots, x_n]]$ <u>is surjective</u>.

One of the main classification results is the

<u>Theorem (Mather)</u> <u>Let</u> $f : (N, x) \to (P, y)$ <u>be a stable map germ. Then</u> f <u>is determined up to equivalence by the isomorphism class of the local algebra</u> $Q_{p+1}(f)$.

For a sketch of the proof, see lecture 5.

Given a local algebra $\hat{Q}$, we can find corresponding maps by choosing generators $x_1, \ldots, x_n$ (in the maximal ideal $\mathfrak{m}$) and functions $r_1, \ldots, r_p$ of the x_i which give defining relations; if these are smooth functions, they define a smooth map germ $\mathbb{R}^n \xrightarrow{f} \mathbb{R}^p$ with local algebra $\hat{Q}$.

It can also be shown (using Nakayama's lemma : see lecture 4) that $x_1, \ldots, x_n \in \mathfrak{m}$ generate $\hat{Q}$ if and only if they span $\mathfrak{m}$ mod $\mathfrak{m}^2$, so a minimal generating set has just $\dim_{\mathbb{R}}(\mathfrak{m} / \mathfrak{m}^2)$ elements. Choosing such a set gives a map f with $T_0 f = 0$. In general given a map-germ $f : (N, x) \to (P, y)$ with $T_x f \neq 0$ one can make a preliminary simplification of local coordinates as follows. Let rank $(T_x f) = r$; choose local coordinates $y_1, \ldots, y_p$ for P at y such that $d(y_1 \circ f), \ldots, d(y_r \circ f)$ are a base of the image of the dual of T_x, and $d(y_i \circ f) = 0$ for $i > r$. Since $x_1 = y_1 \circ f, \ldots, x_r = y_r \circ f$ have linearly independent differentials at x, they form part of a local coordinate system $x_1, \ldots, x_n$ for N at x. Such local coordinates are said to be linearly adapted to f ; we shall in general try to use such coordinates in examples (and for theory).

The idea to obtain stable germs is to hope that this has finite codimension in the appropriate sense (see later) and take the universal unfolding. An unfolding is a q-parameter family $F_q : \mathbb{R}^n \times \mathbb{R}^q \to \mathbb{R}^p \times \mathbb{R}^q$ whose restriction to $0 \in \mathbb{R}^q$ gives f ; universality means roughly speaking, that for any other such unfolding F_r, there are smooth map-germs $G_n : \mathbb{R}^n \times \mathbb{R}^r \to \mathbb{R}^n \times \mathbb{R}^q$, $G_p : \mathbb{R}^p \times \mathbb{R}^r \to \mathbb{R}^p \times \mathbb{R}^q$ inducing a commutative diagram $G_p \circ F_r = F_q \circ G_n$. One cannot require G_n, G_p to be uniquely determined, but their first partial derivatives (or at any rate, most of them) are. The theory of unfoldings has not yet had a complete exposition: there are several not quite equivalent definitions. There are unpublished versions of Brieskorn (to appear in Shih Weishu seminar, I.H.E.S., 1970) and Mather (to appear in Thom's book); see also Turina's paper listed in the references.

Computationally, we proceed as follows. Write A for the ring $\mathbb{R}[[x_1, \ldots, x_n]]$; $\mathcal{M}$ for its maximal ideal, A^p for a free module of rank p. Let I be the ideal generated by the relators $r_1, \ldots, r_p$ (which we assume to belong to $\mathcal{M}^2$), and let L be the submodule of $\mathcal{M} A^p$ generated by IA^p and by the vectors

$$\partial_i r = \left(\frac{\partial r_1}{\partial x_i}, \ldots, \frac{\partial r_p}{\partial x_i}\right) \qquad (1 \leq i \leq n) .$$

Then $q = \dim_{\mathbb{R}}(\mathcal{M} A^p/L)$ is the codimension of f, and we assume this finite and choose $v_1, \ldots, v_q \in \mathcal{M} A^p$ whose classes mod L form a basis. A universal unfolding is then given by

$$\begin{cases} y_i \circ f = r_i(x_1, \ldots, x_n) + \sum_{j=1}^{q} x_{n+j} v_{j,i}(x_1, \ldots, x_n) & (1 \leq i \leq p) \\ y_{p+j} \circ f = x_{n+j} & (1 \leq j \leq p) \end{cases}$$

Any stable map-germ can be so obtained (up to choice of local coordinates), and any such map-germ is stable.

Example 1 $\quad Q = \mathbb{R}[x/x^{m+1} = 0]$

If we just take the one relator $r = x^{m+1}$, we have $p = 1$, $I = \langle x^{m+1} \rangle$, $\partial r = (m+1)\,x^m$ and so $L = x^m\,\mathbb{R}[[x]] = x^m A$.

We can take $v_i = x^i$ $(1 \leq i \leq m-1)$, so the codimension is $m-1$ and the universal unfolding (writing x_m for x)

$$y_i = x_i \qquad\qquad 1 \leq i \leq m-1$$

$$y_m = x_m^{m+1} + \sum_{i=1}^{m-1} x_i\, x_m^i \;.$$

The cases $m = 2,3,4,5,6$ appear on Thom's list of catastrophes, entitled re pectively : simple minimum, fold, cusp, dovetail (the translation swallow's tail seems to be misleading) and butterfly.

If we take p relators of which $(p-1)$ are zero and the last, $r_p = x^{m+1}$ then $L = x^{m+1}A^p + \mathbb{R}(0,\ldots,0,x^m)$ and for v_j we can choose the vectors $x^k e_i$ (e_i the unit vector in i^{th} place) $1 \leq k \leq m$, $1 \leq i \leq p$, $(k,i) \neq (m,p)$. The unfolding here is Morin's canonical form for the singularity $\sum^{1,1,\ldots,1}$ (m times) .

Example 2 The other cases on Thom's list are :

Generators x, y; relator $x^3 + y^3$ (hyperbolic umbilic), $x^3 - 3xy^2$ (elliptic umbilic) or $x^2y + y^4$ (parabolic umbilic). We have

$$\partial_1 r = 3x^2 \quad \text{resp.} \quad 3(x^2 - y^2) \quad \text{resp.} \quad 2xy$$

$$\partial_2 r = 3y^2 \quad \text{resp.} \quad -6xy \qquad \text{resp.} \quad x^2 + 4y^3$$

so for v_j we can choose x, y and xy resp. $x^2 + y^2$ resp. x^2, y^2 .

Example 3 Mather has shown (see Mather VI §7 below) that a germ representing a singularity of type $\sum^{2,0}$ which is stable (or at least finitely $\mathcal{K}$ -determined) has local algebra equivalent to one of

$I_{a,b}$ $\quad R[[x, y]]/\{xy, x^a + y^b\}$, $b \geq a \geq 2$,

$II_{a,b}$ $\quad R[[x, y]]/\{xy, x^a - y^b\}$, $b \geq a \geq 2$ even

$III_{a,b}$ $\quad R[[x, y]]/\{x^a, y^b, xy\}$, $b \geq a$,

IV_a $\quad R[[x, y]]/\{x^2 + y^2, x^a\}$, $a \geq 3$,

V_a $\quad R[[x, y]]/\{x^2 + y^2, x^a, yx^{a-1}\}$, $a \geq 3$.

Exercise Find universal unfoldings .

Example 4 Generators $x_1, \ldots, x_{n+1}$; relator $- \sum_1^\lambda x_i^2 + \sum_{\lambda+1}^n x_i^2 + x_{n+1}^{r+1}$.

This recovers another normal form due to Morin, (and generalising the Morse Lemma).

Lecture 3 : Transversality

Transversality is the analogue for differentiable mappings of 'general position' arguments in algebraic geometry, and is indeed not unrelated to them. The notion is due (like so much else) to Thom. Before I indicate the proof, let us compute the dimensions of some algebraic varieties.

Example 1 Consider linear maps $\mathbb{R}^n \to \mathbb{R}^p$. Dim = np. Those of rank r form a submanifold. We compute its codimension using the

$$\begin{array}{c c} & \begin{array}{cc} r & n-r \end{array} \\ \begin{array}{c} r \\ p-r \end{array} & \begin{pmatrix} A & B \\ C & D \end{pmatrix} \end{array}$$

matrix representation, partitioned in blocks. For an open subset of matrices, A is nonsingular. Subject to this, our manifold is defined by the $q = (p - r)(n - r)$ independent conditions $D = CA^{-1}B$. Its dimension is $pn - q = r(n + p - r)$.

Note for an $n \times n$ symmetric matrix to have rank $r = n - k$ we can use the same notation ($CA^{-1}B$ is automatically symmetric therefore OK) ; the codimension is again the dimension of the space of D : $\frac{1}{2}k(k + 1)$. The same is valid in the skew-symmetric case, except that the rank r must be even; the codimension is $\frac{1}{2}k(k-1)$.

Example 2 Denote by $G_k(\mathbb{R}^n)$ the Grassmannian of k-dimensional subspaces of $\mathbb{R}^n$. If A is such a subspace, B a complement of it, then subspaces near A also complement B, hence are graphs of linear maps $A \to B$. The dimension we want is thus $k(n-k)$.

Example 3 Let $\mathbb{R}^p \subset \mathbb{R}^n$: then a k-dimensional subspace A will 'generally' satisfy

$$\mathbb{R}^p \cap A = \{0\} \text{ (if } n \geq k+p) \text{ or } \mathbb{R}^p + A = \mathbb{R} \text{ (if } n \leq k+p)$$

I leave you as an exercise to show that the codimension in $G_k(\mathbb{R}^n)$ of the space of A with $\dim(\mathbb{R}^p \cap A) = r$ is

$$r(n-k-p+r) = \dim(\mathbb{R}^p \cap A) \ \text{codim} \ (\mathbb{R}^p + A) .$$

We call two subspaces of a vector space transverse if they span the whole space. [More generally, $W_1, \ldots, W_r \subset V$ are said to be transverse if the natural map $V \to \oplus V/W_i$ is surjective.]

Let now V be a smooth manifold, W a submanifold, $f : M \to V$ a smooth map. Suppose $x \in M$ has $f(x) \in W$. Then f is said to meet W transversaly at x if $T_x f(T_x M)$ and $T_{fx} W$ are transverse :

$$T_x f\,(T_x M) + T_{fx} W = T_{fx} V .$$

If this holds for all $x \in M$ with $f(x) \in W$, we say f is transverse to W, and write $f \pitchfork W$. Similarly one can define transversality of two maps.

Lemma If $f \pitchfork W$, then $N = f^{-1}(W)$ is a smooth submanifold of M, and its normal bundle in M is induced (by f) from that of W in V. In particular, codim N = codim W.

Proof. Let $f_1 \ldots f_q$ be smooth functions on V at $y = f(x)$, with linearly independent derivatives, such that W is defined by $f_1 = \ldots = f_q = 0$. By transversality, the $f_i \circ f$ have linearly independent derivatives; since $f^{-1}W$ is defined by the $f_i \circ f = 0$ it is a submanifold, and the $T_x(f_i \circ f)$ span its normal bundle. The rest is clear.

Note an important special case : if $\dim M < \text{codim } W$, then $f \pitchfork W \Leftrightarrow f(M) \cap W = \phi$. This is the most useful case for applications; this is why

I computed some codimensions above.

The following is the simplest case of the transversality theorem :

Theorem Let V be a smooth manifold, W a submanifold of codimension q ; M a compact smooth manifold. Then $\{f : M \to V : f \pitchfork W\}$ is a dense open subset of $C^\infty(M, V)$.

The openness is clear since transversality is a condition on 1-jets defining an open subset of $J^1(M, V)$, so we have an open set even in the C^1-topology.

For density, first suppose constructed a map

$$F : M \times \mathbb{R}^k \to V \qquad \text{(some } k)$$

such that $F(m, 0) = f(m)$ for all m , and TF is surjective on each fibre (F is a submersion). Clearly $F \pitchfork W$, so $F^{-1}(W)$ is a submanifold of $M \times \mathbb{R}^k$. We will deform f to f_u , where $f_u(m) = F(m, u)$. Then

$$f_u \pitchfork W \iff M \times u \pitchfork F^{-1} W \quad \text{(since the normal spaces to } W \text{ and } F^{-1}(W) \text{ are 'the same')}$$

$$\iff \pi \pitchfork u \qquad \text{(by the same argument)} ,$$

where π is the projection of $F^{-1}W \subset M \times \mathbb{R}^k$ on $\mathbb{R}^k$. But by Sard's theorem, this is true for almost all $u \in \mathbb{R}^k$.

One can find such an F locally without trouble: global density can then be proved by piecing together using bump functions and a Baire category argument. It is simpler to use an embedding $V \subset \mathbb{R}^k$ for some k ; now if N is a neighbourhood of 0 in $\mathbb{R}^k$,

$$M \times N \to \mathbb{R}^k$$

by $(x, y) \to f(x) + y$; this is a submersion, and if we compose with a retraction on V of a neighbourhood in $\mathbb{R}^k$ we get a submersion

$$F : M \times N \to V$$

which suffices for the argument.

Using essentially the same techniques, several extensions of this result can be proved.

(A) For any submanifold S of $J^r(M, V)$, the set of $f \in C^\infty(M, V)$ such that $j^r f \pitchfork S$ is a dense subset, open in the C^{r+1} topology.

<u>Example</u> Let $S = \Sigma^r \subset J^1(M, V)$ be the set of 1-jets corresponding to linear maps $T_xM \to T_yV$ of kernel rank r, i.e. rank $m - r$:
this is a smooth submanifold with codimension $r(v - m + r)$ (Example 1). For a dense open set of maps f, $\Sigma^r f = (j^1 f)^{-1}\Sigma^r$ will be a smooth submanifold of M with this codimension : in particular, it will be empty if $m < r(v - m + r)$. We say that <u>generically</u> f has this property. In particular if $v \geq 2m$ a generic map has $\Sigma^r = \phi$ for all $r \geq 1$; i.e. it is an immersion. This result is due to Whitney.

Thom calls a map f <u>correct</u> if it satisfies "all natural" transversality conditions of this type.

(B) Multiply s copies of $J^r(M, V)$: there is a natural projection π on M^s. If $M^s - \Delta$ denotes the subset of M^s consisting of s-tuples of <u>distinct</u> points of M, write (following Mather)

$$ {}_sJ^r(M, V) = \pi^{-1}(M^s - \Delta) . $$

Now for any smooth $f : M \to V$ its r-jet induces a section

$$ {}_sj^r(f) : M^s - \Delta \to {}_sJ^r(M, V) . $$

The transversality theorem extends to show that

${}_sj^r(f)$ <u>is generically transverse to any given submanifold</u> T <u>of</u> ${}_sJ^r(M, V)$.

<u>Example</u> Let T be the submanifold of ${}_2J^0(M, V)$ consisting of pairs of jets with the same image. Then f is transverse to this iff for all $x_1, x_2 \in M$ such that $f(x_1) = f(x_2) = y$, say, in V we have

$$ df_{x_1}(M_{x_1}) + df_{x_2}(M_{x_2}) = V_y \quad - $$

f is 'transverse to itself'. For example, if $v \geq 2m + 1$, this implies that f is injective.

<u>Problem.</u> Is there any natural way to 'fill in the hole along the diagonal' and get a transversality theorem in the completed space ? (so one could, for example, combine injectivity here with the immersion result above) ? A method for doing this in the case $V = \mathbb{R}$ is suggested by recent work of Georges Glaeser.

(C) Using the Baire category theorem, we see that we can make f satisfy a countable number of transversality conditions of the type listed above. For uncountable numbers of conditions, the direct analogue is false; for example, just try to make $f : S^1 \to \mathbb{R}^2$ transverse to each horizontal line (the function $y \circ f$ must have a maximum somewhere on S^1). There is, however, a result applicable to such problems, formulated recently in general terms by Gromov: this will be discussed in André Haefliger's lectures. I will close with an application (due to Mather) to stability.

There is a natural action of Diff N × Diff P on ${}_rJ^k(N, P)$, induced by composition. When I speak of orbits in ${}_rJ^k(N, P)$, I mean orbits of this action: clearly, they can also be locally defined.

<u>Theorem</u>. <u>If</u> $f : N \to P$ <u>is stable</u>, ${}_rj^kf$ <u>is transversal to every orbit in</u> ${}_rJ^k(N,P)$.

For if S is such an orbit, we can approximate f by g with $g \pitchfork S$. Since f is stable, any close enough approximation is equivalent to f under the action of Diff N × Diff P. It follows that $f \pitchfork S$.

For N compact, the converse is also true, (provided $r \geq p + 1$ and $k \geq p$): see lecture 5. Note that we have here an uncountable number of transversality conditions.

<u>Lecture 4</u> : <u>Stability and sufficiency of germs</u>

<u>Note</u> The lecture actually given (though better than the notes handed out) was wrong in several particulars; this is a considerably expanded version.

Recall the terminology of lecture 1 : we had a set S, a topology $\mathcal{J}$, and an equivalence relation e. Let S be the set of C^∞ map-germs $(\mathbb{R}^n, 0) \to (\mathbb{R}^p, 0)$. We have equivalence relations corresponding to various group actions :

$\mathcal{R}$ = germs at 0 of diffeomorphisms of $(\mathbb{R}^n, 0)$
(these act by composition on the right),

$\mathcal{L}$ = germs at 0 of diffeomorphisms of $(\mathbb{R}^p, 0)$
(which act by composition on the left)

$\mathcal{C}$ = germs at 0 of diffeomorphisms of $(\mathbb{R}^n \times \mathbb{R}^p, 0)$ of the form $H(x,y) = (x, h_x(y))$ with $h_x(0) = 0$ - i.e. n-parameter families of diffeo - morphism of $(\mathbb{R}^p, 0)$. These act on map-germs via a natural action on the graph :

$$f^H(x) = h_x(f(x)) .$$

We also have actions of the products (note $\mathcal{L} \subset \mathcal{C}$ as constant families) $\mathcal{A} = \mathcal{R} \times \mathcal{L}, \mathcal{K} = \mathcal{R} . \mathcal{C}$ (a semi-direct product). Each gives rise to an equivalence relation. $\mathcal{K}$-equivalence is also (I don't know why) called contact equivalence. The notion previously contemplated was $\mathcal{A}$-equivalence. The other most natural one is $\mathcal{R}$-equivalence in the case $(p = 1)$ when the target space is $\mathbb{R}$, so we are classifying singularities of real valued functions.

For each group $\mathcal{G}$ of diffeomorphisms, we can also consider the group of r-jets of members of $\mathcal{G}$: denote it by $\mathcal{G}^r$. Since the r-jet of the composite of two maps is determined by their r-jets, $\mathcal{G}^r$ acts on the space of r-jets.

Although S inherits a topology from $C^\infty(\mathbb{R}^n, \mathbb{R}^p)$, this does not lead to the right formulation for studying stability of map germs, since on a small deformation of a stable map, although a given singularity type will remain, in general its source and target will move. It is not hard to formulate a definition which allows for this. However, we also have the $\mathfrak{m}$-adic topology, defined by the ultrapseudometric

$$\|f - g\|_{\mathfrak{m}} = \inf \{e^{-r} : f \sim_r g\} .$$

Thus the neighbourhoods of f are (for various r) the sets of map germs with the same r-jet as f. If f is stable (for a group action by $\mathcal{G}$) in this topology, f is said to be <u>finitely $\mathcal{G}$ - determined</u>. More precisely, if $\|f - g\|_{\mathfrak{m}} \leq e^{-r}$ (i.e. $f \sim_r g$) implies $g \in f\mathcal{G}$, we say that f is $r - \mathcal{G}$ - determined, or that $j^r f$ is <u>$\mathcal{G}$-sufficient</u>. Other related questions : C^0-sufficiency and v-sufficiency of jets : were treated in the lectures by Kuo (see preceding paper).

The groups $\mathcal{C}$ and $\mathcal{K}$ are useful in effecting the transition from geometry ($\mathcal{A}$-equivalence) to algebra. The relation to algebra is given by the

<u>Proposition</u> <u>Map-germs</u> f <u>and</u> g <u>are</u> <u>$\mathcal{C}$-equivalent iff</u> $f^*\mathfrak{m}_y$ <u>and</u> $g^*\mathfrak{m}_y$ <u>generate the same ideal in</u> $C_x^\infty(N)$.
<u>Their k-jets are $\mathcal{K}$-equivalent iff the local algebras</u> $Q_k(f)$ <u>and</u> $Q_k(g)$ <u>are isomorphic.</u>

<u>Proof</u> <u>$\mathcal{E}$=></u> Let Z_f be the ideal in $C^\infty_{(x,y)}(N \times P)$ of functions vanishing on the graph of f. Take local coordinates (x_i) on N with origin x; (y_i) on P with origin y. Then Z_f is generated by $y_i - f_i(x_1, \ldots, x_n)$ for appropriate C^∞ functions f_i. Let $i : N \to N \times P$ be the inclusion as $N \times y$: then $i^* Z_f$ is the ideal generated by the f_i, hence also by $f^* \mathfrak{m}_y$.

If now $f \sim_{\mathcal{E}} g$, there is an automorphism H of $N \times P$ taking graph f to graph g, and hence Z_f to Z_g, and leaving $i(N)$ fixed. Hence

$$\langle f^* \mathfrak{m}_y \rangle = i^* Z_f = i^* H^* Z_g = i^* Z_g = \langle g^* \mathfrak{m}_y \rangle .$$

<u>$\mathcal{E}$<=</u> Our hypothesis is that the $y_i \circ f$ and the $y_i \circ g$ span the same ideal (= submodule) of $C^\infty_x(N)$: write

$$y_i \circ f = \Sigma\, w_{ij} . (y_j \circ g) ,$$
$$y_i \circ g = \Sigma\, v_{ij} . (y_j \circ f) ,$$

for suitable $w_{ij}, v_{ij} \in C^\infty_x(N)$. Taking values at x, we get real matrices W, V: choose C so that $C(I - VW) + W$ is invertible. [In general, given endomorphisms α, β of a vector space, there exists γ with $\gamma\alpha + \beta$ invertible iff $\operatorname{Ker} \alpha \cap \operatorname{Ker} \beta = \{0\}$: here, if $Wx = 0$ then $(I - VW)x = x$.] We can thus replace w_{ij} by

$$w_{ij} + c_{ij} - \Sigma_{k,\ell}\, c_{ik}\, v_{k\ell}\, w_{\ell j}$$

and so suppose w_{ij} an invertible matrix at, and hence near x. The required automorphism H is now given by

$$x_i \circ H = x_i$$
$$y_i \circ H = \Sigma_j\, w_{ij}(x_1, \ldots, x_n)\, y_j .$$

<u>$\mathcal{K}$ =></u> If $j^k g \in (j^k f)\, \mathcal{K}^k$, we can find $f' \sim_k f$ with $g \in f' \mathcal{K}$. Then $Q_k(f) \cong Q_k(f')$ (see lecture 2); the action of $\mathcal{R}$ clearly preserves the local algebras, and we have just shown that $\mathcal{E}$ does, hence so does $\mathcal{K} = \mathcal{R}.\,\mathcal{E}$.

<u>$\mathcal{K}$ <=</u> If $k = 0$, the result is trivial; otherwise $Q_k(f)$ - with unique maximal ideal $\mathfrak{m}$ - determines $Q_1(f) = Q_k(f)/\mathfrak{m}^2$, and hence $\operatorname{rank}(T_x f) = \dim_{\mathbb{R}} Q_1(f) - 1 = r$, say.

Choose linearly adapted charts (see lecture 2) with coordinates x_i, y_j for f

and x_i', y_j' for f' : thus

$$y_i \circ f = x_i \qquad y_i' \circ f' = x_i' \qquad \text{for } i \leq r ,$$

$$d(y_i \circ f) = d(y_i' \circ f') = 0 \qquad r < i \leq p ;$$

in particular $x_{r+1}, \ldots, x_n$ form an (irredundant) set of generators for $Q_k(f)$. For each i, $r < i \leq n$ choose a polynomial $\phi_i(x_{r+1}, \ldots, x_n)$ whose class in $Q_k(f)$ corresponds under the given isomorphism to the class of x_i' in $Q_k(f')$.

By looking at $Q_1(f)$ we see that the jacobian of the ϕ_i with respect to the x_i $(r < i \leq n)$ is nonzero, hence we can take these as new coordinates - this means using an element of $\mathcal{R}$. But in the new chart, the classes of x_i, x_i' correspond in $Q_k(f)$, thus $f^*\mathfrak{m}_y$ and $f'^*\mathfrak{m}_y$ generate the same ideal in $C_x^\infty(N)/\mathfrak{m}_x^{k+1}$. The argument of $\mathcal{C} \Leftarrow$ now shows that j^kf and j^kf' are $\mathcal{C}$-equivalent.

I now ask how to recognise finitely determined germs. Following lecture 1, we expect to use a tangent space to the "manifold of map-germs" ; recall that for $f : N \to P$ we defined $\theta(f)$ as the space of C^∞-sections of $f^*(TP)$. For a map-germ $f : (\mathbb{R}^n, 0) \to (\mathbb{R}^p, 0)$, define $\theta(f)$ as the space of germs of C^∞-sections. Since the bundle in question is trivial, we have a free module of rank p over $C_0^\infty(\mathbb{R}^n)$. For map germs preserving 0, the appropriate space is ${}_x\theta(f)$. Now the tangent space to $\mathcal{R}$ is $\theta(\mathbb{R}^n)$ (Mather calls it B) ; the action of $\mathcal{R}$ induces $tf : \theta(\mathbb{R}^n) \to \theta(f)$ as in lecture 1 by composing $\phi : \mathbb{R}^n \to T\mathbb{R}^n$ with Tf. Clearly, tf is a map of $C_0^\infty(\mathbb{R}^n)$ - modules; its matrix with respect to the obvious bases is just the Jacobian matrix of f.

Similarly, the tangent space to $\mathcal{L}$ is $\theta(\mathbb{R}^p)$, and the action of $\mathcal{L}$ induces $\omega f : \theta(\mathbb{R}^p) \to \theta(f)$ by composition with f. Algebraically, however, this is not analogous : $\theta(\mathbb{R}^p)$ is only a $C_0^\infty(\mathbb{R}^p)$ - module, and though we can regard ωf as a morphism of modules over $C_0^\infty(\mathbb{R}^p)$ using $f^* : C_0^\infty(\mathbb{R}^p) \to C_0^\infty(\mathbb{R}^n)$ to consider the latter ring as algebra over the former, this is a distinctly more complex situation. Again, for germs preserving 0, one must restrict to $\mathfrak{m}_x\theta(\mathbb{R}^n)$ and $\mathfrak{m}_y\theta(\mathbb{R}^p)$.

Finally, the tangent space to $\mathcal{C}$ is $\pi_2^*\mathfrak{m}_y\theta(\pi_2)$, where $\pi_2 : \mathbb{R}^n \times \mathbb{R}^p \to \mathbb{R}^p$ is the projection (for maps may only be deformed parallel to $\mathbb{R}^p$, and $y = 0$ must remain fixed). It follows that the image in $\theta(f)$ is just

$f^*\mathfrak{m}_y\,\theta(f)$.

We define $\tau(\mathcal{G})f \subset \mathfrak{m}_x\theta(f)$ to be

$tf(\mathfrak{m}_x\,\theta(\mathbb{R}^n))$ for $\mathcal{R}$

$\omega f(\mathfrak{m}_y\,\theta(\mathbb{R}^p))$ for $\mathcal{L}$

$f^*\mathfrak{m}_y\,\theta(f)$ for $\mathcal{C}$

and for $\mathcal{A} = \mathcal{R} \times \mathcal{L}$, $\mathcal{K} = \mathcal{R} \cdot \mathcal{C}$ just add .

Theorem For $\mathcal{G} = \mathcal{R}, \mathcal{L}, \mathcal{C}, \mathcal{A}, \mathcal{K}$ we have

(i) If f is r - $\mathcal{G}$ - determined, $\tau(\mathcal{G})f \supset \mathfrak{m}_x^{r+1}\,\theta(f)$.

(ii) For any r there exists $\ell(n, p, r, \mathcal{G})$ such that if $\tau(\mathcal{G})f \supset \mathfrak{m}_x^{r+1}\theta(f)$ then f is ℓ - $\mathcal{G}$ - determined. For $\mathcal{G} = \mathcal{R}$, $\mathcal{C}$ or $\mathcal{K}$ we can take $\ell = r + 1$.

(iii) Thus f is finitely $\mathcal{G}$-determined iff $\tau(\mathcal{G})f$ contains $\mathfrak{m}_x^{k+1}\theta(f)$ for some k . This is equivalent to saying that the codimension

$$d(f, \mathcal{G}) = \dim_{\mathbb{R}}(\mathfrak{m}_x\theta(f)/\tau(\mathcal{G})f) < \infty .$$

Proof (in part). (i) The space $J^\ell(n, p)$ of jets is a quotient of the space of germs. Using our infinitesimal deformation argument gives a map from $\mathfrak{m}_x\theta(f)$ to the tangent space at $j^\ell f$ to $J^\ell(n, p)$. It is easy to see that this is surjective, with kernel $\mathfrak{m}_x^{\ell+1}\,\theta(f)$, so we can identify the tangent space to $J^\ell(n, p)$ accordingly. Clearly the submanifold consisting of those ℓ-jets whose r-jet is $j^r f$ has tangent space $\mathfrak{m}_x^{r+1}\theta(f)/\,\mathfrak{m}_x^{\ell+1}\theta(f)$. By hypothesis, $\mathcal{G}^\ell$ acts transitively here. We deduce that

$$\tau(\mathcal{G})f + \mathfrak{m}_x^{\ell+1}\theta(f) \supset \mathfrak{m}_x^{r+1}\theta(f) .$$

To conclude (i), and even more for the rest of the proof, some algebraic machinery is needed, which is based on the preparation theorem. We now develop the necessary algebra in seven lemmas of increasing difficulty.

Lemma 1 (Nakayama) Let R be a ring, $\mathfrak{m}$ an ideal such that for $x \in \mathfrak{m}$, $1 + x$ is invertible, and $\alpha : A \to C$ a map of R-modules, with C finitely generated. Then

$$\alpha A + \mathfrak{m} C = C \quad \text{implies} \quad \alpha A = C .$$

If elements c_i generate C , we can write

$$c_i = \alpha(a_i) + \Sigma m_{ij}\, c_j \qquad (a_i \in A,\ m_{ij} \in \mathfrak{m}) .$$

But we can solve these equations for the c_i in terms of the $\alpha(a_j)$ since the determinant is invertible.

Lemma 2. R, $\mathfrak{m}$ as before; A a submodule of the finitely generated R-module C. Then

$$\dim_{\mathbb{R}} (C/\mathfrak{m}^{\ell+1} C + A) \leq \ell \text{ implies } \mathfrak{m}^{\ell} C \subset A .$$

We must have $\mathfrak{m}^{i} C + A = \mathfrak{m}^{i+1} C + A$ for some i, $0 \leq i \leq \ell$ (if these were strictly decreasing, the quotient dimension would be $\geq \ell + 1$). By Lemma 1 (with $\alpha = 0$ and $(\mathfrak{m}^{i} C + A)/A$ for C), $\mathfrak{m}^{i} C + A = A$, so $A \supset \mathfrak{m}^{i} C \supset \mathfrak{m}^{\ell} C$.

Lemma 3 Hypotheses of Lemma 2 ; suppose also R an algebra over a field F, and $\dim_F (R/\mathfrak{m}^{i} R) < \infty$ for all i. Then there is a bound for $\dim_F (A/\mathfrak{m} A)$ (and hence, by Lemma 1, for the number of generators for A) depending only on ℓ, R, and the number of generators of C.

For
$$\dim_F (A/\mathfrak{m} A) \leq \dim_F (A/\mathfrak{m}^{\ell+1} C)$$
$$\leq \dim_F (C/\mathfrak{m}^{\ell+1} C)$$

and the result follows since this is smaller than the corresponding number for a free module mapping onto C.

Lemma 4. (Preparation theorem) Let $\phi : R \rightarrow S$ be the map $f^* : C_0^{\infty}(\mathbb{R}^p) \rightarrow C_0^{\infty}(\mathbb{R}^n)$ induced by a C^{∞} map-germ f. Let $\alpha : A \rightarrow C$ be a morphism over ϕ of finitely generated modules. Then

$$\alpha A + \phi(\mathfrak{m}_y) C = C \text{ implies } \alpha A = C .$$

For α induces a map of $A/\mathfrak{m}_y A$ onto $C/\phi(\mathfrak{m}_y)C$. The former has finite $\mathbb{R}$-dimension since A is finitely generated, hence so has the latter. By Malgrange's preparation theorem (see e.g. my account of the preparation theorem in this volume), it follows that C is finitely generated as R-module. The result then follows from Lemma 1.

Lemma 5. Notation of lemma 4 ; also $\underline{a} = \dim_{\mathbb{R}} A/\mathfrak{m}_y A$ (the minimum number of generators of A as R-module). Then

$$\alpha A + (\phi(\mathfrak{m}_y) + \mathfrak{m}_x^{a+1}) C = C \text{ implies } \alpha A = C .$$

Here α maps the $\underline{a}$-dimensional space $A/\mathfrak{m}_y A$ onto $C/(\phi(\mathfrak{m}_y) + \mathfrak{m}_x^{a+1}) C$. By Lemma 2, $\mathfrak{m}_x^{a} C \subset \phi(\mathfrak{m}_y) C$. Thus the result follows from lemma 4.

Lemma 6. Notation of lemmas 4 and 5 ; also let C_0 be an S-submodule of C with $\dim_{\mathbb{R}}(C/C_0) = c$. There is a function $\ell(a, p, c)$ such that

$$\alpha A + \mathfrak{m}_x^{\ell+1} C_0 \supset C_0 \quad \text{implies} \quad \alpha A \supset C_0 .$$

Define $A_0 = \alpha^{-1}(C_0)$. We get a bound ℓ for the number of generators of A_0 from Lemma 3, and the result then follows from Lemma 5.

Lemma 7. Notation of lemma 5; also $d = \dim_{\mathbb{R}}(C/\alpha A) < \infty$. There is a function $k(a, d, p)$ such that $\alpha A \supset \mathfrak{m}_x^k C$.

The proof is the same as that of Lemma 2, save that instead of using (with $\mathfrak{m}^i C$ for C_0)

$$A + \mathfrak{m} C_0 \supset C_0 \Rightarrow A \supset C_0 \qquad \text{(Lemma 1)}$$

we now have

$$\alpha A + \mathfrak{m}^{\ell(a,p,c)} C_0 \supset C_0 \Rightarrow \alpha A \supset C_0 \quad \text{(Lemma 6)} .$$

The function k is (roughly) a d-fold iterate of ℓ : the bounds it gives are astronomical.

This concludes our algebraic session: now we can return to the proof of the theorem. Part (i) is concluded using Lemma 1 (for $\mathcal{R}, \mathcal{C}, \mathcal{K}$) or Lemma 6 (for $\mathcal{L}, \mathcal{A}$).

The idea of the proof of (ii) is as follows. If $j^\ell f = j^\ell g$,take a linear homotopy f_t from f to g . We seek to construct a path α_t in $\mathcal{G}$ with $\alpha_0 = 1$ and $f\alpha_t = f_t$: putting $t = 1$ then shows that f and g are equivalent. Differentiating with respect to t , we find that the image of $\dot{\alpha}_t$ by the differential of the action is the tangent space at f_t to the homotopy. The argument above shows that if $\tau(\mathcal{G})f \supset \mathfrak{m}_x^{r+1}\theta(f)$ and $j^r f = j^r g$, we can find a suitable $\dot{\alpha}_0$. To show the existence of a suitable $\alpha_t^{-1}(\dot{\alpha}_t)$, however, one needs $\tau(\mathcal{G})\, f_t \supset \mathfrak{m}_x^{r+1}\theta(f_t)$: we will discuss this in a moment. However, what is really needed is a separate argument (to be found in Mather III pp.131-134, 149-151) which shows that we can choose $\alpha_t^{-1}(\dot{\alpha}_t)$ to depend C^∞ on the parameter t . Solving this ordinary differential equation (we can do this pointwise: no need for function spaces), we obtain the required deformations.

We now show that if $\tau(\mathcal{G})f \supset \mathfrak{m}_x^{r+1}\,\theta(f)$, and $j^\ell g = j^\ell f$ for suitable ℓ, then $\tau(\mathcal{G})g \supset \mathfrak{m}_x^{r+1}\,\theta(g)$. Regarding $\theta(f)$ as the free module over $C_0^\infty(\mathbb{R}^n)$ with base $\partial/\partial y_1, \ldots, \partial/\partial y_p$ we can identify $\theta(f)$ and $\theta(g)$. By Lecture 2, if $j^{r+1}f = j^{r+1}g$ then $f^*\mathfrak{m}_y$, $g^*\mathfrak{m}_y$ are the same modulo $\mathfrak{m}_x^{r+2}$, hence $\tau(\mathcal{C})f = \tau(\mathcal{C})g \bmod \mathfrak{m}_x^{r+2}$. Similarly, since $j^r(Tf) = j^r(Tg)$, $tf(\theta(\mathbb{R}^n))$ and $tg(\theta(\mathbb{R}^n))$ are the same modulo $_x^{r+1}$, so $\tau(\mathcal{R})f = \tau(\mathcal{R})g$ modulo $_x^{r+2}$. Thus when $j^{r+1}f = j^{r+1}g$ and $\mathcal{G} = \mathcal{R}, \mathcal{C}$ or $\mathcal{K}$, we have

$$\tau(\mathcal{G})g + \mathfrak{m}_x^{r+2}\,\theta(g) = \tau(\mathcal{G})f + \mathfrak{m}_x^{r+2}\theta(g) \supset \mathfrak{m}_x^{r+1}\theta(g) ;$$

now by Lemma 1, $\tau(\mathcal{G})g \supset \mathfrak{m}_x^{r+1}\theta(g)$. Finally, it is again immediate that $j^\ell f = j^\ell g$ implies

$$\tau(\mathcal{L})f + \mathfrak{m}_x^{\ell+1}\theta(f) = \tau(\mathcal{L})g + \mathfrak{m}_x^{\ell+1}\,\theta(g)$$

but for $\mathcal{L}$ and $\mathcal{A}$ we must appeal to Lemma 6 (rather than Lemma 1), which is why we need ℓ rather than just $r + 1$. For the case of $\mathcal{A}$, one must take $\alpha = \omega f$, $A = \mathfrak{m}_y\,\theta(\mathbb{R}^p)$, $C = \theta(g)/\tau(\mathcal{R})g$, C_0 = image of $\mathfrak{m}_x^{r+1}\theta(g)$.

Finally for (iii) it is immediate that if, for some k, $\tau(\mathcal{G})f \supset \mathfrak{m}_x^{k+1}\,\theta(f)$ then $d(f, \mathcal{G}) < \infty$; the converse follows from Lemma 2 (for $\mathcal{R}, \mathcal{C}, \mathcal{K}$) or Lemma 7 (for $\mathcal{L}, \mathcal{A}$).

Although I have presented (following Mather) the five cases in parallel, when we come to actually check the conditions we find very different pictures. Perhaps the most striking is the case of $\mathcal{R}$: Mather has shown* that

<u>For</u> $p \geq 2$, f <u>is finitely</u> <u>$\mathcal{R}$-determined only if it is the germ of a submersion.</u>

Thus the only interesting case is $p = 1$. Here $\theta(f)$ is a free module on one generator, so we identify it with $C_0^\infty(N)$. Since $tf(\theta(\mathbb{R}^n))$ was the submodule generated by the

$$\partial/\partial x_i \quad = \quad \partial y/\partial x_i \quad \partial/\partial y$$

it corresponds to the ideal $\Delta(f)$ generated by the $\partial y/\partial x_i$, and $\tau(\mathcal{R})f$ to $\mathfrak{m}_x\Delta(f)$. Thus the theorem reduces to :

$$\mathfrak{m}_x\Delta(f) \supset \mathfrak{m}_x^r \Rightarrow f \text{ is } r\text{-}\mathcal{R}\text{-determined} \Rightarrow \mathfrak{m}_x\Delta(f) \supset \mathfrak{m}_x^{r+1}.$$

The above $d(f, \mathcal{R})$ is closely related to the invariant

$$\mu(f) = \dim_{\mathbb{R}}(C_0^\infty(\mathbb{R}^n)/\Delta(f))$$

*To appear in an appendix to Thom's book.

of Milnor .

In the case $p = 1$, $n > 1$, f is never finitely $\mathcal{L}$-determined. The example $y \circ f = x_1^2 - x_2^2$ shows that it need not be finitely $\mathcal{C}$-determined even if it is for $\mathcal{R}$ (any function in the ideal generated by y vanishes when $x_1 = x_2$). I do not know an example which is finitely $\mathcal{K}$ - determined but not for $\mathcal{R}$. In general this raises the problem of the relation between $\Delta(f)$ and the Jacobian ideal $J(f)$ generated by $\Delta(f)$ and f (which is the one used, for example, by Boardman). That this is not simple is shown by the examples given by Pham in the second volume of these proceedings.

This last point is related to the following conjectures, which arise in the analogous problem with formal power series over $\mathbb{C}$. If $\Delta(f)$ is of finite codimension, one can show that it contains some $\mathcal{M}_x^r$, and hence $(y \circ f)^r$.

<u>Conjecture 1</u> (Mather) $(y \circ f)^2 \in \Delta(f)$.

<u>Conjecture 2</u> (Brieskorn) $y \circ f \in \Delta(f)$ iff, after some change of coordinates, $y \circ f$ becomes a weighted homogeneous polynomial. Sufficiency here follows from Euler's theorem on homogeneous functions.

Apart from this, only the cases $\mathcal{G} = \mathcal{A}$ or $\mathcal{K}$ play any serious rôle. Map-germs which are finitely $\mathcal{K}$ -determined (this includes all other cases) are called by Mather of <u>finite singularity type</u> : by lecture 2, they are the ones with stable unfoldings, and this plays an important rôle in Mather's work on C^0-stability.

<u>Lecture 5 : C^∞-stable maps</u>

For the result below I will need some of the notation from the previous lecture vamped up so that instead of a point $x \in N$ we have a finite collection S of r points of N with the same image $y \in P$. The previous arguments go through virtually unchanged. Consider $f \in C^\infty(N, P)$ with N compact; let $r \geq p + 1$, $k \geq p$.

<u>Theorem.</u> <u>The following are equivalent</u> :

(a) f <u>is stable</u>

(b) ${}_r j^k f$ <u>is transversal to every A-orbit in</u> ${}_r J^k(N, P)$

(c) ${}_r j^k f$ <u>is transversal to every $\mathcal{K}$-orbit in</u> ${}_r J^k(N, P)$

(d) <u>for every subset</u> S <u>of</u> N <u>with</u> $|S| \leq r$ <u>and</u> $f(S) = y \in P$,

$$tf(\theta_S(N)) + \omega f(\theta_y(P)) + (f^* \mathfrak{m}_y + \mathfrak{m}_S^{k+1})\, \theta_S(f) = \theta_S(f) \qquad (1)$$

(e) f <u>is infinitesimally stable</u> : $tf(\theta(N)) + \omega f(\theta(P)) = \theta(f)$.

We saw in lecture 3 that (a) => (b). Clearly, since $\mathcal{K} \supset A$, (b) => (c). By lecture 4, the tangent space to a $\mathcal{K}$ -orbit in $J^k(N, P)$ when points are held fixed is

$$(tf(\mathfrak{m}_x \theta_x(N)) + f^* \mathfrak{m}_y \theta_x(f) + \mathfrak{m}_x^{k+1} \theta_x(f))/\mathfrak{m}_x^{k+1} \theta_x(f) .$$

Allowing the target to move increases the term $\omega f(\mathfrak{m}_y \theta_y(P))$ (contained in the second term above) to $\omega f(\theta_y(P))$. Adding $T_x j^r f\,(T_x N)$ increases the first term to $tf(\theta_x(N))$. Thus in the case $r = 1$, (1) expresses just the condition (c). The argument for $r > 1$ is virtually the same.

Now I show that (d) => (e). By Lemma 5 from lecture 4, (1) implies

$$tf(\theta_S(N) + \omega f(\theta_y(P)) = \theta_S(f) \qquad (2)$$

giving the local form of what we want globally. Let

$$\Sigma = \{x \in N : T_x f \text{ is } \underline{\text{not}} \text{ surjective}\} .$$

Now if $y \in N$ and $\Sigma_y = \Sigma \cap f^{-1}(y)$ has more than p points, pick a subset S containing $p + 1$. Then on one hand, by (2), ωf induces a surjection of $\theta_y(P)/\mathfrak{m}_y \theta_y(P)$ (of dimension p) onto $\theta_S(f)/\mathfrak{m}_S \theta_S(f) + tf(\theta_S(N))$, so it has dimension $\leq p$. On the other hand, this space is a direct sum of the corresponding spaces for the points $x \in S$; now $\theta_x(f)/\mathfrak{m}_x \theta_x(f)$ can be identified with the tangent space $T_y P$ to the 0-jets with fixed source but not fixed target, and hence our space with $T_y N/T_x f(T_x P)$, which is nonzero since $x \in \Sigma$. Summing over x , the space has dimension $\geq (p + 1)$: a contradiction.

Hence $|\Sigma_y| \leq p$ for all y . Now given $\chi \in \theta(f)$, by (2) we can, for each $y \in P$, choose a germ $\psi_y \in \theta_y(P)$ such that $\chi - \omega f(\psi_y) \in tf(\theta_{\Sigma_y}(N))$. Using a partition of unity on P , and a little topology (the fact that N is compact - or f proper - is used here), we obtain $\psi \in \theta(P)$ such that for each $x \in \Sigma$ there is a germ $\phi_x \in \theta_x(N)$ with $(\chi - \omega f(\psi))_x = tf(\phi_x)$. But there certainly exist such ϕ_x for $x \notin \Sigma$ since then Tf is surjective. Using a partition of unity on N, we obtain $\phi \in \theta(N)$ with $\chi - \omega f(\psi) = tf(\phi)$, as required.

The key step in the theorem is of course (e) => (a). The argument is similar to the one needed for finite determination of map germs in lecture 4. I will not discuss it further here.

It seems appropriate to mention at this point Mather's global version of finite singularity type. If f satisfies the condition locally, the multiplicities

$$q_x = \dim_{\mathbb{R}} (\theta_x(f)/tf(\theta_x(N)) + f^* \quad {}_y\theta_x(f))$$

are finite. Then with the notation above, f is said to be of finite singularity type if

i) $f|\Sigma$ is proper ,

ii) the number of points in Σ_y is bounded,

iii) q_x is bounded for $x \in \Sigma$;

we can also combine ii) and iii) to state that the total multiplicity of Σ_y is bounded. Mather shows that this is equivalent to $\theta(f)/tf(\theta(N))$ being a finitely generated $C^\infty(P)$-module, or $C^\infty(\Sigma(f))$ being one, or to f having a stable unfolding. The proof will appear in his book on C^0-stability.

Now I will sketch the proof that stable map germs are classified by local algebras $Q_{p+1}(f)$. First we show that stable map germs are (p+1)-determined (for $\mathcal{A}$). If f is a stable germ with source x and target y then (by the theorem) (2) holds, with $S = \{x\}$. Multiply by $\mathfrak{m}_y$: since $f^*\mathfrak{m}_y \subset \mathfrak{m}_x$, we have

$$f^*\mathfrak{m}_y\theta_x(f) \subset tf(\mathfrak{m}_x\theta_x(N)) + \omega f(\mathfrak{m}_y\theta_y(P)) \qquad (3) .$$

Now (2) also implies that $tf(\theta_x(N)) + f^*\mathfrak{m}_y\theta_x(f)$ has codimension $\leq p$ in $\theta_x(f)$ [for it contains $tf(\theta_x(N) + \omega f(\mathfrak{m}_y\theta_y(P))$, and $\mathfrak{m}_y\theta_y(P)$ has codimension p in $\theta_y(P)$], so by Lemma 2 of lecture 4, it contains $\mathfrak{m}_x^p\theta_x(f)$. Thus

$$\begin{aligned} tf(\mathfrak{m}_x\theta_x(N)) + \omega f(\mathfrak{m}_y\theta_y(P)) &= tf(\mathfrak{m}_x\theta_x(N)) + f^*\mathfrak{m}_y\theta_x(f) \quad (4) \text{ (by (3))} \\ &\supseteq \mathfrak{m}_x(tf(\theta_x(N)) + f^*\mathfrak{m}_y\theta_x(f)) \\ &\supseteq \mathfrak{m}_x \cdot \mathfrak{m}_x^p\,\theta_x(f) \qquad (5) \end{aligned}$$

so by the theorem of lecture 4, f is finitely $\mathcal{A}$-determined. Our previouse argument also showed that the crucial point now needed to establish that f is $(p+1)$-determined is that $j^{p+1}f' = j^{p+1}f$ implies that (5) holds for f' : by the above it is enough to deduce that f' is stable. But

$$tf'(\theta_x(N)) = tf(\theta_x(N)) \mod \mathcal{M}_x^{p+1}, \text{ and}$$

$$\omega f'(\mathcal{M}_y\theta_y(P)) = \omega f(\mathcal{M}_y\theta_y(P)) \mod \mathcal{M}_x^{p+2} .$$

It thus follows (using (2)) that

$$tf'(\theta_x(N)) + \omega f'(\theta_y(P)) + \mathcal{M}_x^{p+1}\,\theta_x(f') \supseteq \theta_x(f') .$$

By Lemma 5 of lecture 4 we obtain (2) for f', so f' is stable.

Hence to classify stable map germs it suffices to classify their $(p+1)$-jets for $\mathcal{A}$. We have already seen that $Q_{p+1}(f)$ gives the classification for $\mathcal{K}$. Write St for the set of stable $(p+1)$-jets : clearly open, since its complement is defined by algebraic equations. It remains to show that if $z \in \mathrm{St}$, then $z\mathcal{A} = z\mathcal{K} \cap \mathrm{St}$. Now (4) shows that if $y \in z\mathcal{K} \cap \mathrm{St}$, then $z\mathcal{A}$ and $z\mathcal{K}$ have the same tangent space at y. Thus each orbit $y\mathcal{A}$ is open, hence also each is closed, in $z\mathcal{K} \cap \mathrm{St}$. From this to the desired result is non-trivial: see Mather IV, section 5 - though the complex analogue is trivial ($z\mathcal{K} \cap \mathrm{St}$ is then connected). The proof includes the derivation of the normal form for stable germs.

<u>Example.</u> $(n = p = 2)$. All the 3-jets

$$y_1 = x_1^3 + \lambda x_1 x_2$$

$$y_2 = x_2$$

are in the same $\mathcal{K}$-orbit. Those with $\lambda \neq 0$ are stable and lie in an $\mathcal{A}$-orbit.

Finally a brief indication of how to show that stable maps are dense in some dimensions. Suppose there is a $\mathcal{K}$-invariant subset X of $J^p(n, p)$, of codimension $> n$, whose complement contains only a finite number of contact classes. By the transversality theorem we can approximate any $f : N^n \to P^p$ by one whose p-jet at any point avoids X. For stability, by (c) of the Theorem above, it suffices for ${}_{p+1}j^p(f)$ to be transverse to all $\mathcal{K}$-orbits. But for multijets, $\mathcal{K}$ operates independently at the points of S, so a $\mathcal{K}$-orbit here is a collection of $(p+1)$ $\mathcal{K}$-orbits in $J^p(n, p)$. If we continue to avoid X, there are only finitely many such orbits. A further application of the transversality theorem now allows us to impose the required condition after perturbing f .

By Lecture 3, if $p \geq 2n$, a map $N^n \to P^p$ is generically an immersion, so stable. If $2p \geq 3(n-1)$, then generically $\Sigma^2 = \emptyset$. We can thus take X to be the jets with kernel rank ≥ 2. For the rest, $Q_{p+1}(f)$ has one generator, so is isomorphic to one of the (finite) number of algebras $\mathbb{R}[x/x^m = 0]$, $1 \leq m \leq p+1$. Thus in this range too, stable map-germs are dense.

If $3p \geq 4(n-2)$, we can take X to be the jets with kernel rank ≥ 3. For here again the number of algebras on two generators with given degree p of nilpotency is finite : the list for $\Sigma^{2,0}$ was given in lecture 2. For other cases, see the paper below.

REFERENCES

Note The references given here are the classics of the subject of singularities.

V.I. Arnold, Singularities of smooth mappings, Russian Math. Surveys (1969) 1 - 43. (Translated from Uspehi Math. Nauk 23 (1968) 3 - 44).

J.M. Boardman, Singularities of differentiable maps, Publ. Math. I.H.E.S. 33 (1967) 21 - 57.

B. Malgrange, Ideals of differentiable functions, Oxford University Press, 1966.

J.N. Mather, Stability of C^∞ mappings.

I The division theorem. Ann. of Math.87 (1968) 89-104.

II Infinitesimal stability implies stability. Ann. of Math.

III Finitely determined map-germs. Publ. Math. I.H.E.S. 35 (1968) 127-156.

IV Classification of stable germs by R-algebras. Publ.Math. I.H.E.S. 37 (1969) 223 -

V Transversality. Advances in Math.

VI The nice dimensions. This volume, pp.207-253.

J.N. Mather, Notes on topological stability. Lecture notes, Harvard University 1970.

J.W. Milnor, Singular points of complex hypersurfaces, Ann. of Math. Study 61, Princeton University Press, 1968.

B. Morin, Formes canoniques des singularités d'une application différentiable, Comptes Rendus Acad. Sci. Paris 260 (1965) 5662 - 5665 and 6503 - 6506.

R. Thom, Stabilité structurelle et morphogenèse, Benjamin, to appear.

G.N. Turina, On flat semi-universal deformations of analytic sets with isolated singularities (in Russian) Izv. Akad. Nauk, S.S.S.R. ser. mat. 33 (no. 5) (1969) 1026-1058.

STABILITY OF C^∞ MAPPINGS : VI

THE NICE DIMENSIONS

J.N. Mather

A smooth (i.e. C^∞) mapping f from one finite dimensional manifold to another is said to be stable if any sufficiently close approximation g to f is equivalent to f in the sense that there exist smooth diffeomorphisms of the source and the target onto themselves which transform f into g . This is the basic notion which we have studied in this series of papers [3] . (See II in this series for a precise definition of stable mappings.)

In V of this series, we gave a characterization of stable mappings, which shows that at least for proper mappings, whether a mapping is stable or not depends only on local properties of the mapping. On the other hand, one can find mappings which are not proper and fail to be stable (for essentially pathological reasons) even though they satisfy the local criteria for stability. For this reason, the notion of "stable mappings" in the non proper case is of little interest. Thus, we are led to formulate the following question. Let N^n and P^p be manifolds of dimensions n and p respectively.

Question. Are the proper stable mappings of N into P dense in the set of proper mappings of N into P ?

If there are no proper mappings of N into P , the answer is yes, vacuously. Thus, we are led to impose the hypothesis that there is at least one proper mapping of N into P . In V , we showed that under this hypothesis the answer depends only on the dimensions n and p . More precisely, we defined an integer valued function $\sigma(n, p)$ of two variables n and p ranging over all positive integers and showed that the proper stable mappings are dense in the proper mappings of N^n into P^p if and only if $n < \sigma(n, p)$.

In this paper, we will outline the computation of $\sigma(n, p)$. Our computations are divided into two broad cases.

Case I. $n \leq p$. We have

$$\sigma(n, p) = 6(p - n) + 8 \quad \text{if} \quad p - n \geq 4 \quad \text{and} \quad n \geq 4$$

$$= 6(p - n) + 9 \begin{cases} \text{if} \quad 3 \geq p - n \geq 0 \quad \text{and} \quad n \geq 4 \\ \text{or if} \quad n = 3 \end{cases}$$

$$= 7(p - n) + 10 \quad \text{if} \quad n = 2$$

$$= \infty \quad \text{if} \quad n = 1$$

Case II. $n > p$. We have

$$\sigma(n, p) = 9 \quad \text{if} \quad n = p + 1$$

$$= 8 \quad \text{if} \quad n = p + 2$$

$$= n - p + 7 \quad \text{if} \quad n \geq p + 3$$

From this calculation, it then follows that the stable proper mappings are dense in the proper mappings if and only if the pair (n, p) satisfies one of the following conditions.

$$n < \frac{6}{7} p + \frac{8}{7} \quad \text{and} \quad p - n \geq 4$$

$$n < \frac{6}{7} p + \frac{9}{7} \quad \text{and} \quad 3 \geq p - n \geq 0$$

$$p < 8 \quad \text{and} \quad p - n = -1$$

$$p < 6 \quad \text{and} \quad p - n = -2$$

$$p < 7 \quad \text{and} \quad p - n \leq -3$$

Any pair (n, p) for which one of these conditions hold will be called nice. In figure 1, we have sketched the boundary of the nice region in the (n, p) plane.

Figure 1

The nice region

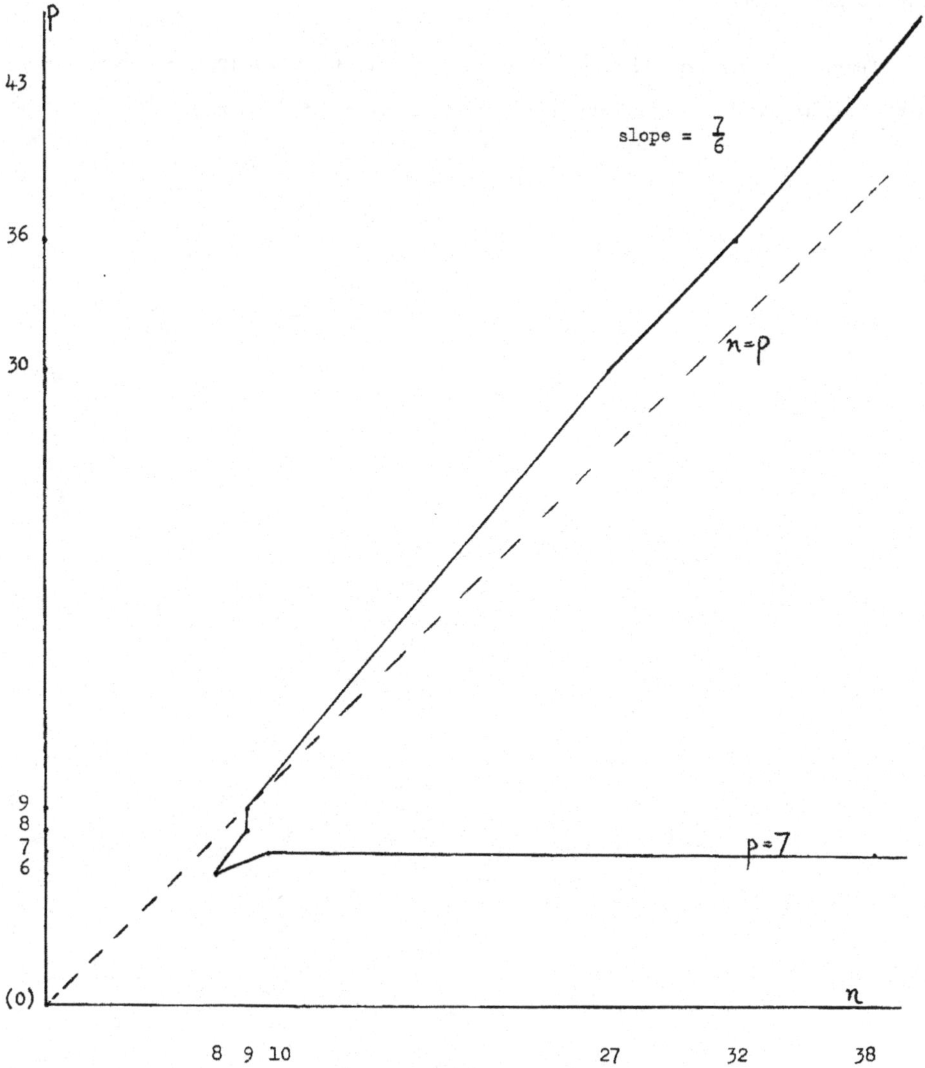

The nice region is theregion to the left of and below the solid broken line.

The arguments of this paper can be used to give a complete classification of stable map - germs in the nice region. We give the details of this classification in some cases in this paper. However, we will postpone the complete classification until a later paper.

Complete proofs are given in §§1 - 7 . However, in §§8 - 12, proofs are given only in outline form. Details will be given in a later paper.

CONTENTS

§1 The scheme of the calculation

In this section, we will outline how we will go about calculating $\sigma(n, p)$.

To begin with, we recall the definition of $\sigma(n, p)$ (compare V §7). Let k be a positive integer. We defined in IV a group $\mathcal{K}^k$ which acts on the space $J^k(n, p)$ of k jets of mappings of $\mathbb{R}^n$ into $\mathbb{R}^p$ whose source and target are both 0. Two k jets have isomorphic associated $\mathbb{R}$ algebras if and only if they are in the same orbit of $\mathcal{K}^k$. The group $\mathcal{K}^k$ is an algebraic group. It follows that there is a smallest closed algebraic $\mathcal{K}^k$-invariant subset $\Pi^k(n, p)$ of $J^k(n, p)$ with the property that only finitely many orbits of $\mathcal{K}^k$ lie outside $\Pi^k(n, p)$. (An explicit construction of $\Pi^k(n, p)$ was given in V §7.) We let $\sigma^k(n, p)$ denote the codimension of $\Pi^k(n, p)$ in $J^k(n, p)$ if $\Pi^k(n, p)$ is non-empty and let $\sigma^k(n, p) = \infty$ if $\Pi^k(n, p)$ is empty.

Clearly $\Pi^{k+1}(n, p)$ contains the inverse image of $\Pi^k(n, p)$ under the projection $J^{k+1}(n, p) \to J^k(n, p)$. Hence $\sigma^{k+1}(n, p) \leq \sigma^k(n, p)$. We let $\sigma(n, p) = \inf_k \sigma^k(n, p)$.

To compute $\sigma(n, p)$, we will need to introduce certain other functions $\sigma_r(n, p)$. Let Σ_r^k denote the set of $z \in J^k(n, p)$ whose kernel rank is r. In other words, a k-jet z is in Σ_r^k if and only if for one (and therefore for every) representative f of z, we have $\dim \ker df_0 = r$. Clearly $J^k = \cup \{\Sigma_r^k : r \geq \max(0, n - p)\}$. The set Σ_r^k is invariant under the action of $\mathcal{K}^k$. It follows that there is a smallest closed algebraic $\mathcal{K}^k$-invariant subset $\Pi_r^k(n, p)$ of Σ_r^k with the property that only finitely many of the orbits of $\mathcal{K}^k$ lie in $\Sigma_r^k - \Pi_r^k(n, p)$. We set $\sigma_r^k(n, p)$ equal to the codimension of $\Pi_r^k(n, p)$ in $J^k(n,p)$, if $\Pi_r^k(n, p)$ is non-empty, and we set $\sigma_r^k(n, p) = \infty$ otherwise. Clearly

$$\sigma^k(n, p) = \inf_r \sigma_r^k(n, p) .$$

We set $\sigma_r(n, p) = \inf_k \sigma_r^k(n, p)$. (Note that $\sigma_r^{k+1}(n, p) \leq \sigma_r^k(n, p)$.) From the above formula it follows that

$$\sigma(n, p) = \inf_r \sigma_r(n, p) = \inf_{r,k} \sigma_r^k (n, p) .$$

Thus it is enough to compute $\sigma_r(n, p)$.

It is clear that $\sigma_r^1(n, p) = \infty$. On the other hand $\sigma_r^2(n, p)$ is often finite. Sections 2 and 3 will be devoted to the calculation of $\sigma_r^2(n, p)$. In many cases this suffices to determine $\sigma_r(n, p)$.

§2 The structure of $\mathcal{K}^2$ orbits.

The purpose of this section is to describe the structure of $\mathcal{K}^2$ orbits in $\Sigma_r^2(n, p)$, where r is a fixed integer with $n \geq r \geq n - p$. For this, we need I. Porteous' notion of the intrinsic second derivative of a mapping, which we describe below.

If f is a C^2 mapping from an open set U in $\mathbb{R}^n$ to $\mathbb{R}^p$ and $x \in U$, then the second derivative D^2f_x of f at x is the bilinear symmetric mapping $\mathbb{R}^n \times \mathbb{R}^n \to \mathbb{R}^p$ defined by $D^2f_x(v, w) = D(Df)_x (v)(w)$. The second derivative obeys the following law of coordinate changes. If $\phi : U' \to U$ is a diffeomorphism, where U' is a second open set in $\mathbb{R}^n$ and ψ is a diffeomorphism of an open neighbourhood of $f(x)$ in $\mathbb{R}^p$ onto a second open set in $\mathbb{R}^p$, then

$$\begin{aligned} D^2(\psi f \phi)_{\phi^{-1}(x)} &= D^2\psi_{f(x)} \cdot (Df_x \cdot D\phi_{\phi^{-1}(x)}, Df_x \cdot D\phi_{\phi^{-1}(x)}) \\ &+ D\psi_{f(x)} \cdot D^2f_x \cdot (D\phi_{\phi^{-1}(x)}, D\phi_{\phi^{-1}(x)}) \\ &+ D\psi_{f(x)} \cdot Df_x \cdot D^2\phi_{\phi^{-1}(x)} . \end{aligned}$$

Let $K = K_{x,f} = \ker (Df_x : \mathbb{R}^n \to \mathbb{R}^p)$ and $C = C_{x,f} = \operatorname{cok} (Df_x : \mathbb{R}^n \to \mathbb{R}^p)$. The second derivative induces a mapping

$$\underset{\sim}{D}^2 f_x : \mathbb{R}^n \times K \to C$$

called the intrinsic second derivative of f at x. The intrinsic second derivative obeys the following law of coordinate changes :

$$\underset{\sim}{D}^2(\psi f \phi)_{\phi^{-1}(x)} = \underset{\sim}{D}\psi_{f(x)} \cdot \underset{\sim}{D}^2 f_x \cdot (D\phi_{\phi^{-1}(x)}, D\phi_{\phi^{-1}(x)}),$$

where $\underset{\sim}{D}\psi_{f(x)} : C_{x,f} \to C_{x,\psi f(x)}$ is the mapping induced by $D\psi_{f(x)}$.

It follows from this law of coordinate changes that the intrinsic second derivative can be defined for mappings from one manifold N into a second P (whereas, the second derivative cannot be defined in this generality). Thus if $f : N \to P$ is C^∞, $x \in N$, and $K = K_{x,f} = \ker (Df_x : TN_x \to TP_{f(x)})$ and $C = C_{x,f} = \operatorname{cok} (Df_x : TN_x \to TP_{f(x)})$, the intrinsic second derivative

$$\underset{\sim}{D}^2 f_x : TN_x \times K \to C$$

is defined. (For a different way of defining the intrinsic second derivative, see [1, §3].)

In what follows we will be interested only in the restriction $d^2 f_x = \underset{\sim}{D}^2 f_x | K \times K$. Note that $d^2 f_x$ depends only on the 2-jet of f at x ; if z is any 2-jet of f at x we will write $d^2 z = d^2 f_x$ where f is any representative of z.

Clearly $d^2 f_x$ is a symmetric bilinear mapping. We define an equivalence relation on symmetric bilinear mappings as follows. If $\phi : K \times K \to C$ and $\phi_1 : K_1 \times K_1 \to C_1$ are symmetric bilinear mappings of $\mathbb{R}$ vector spaces, we say ϕ and ϕ_1 are <u>equivalent</u> and write $\phi \sim \phi_1$ if there exist $\mathbb{R}$ vector space isomorphisms $\alpha : K \to K_1$ and $\beta : C \to C_1$ such that $\phi_1 = \beta \cdot \phi \cdot (\alpha^{-1} \times \alpha^{-1})$.

If K is a vector space, let S^2K denote the symmetric product of K with itself. The symmetric bilinear mappings $K \times K \to C$ are in natural one-one correspondence with the linear mappings $S^2K \to C$. If $\phi : K \times K \to C$ is a symmetric bilinear mapping, we denote the corresponding linear mapping $S^2K \to C$ by the same letter, and let $\phi^* : C^* \to S^2K^*$ denote its dual.

Let $SK^* = \sum_{i \geq 0} S^i K^*$ denote the symmetric algebra on K^*. If $\phi : K \times K \to C$ is a symmetric bilinear mapping, we define its <u>associated</u> $\mathbb{R}$ <u>algebra</u> $q(\phi)$ by

$$q(\phi) = \frac{SK^*}{\phi^*(C^*) \oplus \sum_{i \geq 3} S^i K^*}$$

If K and K_1 are two $\mathbb{R}$ vector spaces, and $W \subseteq S^2K$, $W_1 \subseteq S^2K_1$ are vector subspaces, we say (K, W) and (K_1, W_1) are <u>equivalent</u> and write $(K, W) \sim (K_1, W_1)$ if there exists an isomorphism $\alpha : K \to K_1$ such that $S^2\alpha(W) = W_1$.

<u>Lemma 2.1.</u> *If* $\phi_i : K_i \times K_i \to C_i$, $i=1, 2$ *are two symmetric bilinear mappings of finite dimensional* $\mathbb{R}$ *vector spaces and* $\dim K_1 = \dim K_2$, *then the following conditions are equivalent* :

a) $\phi_1 \sim \phi_2$

b) $(K_1, W_1) \sim (K_2, W_2)$ *where* $W_i = \ker(\phi_i : S^2K_i \to C_i)$

c) $(K_1^*, W^\perp) \sim (K_2^*, W^\perp)$ *where* $W^\perp$ *is the annihilator of* W_2; *that is* $W^\perp = \mathrm{im}(\phi_i^* : S^2K_1^* \to C_i^*)$.

d) $q(\phi_1) \approx q(\phi_2)$ (*where* $\approx$ *denotes* $\mathbb{R}$ *algebra isomorphism*).

<u>Proof</u> Obvious.

Let N and P be manifolds and let $z \in J^2(N, P)$. It is easily verified that

$$q(d^2 z) \approx Q(z).$$

Since two jets are contact equivalent if and only if their associated $\mathbb{R}$ algebras are isomorphic, lemma 2.1 and the above formula imply the following result.

<u>Proposition 2.2.</u> *Let* N *and* P *be manifolds and let* $z, z' \in J^2(N, P)$. *Then* z *and* z' *are contact equivalent if and only if* d^2z *and* d^2z' *are equivalent as bilinear mappings.*

Now we apply this result to the study of the orbits of $\mathcal{K}^2$ in $\Sigma_r^2(n, p)$. Let $S = S^2L(r, p-n+r)$ denote the $\mathbb{R}$ vector space of symmetric bilinear mappings of $\mathbb{R}^r \times \mathbb{R}^r$ into $\mathbb{R}^{p-n+r}$. Let $G = G\ell(r) \times G\ell(p-n+r)$ act on S in the obvious way: namely, if $h \in G\ell(r)$, $h' \in G\ell(p-n+r)$ and $\phi \in S$ then

$$(h, h').\phi = h' \circ \phi \circ (h^{-1}, h^{-1}).$$

The following is an immediate consequence of proposition 2.2 and the fact that two elements of $J^r(n, p)$ are contact equivalent if and only if they are in the same orbit of $\mathcal{K}^r$.

Corollary 2.3. There is a unique one-one correspondence θ *between the orbits of the action of* $\mathcal{K}^2$ *on* $\Sigma_r^2(n, p)$ *and the orbits of the action of* $G\ell(r) \times G\ell(p-n+r)$ *on* $S^2L(r,p-n+r)$ *such that if* O *is an orbit of* $\mathcal{K}^2$ *and* O' *the corresponding orbit of* $G\ell(r) \times G\ell(p-n+r)$ *then for any* $z \in O$ *and any* $\phi \in O'$, *we have*

$$Q(z) \sim q(\phi).$$

Now we want to analyze θ. In what follows, we set $\Sigma = \Sigma_r^2(n, p)$, $\mathcal{K} = \mathcal{K}^2$, $S = S^2L(r,p-n+r)$, and $G = G\ell(r) \times G\ell(p-n+r)$. We provide Σ and S with the Zariski topology and the orbit spaces $\Sigma/\mathcal{K}$ and S/G with the quotient topology.

Proposition 2.4. The mapping $\theta : \Sigma/\mathcal{K} \to S/G$ *is a homeomorphism.*

Proof Let Λ denote the subset of Σ consisting of all 2-jets of mappings f satisfying

$$(*) \qquad \begin{cases} y_i \circ f = x_i & 1 \leq i \leq n - r \\ d(y_i \circ f)(0) = 0 & n - r + 1 \leq i \leq p. \end{cases}$$

Clearly Λ is a Zariski closed subset of Σ. The composition

$$\Lambda \hookrightarrow \Sigma \to \Sigma/\mathcal{K}$$

is onto because any germ of a mapping can be put in the form $(*)$. Moreover, this composition is an identification mapping (in the sense that a subset of $\Sigma/\mathcal{K}$ is closed if and only if its inverse image in Λ is closed) by V, proposition 9.1.

For any f of the form $(*)$, the kernel of df_0 is $\{dx_1 = \dots = dx_{n-r} = 0\}$ and the image is $\{dy_{n-r+1} = \dots dy_p = 0\}$. If we identify the kernel with $\mathbb{R}^r$ and the cokernel with $\mathbb{R}^{p-n+r}$ we may consider d^2f_0 as a symmetric bilinear mapping $\mathbb{R}^r \times \mathbb{R}^r \to \mathbb{R}^{p-n+r}$. Thus $d^2f_0 \in S$. Therefore $d^2z \in S$ for any $z \in \Lambda$. Now consider the commutative diagram

$$\begin{array}{ccccc} S & \xleftarrow{\;d^2\;} & \Lambda & \hookrightarrow & \Sigma \\ \downarrow & & & & \downarrow \\ S/G & & \xleftarrow{\quad\theta\quad} & & \Sigma/\mathcal{K} \end{array}$$

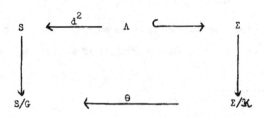

Since the composition $\Lambda \hookrightarrow \Sigma \to \Sigma/\mathcal{K}$ is an identification mapping and Θ is a bijection (corollary 2.3), it is enough to prove that the composition $\Lambda \to S \to S/G$ is an identification mapping. Since $S \to S/G$ is an identification mapping it is enough to show that $\Lambda \to S$ is an identification mapping. But $\Lambda \to S$ is equivalent to a projection $\mathbb{R}^{\frac{n(n+1)}{2}(p-n+r)} \to \mathbb{R}^{\frac{(n-r)(n-r+1)}{2}(p-n+r)}$ and this is clearly an identification mapping (with respect to the Zariski topologies). Q.E.D.

If X is a Zariski closed subset of Σ (or S) then the codimension of X in Σ (or in S) is defined. Therefore if Y is a closed subset of $\Sigma/\mathcal{K}$ (or S/G) we may define its codimension by

$$\mathrm{cod}\, Y = \mathrm{cod}\, \pi^{-1} Y$$

where $\pi : \Sigma \to \Sigma/\mathcal{K}$ (or $\pi : S \to S/G$) is the identification mapping and the right hand side denotes codimension relative to Σ (or S).

Proposition 2.5. If Y is a closed subset of $\Sigma/\mathcal{K}$ then

$$\mathrm{cod}_{S/G}\, \Theta(Y) = \mathrm{cod}_{\Sigma/}\, Y.$$

Proof The inverse image Y' of Y under the composition $\Lambda \hookrightarrow \Sigma \to \Sigma/\mathcal{K}$ is the same as the inverse image of $\Theta(Y)$ under the composition $\Lambda \xrightarrow{d^2} S \to S/G$. Since d^2 is equivalent to a projection of real number spaces, $\mathrm{cod}_\Lambda\, Y' = \mathrm{cod}$ $\mathrm{cod}_{S/G}\, \Theta(Y)$. From V, proposition 9.1, it follows that $\mathrm{cod}_\Lambda\, Y' = \mathrm{cod}_{\Sigma/\mathcal{K}}\, Y$. Combining these two formulae, we get the desired result. Q.E.D.

The mapping Θ induces a one-one correspondence between subsets of $\Sigma/\mathcal{K}$ and S/G. Since the subsets of $\Sigma/\mathcal{K}$ (or S/G) are in one-one correspondence with the invariant subsets of Σ (or S) we get (by composing these one-one correspondences) a one-one correspondence θ between invariant subsets of Σ and invariant subsets of S. Then propositions 2.4 and 2.5 can be reformulated as follows.

Proposition 2.4'. Let X be an invariant subset of Σ. Then X is closed (in the Zariski topology) if and only if $\theta(X)$ is closed.

Proposition 2.5'. Let X be a closed invariant of Σ. Then $\mathrm{cod}_\Sigma X = \mathrm{cod}_S \theta(X)$.

From proposition 2.4' we deduce:

Proposition 2.6. $\theta(\pi_r^2(n, p))$ is the smallest Zariski closed invariant subset of the vector space $S^2L(r, p-n+r)$ whose complement contains only finitely many orbits of $G\ell(r) \times G\ell(p-n+r)$.

Combining this with proposition 2.5' and the observation that the codimension of $\Sigma_r^2(n, p)$ in $J^2(n, p)$ is $r(p-n+r)$, we obtain:

Proposition 2.7. $\sigma_r^2(n, p) = r(p-n+r) + \alpha_{r,n,p}$ where $\alpha_{r,n,p}$ is the codimension in $S = S^2L(r, p-n+r)$ of the smallest closed $G = G\ell(r) \times G\ell(p-n+r)$-invariant subset of S whose complement contains only finitely many G-orbits.

§3 Calculation of $\sigma_r^2(n, p)$.

We divide the calculation of $\sigma_r^2 = \sigma_r^2(n, p)$ into two cases.

Case I. $n \leq p$. We will show

a) $\sigma_r^2 = \infty$ if $r = 1$ or 2

b) $\sigma_3^2 = 6(p-n) + 9$

c) $\sigma_r^2 = r(p-n+r)$ if $r \geq 4$ and $p-n \leq \frac{r(r-1)}{2} - 2$

d) $\sigma_r^2 = (r+2)(p-n+1) + 2$ if $r \geq 4$ and $p-n \geq \frac{r(r-1)}{2} - 2$

Case II. $n > p$

a) $\sigma_r^2 = r(p-n+r)$ $\begin{cases} \text{if } n-p \geq 2 \text{ and } r \geq n-p+2 \\ \text{or } n-p=1 \text{ and } r \geq 4 \end{cases}$

b) $\sigma_r^2 = \infty$ $\begin{cases} \text{if } r=n-p+1 \\ \text{or } n-p=1 \text{ and } r=3 \end{cases}$

These formulae will be obtained by an application of proposition 2.7. (Recall that we have made the convention that the empty set has codimension ∞.)

Let the kernel rank of a symmetric bilinear mapping $\phi : K \times K \to C$ be defined to be the dim of $\ker(\phi : S^2K \to C)$. For any non-negative integer s, let $S(s) = S^2L(r, p-n+r)_s$ denote the set of $\phi \in S$ for which $\ker \text{rank}\ \phi = S$. Clearly $S(s)$ is empty unless

$$\max(0, \frac{r(r-1)}{2} - p+n) \leq s \leq \min(\frac{r(r+1)}{2}, p-n+r),$$

and it is non-empty when this inequality is satisfied.

Furthermore $\overline{S(s)} = \bigcup_{s' \geq s} S(s')$, and $\operatorname{cod}_S \overline{S(s)} = s(p-n - \frac{r(r-1)}{2} + s)$. Finally $S(s)$ is invariant under the action of G.

Let $Gr(s, r)$ denote the Grassmannian of s planes in $S^2\mathbb{R}^r$. The projective linear group $PG\ell(r)$ acts on $Gr(s, r)$, and there are natural mappings $S(s) \to Gr(s, r)$ and $S(s) \to Gr(\frac{r(r+1)}{2} - s, r)$, defined respectively by $\phi \longmapsto \ker(\phi : S^2\mathbb{R}^r \to \mathbb{R}^{p-n+r})$ and $\phi \longmapsto \operatorname{im}(\phi^* : \mathbb{R}^{p-n+r} \to S^2\mathbb{R}^r)$. In what follows, we set $s' = \frac{r(r+1)}{2} - s$. By lemma 2.1 these mappings induce bijections of orbit spaces:

$$Gr(s', r)/PG\ell(r) \underset{\sim}{\leftarrow} S(s)/G \underset{\sim}{\rightarrow} Gr(s,r)/PG\ell(r),$$

in case $S(s)$ is non-empty.

It is easily seen that these bijections are homeomorphisms and preserve codimension of closed subsets (where the codimension of a closed set in an orbit space is defined as the codimension of the union of its orbits.)

We will call the homeomorphism

$$Gr(s', r)/PG\ell(r) \approx Gr(s, r)/PG\ell(r)$$

given above the <u>duality principle</u>. The duality principle can also be given the following form: <u>If $S(s)$ and $S(s')$ are both non-empty then there is a homeomorphism $S(s)/G \sim S(s')/G$ which preserves codimension (of closed sets)</u>.

Now we consider case IIa. Since $n > p$, we have $\frac{r(r-1)}{2} - p+n > 0$, and this number is the smallest value of s for which $S(s)$ is non-empty. Hence $S(s)$ is open and dense in S for this value of s. We have

$$\dim Gr(s, r) = s(\frac{r(r+1)}{2} - s) = (\frac{r(r-1)}{2} - p+n)(p-n+r)$$

$$\dim PG\ell(r) = r^2 - 1$$

from which it follows easily that

$$\dim Gr(s, r) > \dim PG\ell(r) \quad \text{for} \quad s = \frac{r(r-1)}{2} - p+n.$$

It follows that there are no finite open sets in the orbit space $Gr(s, r)/PG\ell(r) \approx S(s)/G$. Since $S(s)/G$ is dense in S/G, there are no finite open sets in S/G. By proposition 2.7, it then follows that $\sigma_r^2 = r(p-n+r)$ in case IIa.

Next, we consider cases Ic and Id. It is easily seen that

$$\dim Gr(s, r) > \dim PG\ell(r)$$

when $n \leq p$, $r \geq 4$, and $2 \leq s \leq \frac{r(r+1)}{2} - 2$. In case Ic, the smallest value of s for which $S(s)$ is non-empty lies in this range; hence $\sigma_r^2 = r(p-n+r)$ in this case, by the same reasoning as in case IIa.

In case Id, the smallest value of s for which $S(s)$ is non-empty is ≤ 2. Furthermore, the elements of $S(1)/G \approx Gr(1, r)/G\ell(r)$ are in one-one correspondence with equivalence classes of real non-zero quadratic forms in r variables, and hence they form a finite set. Furthermore $S(0)/G \approx G\ell(0, r)/G\ell(r)$ obviously contains only one point.

Hence the smallest closed invariant subset of S whose complement contains only finitely many orbits in $\overline{S(2)}$. From this it follows easily that σ_r^2 has the value given in the table in case Id.

Next, we consider the case Ia. When $r = 1$, it is trivial that S/G has exactly two points $S(1)/G$ and $S(0)/G$; hence $\sigma_1^2 = \infty$. When $r = 2$, only $S(0)$, $S(1)$, $S(2)$, and $S(3)$ are non-void (and $S(0)$ is empty when $n = p$). Clearly $S(0)/G$ and $S(3)/G$ each have only one point, and $S(1)/G \approx Gr(1, r)/G\ell(r)$ is in one-one correspondence with the set of equivalence classes of non-zero real quadratic forms in two variables. Furthermore by the duality principle $S(2)/G$ is also in one-one correspondence with the set of non-zero real quadratic forms in two variables. Hence all $S(i)/G$ are finite. Hence $\sigma_2^2 = \infty$.

Now we consider case Ib. Since $\dim Gr(3, 3) = 9$ and $\dim PG\ell(3) = 8$, we have that $S(3)/G \approx Gr(3, 3)/PG\ell(3)$ has no finite open subsets, it follows that $\overline{S(3)}$ is contained in any invariant subset of S whose complement contains only finitely many orbits.

By arguments given above $S(0)/G$ in one point and $S(1)/G$ is a finite set. We will show below that $S(2)/G \approx Gr(2, 3)/G\ell(3)$ is also a finite set, containing exactly 13 points. Then it will follow that $\overline{S(3)}$ is the smallest closed invariant subset of S whose complement contains only finitely many orbits, and hence $\sigma_3^2 = 6(p-n) + 9$, by proposition 2.7.

We now begin the proof that $Gr(2, 3)/G\ell(3)$ contains exactly 13 points. We adopt the following notational conventions. We set $X = Gr(2, 3)/G\ell(3)$. We let $Gr(2, 3; \mathbb{C})$ denote the Grassmannian of complex 2-planes in $S^2 \mathbb{C}^3$, provided with the Zariski topology. We let $X_{\mathbb{C}}$ denote the orbit space $Gr(2, 3; \mathbb{C})/G\ell(3)$, provided with the quotient topology. The embedding $Gr(2, 3) \hookrightarrow Gr(2, 3; \mathbb{C})$ defined by "complexification" induces a mapping $X \to X_{\mathbb{C}}$. Clearly a set in X is open if and only if it is the inverse image by this mapping of an open set in $X_{\mathbb{C}}$. Thus to describe X as a topological space, it is enough to describe $X_{\mathbb{C}}$ as a topological space, and say how many points there are in the inverse image of each point of $X_{\mathbb{C}}$, i.e., how many "real forms" correspond to each point of $X_{\mathbb{C}}$.

We now begin the study of $X_{\mathbb{C}}$. We will show that $X_{\mathbb{C}}$ contains exactly eight points and describe the topology of $X_{\mathbb{C}}$ explicitly.

A point in $Gr(2, 3; \mathbb{C})$ is a complex 2-plane in $S^2\mathbb{C}^3$. Since $S^2\mathbb{C}^3$ can be identified with the set of homogeneous quadratic polynomials in 3 variables, a line in $S^2\mathbb{C}^3$ corresponds to a pencil of curves ω in $\mathbb{C}P^2$.

Let γ_1 and γ_2 be two distinct curves in ω, and let $\Gamma = \gamma_1 \cap \gamma_2$. One calls Γ the set of <u>base points</u> of ω. Clearly Γ is independent of the choice of γ_1 and γ_2. We consider the following cases:

(i) $\dim \Gamma = 0$

(ii) $\dim \Gamma = 1$.

In case (i), it follows from Bezout's theorem that Γ has four points with multiplicities counted. It follows that the only possibilities are:

(α) Γ consists of four distinct points,

(β) Γ consists of a point of multiplicity 2 and two points of multiplicity 1,

(γ) Γ consists of two points of multiplicity 2,

(δ) Γ consists of a point of multiplicity 3 and a point of multiplicity 1,

and

(ϵ) Γ consists of a single point of multiplicity 4.

(See figure 2)

We will denote the corresponding subsets of X (resp. $X_{\mathbb{C}}$) by $X(i\alpha)$, $X(i\beta)$, etc. (resp. $X_{\mathbb{C}}(i\alpha)$, $X_{\mathbb{C}}(i\beta)$, etc.). Thus $X(i\alpha)$ is the inverse image of $X_{\mathbb{C}}(i\alpha)$ and so on.

Now we consider case $(i\alpha)$. No three points of Γ lie in a line ℓ, for otherwise we would have $\ell \subseteq \gamma_1$ and $\ell \subseteq \gamma_2$ (since γ_1 and γ_2 are second order curves) which contradicts the hypothesis $\dim \Gamma = 0$.

It is easily seen that in this case the pencil of curves ω is uniquely determined by Γ. For, the projective space of all quadratic curves on $\mathbb{C}P^2$ is 5 dimensional, and the condition that a curve γ pass through the four points of Γ imposes four independent conditions on γ, so that the set of all curves passing through Γ is reduced to the pencil ω.

Since any four points in $\mathbb{C}P^2$, no three of which are collinear, can be transformed by a linear transformation onto any other set of four points, no three of which are collinear, it follows that $X_{\mathbb{C}}(i\alpha)$ is reduced to a single point. This point has three "real forms" (i.e. $X(i\alpha)$ consists of three points) corresponding to whether 4, 2, or 0 points of Γ are in $\mathbb{R}P^2$.

Similar arguments show that $X_{\mathbb{C}}(i\beta)$, $X_{\mathbb{C}}(i\gamma)$, and $X_{\mathbb{C}}(i\delta)$ each consist of a single point. The corresponding number of real forms are: $X(i\beta)$, two (corresponding to whether the two points in Γ of multiplicity 1 are real); $X(i\gamma)$, two (corresponding to whether the two points in Γ are real); and $X(i\delta)$, one.

In the case ϵ, we have to distinguish two subcases:

(1) a generic curve in ω is non-singular, and

(2) a generic curve in ω is singular.

In case $i\epsilon 1$, one sees easily that W is spanned by ℓ^2 and p where ℓ is a linear form and p is an irreducible quadratic form such that the line $\{\ell=0\}$ is a tangent to the curve $\{p=0\}$. It follows easily that $X_{\mathbb{C}}(i\epsilon 1)$ consists of one point, which has one real form.

In case $(i\epsilon 2)$, any curve γ in the pencil ω is a singular quadratic curve, i.e., the union of two lines, or a single line counted twice. In the former case, the unique singular point of γ must be Γ, because if γ' is a second member of ω, the curves γ and γ' intersect in Γ alone. It follows that if we choose homogeneous linear coordinates x, y, z for $\mathbb{C}P^2$ such that Γ has coordinates $(1, 0, 0)$, then the defining equation for any γ in ω is a form in y and z alone. Therefore W is a two dimensional subspace of the three dimensional space $S^2\mathbb{C}^2$ of all quadratic forms in y and z. It follows from the duality principle that equivalence classes of such subspaces are classified by equivalence classes of one dimensional subspaces of $S^2\mathbb{C}^2$, i.e. equivalence classes of quadratic forms in two variables. Furthermore, it is easily seen that the hypothesis that $\dim \Gamma = 0$ implies that the quadratic form corresponding to W is non-degenerate.

Hence $X_{\mathbb{C}}(i\epsilon 2)$ consists of one point, and it has two "real forms".

This completes the study of case (i).

In case (ii), when $\dim \Gamma = 1$, we have that Γ is a line in $\mathbb{C}P^2$. Any $\gamma \in \omega$ is the union of Γ and another line $\ell(\gamma)$. The family of lines $\{\ell(\gamma) : \gamma \in \omega\}$ forms a pencil of lines in $\mathbb{C}P^2$; hence it consists of all lines passing through a given point $p \in \mathbb{C}P^2$. There are two subcases to consider:

(α) $p \notin \Gamma$

(β) $p \in \Gamma$

In either case p and Γ uniquely determine the pencil ω and hence the subspace W. It follows that $X_{\mathbb{C}}(ii\alpha)$ and $X_{\mathbb{C}}(ii\beta)$ each consists of one point.

Note that since $X_{\mathbb{C}}$ is a finite topological space, its topology is determined by the closure operator applied to points. In the table 1 we describe the topology on $X_{\mathbb{C}}$. We leave to the reader the verification of the information in this table which was not deduced above.

This completes the study of case Ib: we have shown that $\sigma_3^2 = 6(p-n)+9$ when $n \leq p$.

The only remaining case in the computation of σ_r^2 is case IIb. The assertion that $\sigma_r^2 = \infty$ is equivalent to the assertion that $\Sigma_r^2/\mathcal{K}^2$ is a finite set. But $\Sigma_r^2/\mathcal{K}^2 \approx S/G$ by proposition 2.4, where $S = SL^2(r, p-n+r)$ and $G = G\ell(r) \times G\ell(p-n+r)$. In the case $r = n-p+1$, the points of S/G are in one-one correspondence with equivalence classes of real quadratic forms in $n-p+1$ variables, and there are only a finite number of these. Hence $\sigma_r^2 = \infty$ when $r = n-p+1$.

In the remaining case, when $n-p = 1$ and $r = 3$, we have $S = SL^2(3, 2) = S(4) \cup S(5) \cup S(6)$. Clearly $S(6)/G$ is one point and $S(5)/G$ is finite. Furthermore, we have shown above that

$$S(4)/G \approx Gr(2, 3)/PG\ell(3),$$

and we have just finished showing that this space contains exactly 13 points. Hence $\sigma_r^2 = \infty$ in this case too.

This completes the verification of the formulae for σ_r^2 asserted at the beginning of this section.

TABLE 1.

The Structure of $X_{\mathbb{C}}$.

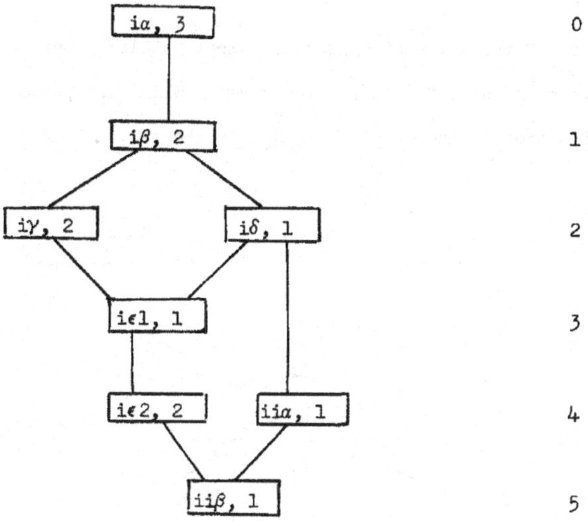

Each box represents a point of $X_{\mathbb{C}}$. The first symbol in each box is the case to which that point corresponds (see text). The second symbol is the number of real forms of the point. If two boxes are connected by a line, the point corresponding to the lower box is in the closure of the point corresponding to the upper box. More generally if p is any point of $X_{\mathbb{C}}$ the closure of p is the set of points whose boxes are connected to the box corresponding to p by a descending chain of lines in the above diagram. Finally all points represented by boxes in the same row have the same codimension in $X_{\mathbb{C}}$, the codimension is indicated by the number in the right hand column.

Figure 2

Examples of pencils of conics

in the plane

Each of the pairs of conics illustrated below generates a pencil of conics. The letter to the left denotes the case to which the pencil belongs (see the classification of the points of $X_{\mathbb{C}}$ in the text).

(iα)

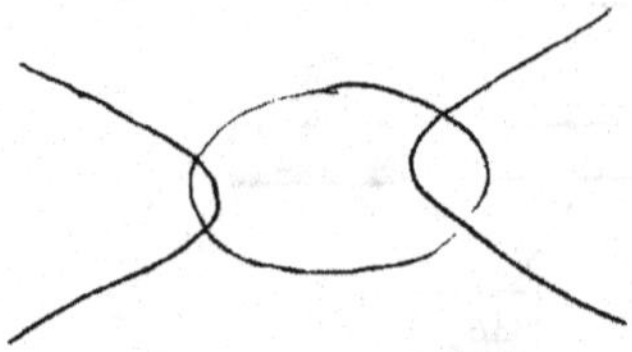

(iβ)

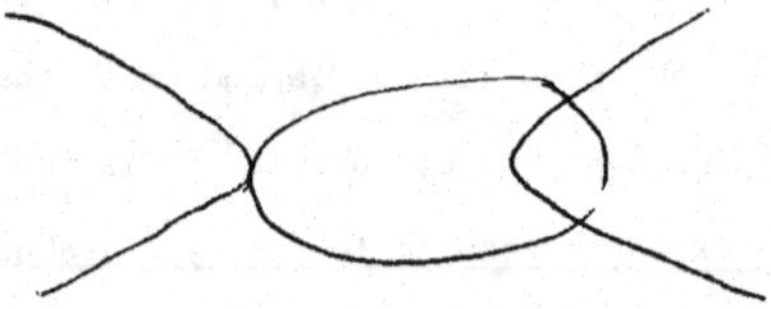

(iγ)

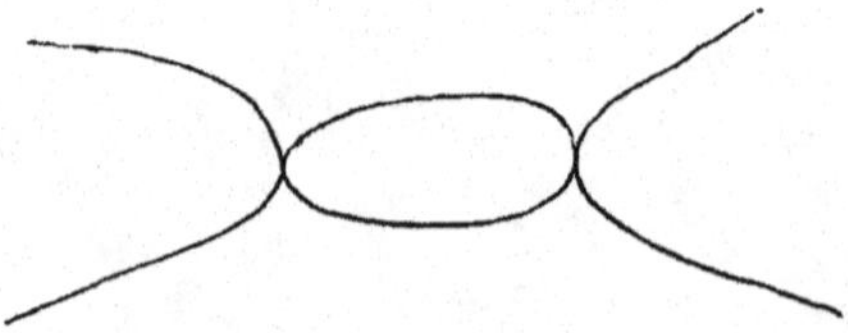

figure 2 (cont.)

(iδ)

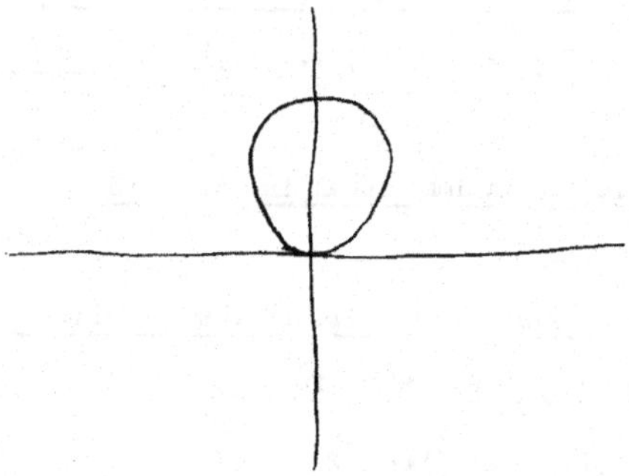

(iε1)

line counted with
multiplicity 2.

§4 A partial classification of certain algebras.

The purpose of this section is to prove the following result.

Proposition 4.1. Let k be a field of characteristic 0. Let I be an ideal in the formal power series ring $A = k[[x_1, \ldots, x_n]]$ such that for some integer p we have

(i) $I \subseteq \mathfrak{m}^p$ (where $\mathfrak{m}$ is the maximal ideal in A), and

(ii) $\dim_k \mathfrak{m}^p/(I + \mathfrak{m}^{p+1}) = 1$.

Then either $\mathfrak{m}^{p+1} \subseteq I$, or there exists a k algebra automorphism of A which carries I onto an ideal of one of the following forms.

$$\text{(a)} \quad \Sigma_{\omega\in\Omega} [(x^\omega + a_\omega x_1^\ell)A] + \Sigma_{\omega\in\Omega'}(x^\omega A),$$

where $\ell>p$, Ω denotes the set of all multi-indices $\omega = (\omega_1, \ldots, \omega_n)$ such that $|\omega| = p$ and $\omega_1 < p-1$, Ω' denotes the set of all multi-indices ω such that $|\omega| = p$ and $\omega_1 = p-1$, and where $a_\omega \in k$ and $a_\omega \neq 0$ for at least one $\omega \in \Omega$.

$$\text{(b)} \ \Sigma_{\omega\in\Omega\cup\Omega'}(x^\omega A) + x^\ell A, \quad \text{where } \ell>p+1.$$

$$\text{(c)} \ \Sigma_{\omega\in\Omega\cup\Omega'}(x^\omega A)$$

Most of the applications of this theorem that we will have occur in the case $p=2$. However, we will use the case $p=3$ of this theorem once in what follows.

For the proof we will need to recall some facts about the symmetric algebra on a vector space. Let V be a finite dimensional vector space over k. Then the symmetric algebra

$$SV = \bigoplus_{i=0} S^iV = k \oplus V \oplus S^2V \oplus \ldots \oplus S^rV \oplus \ldots$$

is defined, and it is the polynomial algebra $k[x_1, \ldots, x_n]$ where $x_1, \ldots, x_n$ is any basis for V. Likewise we may define

$$\hat{S}V = \prod_{i\geq 0} S^iV.$$

If $x_1, \ldots, x_n$ is a basis for V, then $\hat{S}V$ is the formal power series algebra $k[[x_1, \ldots, x_n]]$.

We will regard SV and $\hat{S}V$ as topological algebras, where SV is provided with the discrete topology and $\hat{S}V$ is provided with the Krull topology, i.e. the topology such that $\{\mathfrak{m}^n\}$ is a basis for the family of neighbourhoods of 0, where $= \prod_{i \geq 1} S^i V$ is the unique maximal ideal in $\hat{S}V$.

In what follows we will regard k as a topological field with the discrete topology and V as a topological vector space over k with the discrete topology.

We let V^* be the dual of V. There is a canonical identification of $S^r(V^*)$ with $(S^r V)^*$ defined by the pairing

$$(v_1^* \dots v_r^*, v_1 \dots v_r) = \sum_{\pi} \prod_{i=1}^{r} (v_i^*, v_{\pi i}),$$

where $v_1, \dots, v_r \in V$ and $v_1^*, \dots, v_r^* \in V^*$. This identification commutes with the canonical actions of $\mathrm{Aut}(V)$ on $S^r(V^*)$ and $(S^r V)^*$.

It provides identifications of $(SV)^*$ with $\hat{S}V^*$ $(= \hat{S}(V^*))$ and of $(\hat{S}V)^*$ with SV^* $(= S(V^*))$, where the dual of $\hat{S}V$ means the continuous linear mappings of $\hat{S}V$ into k.

In terms of these identifications, if T is a linear endomorphism of SV, its dual T^* is a continuous linear endomorphism of $\hat{S}V^*$, and if T is a continuous linear endomorphism of $\hat{S}V$, then its dual T^* is a linear endomorphism of SV^*.

Note that if $x_1, \dots, x_n$ is a basis of V and $x_1^*, \dots, x_n^*$ is the dual basis of V^*, then multiplication by x_i and partial differentiation by x_i^* are dual operations. It follows that a vector subspace U of SV^* is the annihilator $I^{\perp}$ of an ideal I in $\hat{S}V$ if and only if it is closed under the operations $\partial/\partial x_i^*$, $i=1, \dots, n$. This suggests the following definition.

<u>Definition</u> A vector subspace U of SV will be said to be a <u>co-ideal</u> if it is closed under the operations $\partial/\partial x_1, \dots, \partial/\partial x_n$, where $x_1, \dots, x_n$ is a basis of V. (It is clear that this condition is independent of the basis chosen.)

It is trivial to verify that if I is an ideal in $\hat{S}V$, then the annihilator $I^{\perp}$ of I is a co-ideal in SV^*; similarly, if U is a co-ideal in SV^* then its annihilator $U^{\perp}$ is an ideal in $\hat{S}V$. It is also trivial to verify that $U^{\perp\perp} = U$. Krull's theorem that any ideal in $\hat{S}V$ is closed implies that $I^{\perp\perp} = I$. Hence the operation $I \to I^{\perp}$ defines a one-one correspondence between ideals in $\hat{S}V$ and co-ideals in SV^*.

Lemma 4.2. Let U be a co-ideal in SV^*. Let $U_r = U \cap S^rV^*$. Suppose for some integer r, $\dim U_r = 1$. Then $\dim U_{r+1} \leq 1$ and if $\dim U_{r+1} = 1$ then there is a basis $y_1^*, \ldots y_n^*$ for V^* such that $U_s = 0$, or is the vector space spanned by y_1^{*s}, for $s \geq r$.

Proof If $\dim U_{r+1} = 0$ there is nothing to prove, so suppose $u \in U_{r+1}$ and $u \neq 0$. Let $x_1^*, \ldots, x_n^*$ be a basis for V^*. Then at least one $\partial u/\partial x_i^* \neq 0$; without loss of generality, we may suppose that $\partial u/\partial x_1^* \neq 0$. Since each $\partial u/\partial x_i^* \in U_r$ and $\dim U_r = 1$, it follows that there exist $\beta_i \in k$ (for $2 \leq i \leq n$) such that

$$\partial u/\partial x_i^* = \beta_i(\partial u/\partial x_1^*).$$

Define the basis $y_1^*, \ldots, y_n^*$ of V^* so that

$$x_1^* = y_1^* - \Sigma_{i=2}^n \beta_i y^*$$

$$x_i^* = y_i^* \qquad i \geq 2.$$

Then for $i \geq 2$, we have

$$\frac{\partial u}{\partial y_i^*} = \sum \frac{\partial u}{\partial x_j^*} \frac{\partial x_j^*}{\partial y_i^*} = 0 .$$

Since u is homogeneous of order $r+1$, it follows that u is a multiple of y_1^{*r+1}. Now if u_1 is any element of $S^{r+1}V^*$ which is independent of u, then $\partial u_1/\partial y_i^*$ is independent of $\partial u/\partial y_1^*$ for some i. If $u_1 \in U_{r+1}$ then both $\partial u_1/\partial y_i^*$ and $\partial u/\partial y_1^*$ are in U_r and we get a contradiction to the hypothesis $\dim U_r = 1$.

Hence U_{r+1} is spanned by y_1^{*r+1}. Since $\partial(y_1^{*r+1})/\partial y_1^* \in U_r$ and $\dim U_r = 1$, it follows that U_r is spanned by y_1^{*r}. Finally it follows by induction that every element of U_s for $s \geq r+1$ is a multiple of y_1^{*s}. QED

Proof of proposition 4.1. We may suppose that $x_1, \ldots, x_n$ is the basis of a vector space V. Then $A = \hat{S}V$. If $\mathfrak{m}^{p+1} \subseteq I$, then the theorem is true, so we may suppose that this is not the case. Then $\dim I_p^\perp = 1$ and $\dim I_{p+1}^\perp \neq 0$, so by lemma 4.2, there exists a basis $y_1, \ldots, y_n$ of V such that $I_s^\perp$ is 0 or is spanned by y_1^{*s} for all $s \geq p$.

In terms of such a basis, I is generated by elements of the form

$$y^{\omega} + f_{\omega} \qquad \text{order } (f_{\omega}) > p$$

where ω ranges over $\Omega \cup \Omega'$, and elements of order $> p$. It is enough to see that we may assume that f_{ω} is a power series in y_1 alone and that by "change of coordinates" we may arrange that $f_{\omega} = 0$ for $\omega \in \Omega'$. Likewise, it is easy to see that if order $(f_{\omega}) = \ell$ for some $\ell > p$ and some $\omega \in \Omega$, then $\mathfrak{M}^{\ell+1} \subseteq I$. Theorem 4.1 follows immediately. Q.E.D.

§5 Calculation of $\sigma_r(n, p)$ for $r \geq 4$ and $n \leq p$.

In this section, we will finish the calculation of σ_r in the range indicated in the title. We will show that $\sigma_r = \sigma_r^2$ in this range. We recall that one always has $\sigma_r \leq \sigma_r^2$. It is clear that $\sigma_r(n, p) \geq r(p-n+r) = \operatorname{cod} \Sigma_r(n, p)$. Hence when $\sigma_r^2 = r(p-n+r)$ we have $\sigma_r = \sigma_r^2$; e.g., in case Ic of §3 we have $\sigma_r = \sigma_r^2$. Hence it remains only to show that in case Id we have $\sigma_r = \sigma_r^2$.

Recall that σ_r^k is defined as the codimension of $\Pi_r^k(n, p)$ in $J^k(n, p)$. Let $\Pi_{k2} : J^k(n, p) \to J^2(n, p)$ denote the projection. We claim that in case Id, we have

$$(1) \qquad \Pi_r^k(n, p) = \Pi_{k2}^{-1}\,(\Pi_r^2(n, p)).$$

This formula clearly implies $\sigma_r = \sigma_r^2$. To prove it, it is enough to show that $Q(z)$ always belongs to one of only finitely many isomorphism classes for $z \in \Sigma_r^k(n, p) - \Pi_{k2}^{-1}\,(\Pi_r^2(n, p))$, by the definition of $\Pi_r^k(n, p)$.

In studying case Id in §3, we showed that $\overline{S(2)}$ is the smallest closed $G = G\ell(r) \times G(p-n+r)$-invariant subset of $S = S^2L(r, p-n+r)$ whose complement contains only finitely many G orbits. In view of proposition 2.6, this means that $\Pi_r^2(n, p)$ is the set of all $z \in \Sigma_r^2(n, p)$ such that ker rank $(d^2z) \geq 2$. Let $\mathfrak{m}(z)$ denote the maximal ideal of $Q(z)$. From the isomorphism $q(d^2z) \approx Q(z)$, it follows that $\dim \mathfrak{m}(z)^2 = \text{ker rank } d^2z$ for any 2-jet z. Hence $\Pi_r^2(n, p)$ is the set of all $z \in \Sigma_r^2(n, p)$ such that $\dim \mathfrak{m}(z)^2 \geq 2$.

Hence, it suffices to show that there are only finitely many isomorphism classes of associated $\mathbb{R}$ algebras $Q(z)$ to $z \in \Sigma_r^k$ such that $\dim \mathfrak{m}(z)^2/\mathfrak{m}(z)^3 = 1$, in order to show that (1) holds.

By an argument which we have already given in §3, there are only finitely many classes of $Q(z)$ for which $\dim \mathfrak{m}(z)^2/\mathfrak{m}(z)^3 = 1$ and $\mathfrak{m}(z)^3 = 0$. Hence, it is enough to study the case $\mathfrak{m}(z)^3 \neq 0$.

In this case $Q(z) = A/I$, where $A = \mathbb{R}[[x_1, \ldots, x_r]]$ and I is an ideal in A satisfying the hypotheses of proposition 4.1. By the conclusion of proposition 4.1, we can assume that I has one of the forms (a), (b), or (c) of proposition 4.1. Form (c) is eliminated by the fact that $\mathfrak{m}(z)^{r+1} = 0$, and there are only finitely many I of the form (b) for which $I \supseteq \mathfrak{m}^{r+1}$.

Hence, all that is left to show is that there are only finitely many equivalence classes of I of the form (a) for which $I \supseteq \mathfrak{m}^{r+1}$, where two ideals in A are said to be *equivalent* if there is an $\mathbb{R}$-algebra automorphism of A which carries one ideal onto the other. The fact that this is so follows immediately from the next result.

Proposition 5.1. Let k *be a field of characteristic* 0. *Let* I *and* I' *be two ideals in* $A = k[[x_1, \ldots, x_n]]$ *which satisfy hypotheses* (i) *and* (ii) *of proposition* 4.1, *for* $p = 2$. *Suppose furthermore that* I *and* I' *both have the form (a) of proposition* 4.1, *for the same value of* ℓ, i.e.,

$$(2) \qquad I = \Sigma_{1<i\leq j\leq n}[(x_i x_j + a_{ij} x_1^{\ell}) A] + \Sigma_{1<i\leq n}(x_1 x_i A)$$

$$I' = \Sigma_{1<i\leq j\leq n}[(x_i x_j + a'_{ij} x_1^{\ell}) A] + \Sigma_{1<i\leq n}(x_1 x_i A)$$

Then the ideals I *and* I' *are equivalent if and only if there exists an invertible matrix* $(c_{ij})_{2\leq i,j\leq n}$ *with entries in* k, *and* $\alpha \neq 0$ *in* k *such that*

$$(3) \qquad a'_{ij}\,\alpha^{\ell} = \sum_{k,\ell\geq 2} c_{ik}\, c_{j\ell}\, a_{k\ell}$$

(where we define $a_{k\ell} = a_{\ell k}$ *if* $k > \ell$).

Proof The ideals I and I' are equivalent if and only if we can find a set of n generators $y_1, \ldots, y_n$ of the maximal ideal of A such that

(4) $$I = \Sigma_{1<i\leq j\leq n} [(y_i y_j + a'_{ij}\, y_1^{\ell})\, A] + \Sigma_{1<i\leq n} (y_1 y_i A).$$

Suppose that such a set of y_i exist. Write

(5) $$y_i = \Sigma\, c_{ij}\, x_j + \text{higher order terms}$$

Comparing the two expressions for I, one sees that the matrix $C = (c_{ij})$ has the form

$$C = \begin{pmatrix} \alpha & 0 \\ 0 & C' \end{pmatrix}$$

where C' is the $(n-1) \times (n-1)$ matrix $(c_{ij})_{2\leq i,j\leq n}$ and $\alpha \in k$. Also $y_i(x_1, 0, \ldots, 0)$ is of order $\geq \ell$, for $i > 1$, for otherwise we would have $x_1 x_i + \beta x_1^m \in I$ for some $m \leq \ell$ and $\beta \neq 0$ by (4), which is clearly impossible by (2).

By substituting (5) in (4), we get $\Sigma_{k,\ell\geq 2}\, c_{ik}\, c_{j\ell}\, x_k\, x_\ell + a'_{ij}\, \alpha^{\ell}\, x_1^{\ell} \in I$. Hence from (2), it follows that (3) holds.

Conversely suppose that (3) holds and define

$$y_1 = \alpha x_1$$

$$y_i = \sum_{j\geq 2} c_{ij}\, x_j$$

Then it is easily seen that (4) holds. Q.E.D.

§6 Classification of Stable Germs of Type Σ_1, for $n \leq p$.

In the previous section, we have computed $\sigma_r(n, p)$ in the case $r \geq 4$ and $n \leq p$. In the case $r=1$ and $n \leq p$, it is trivial that $\sigma_r(n, p) = \infty$, for if $z \in \Sigma_1^k$, then $Q(z) \approx \mathbb{R}[[x]]/(x^\ell)$, where $2 \leq \ell \leq k+1$, so there are only finitely many $\mathcal{K}^k$ orbits in Σ_1^k. For the calculation of $\sigma_r(n, p)$ when $n \leq p$, all that remains to be done are the cases $r=2$ and $r=3$.

Using the classification theorem of IV, it is now easy to classify stable germs of mappings of type Σ_1 from an n-manifold N^n to a p-manifold P^p, in the case $n \leq p$. The purpose of this section is to work out this classification in detail. This is somewhat of a digression from the main line of our argument, whose purpose is to compute $\sigma_r(n, p)$.

The reason for this digression is to recall the way in which classification theorems for $\mathbb{R}$ algebras lead to classification theorems for stable map-germs. In IV, we proved that a stable map-germ $f : (N^n, x) \dashrightarrow (P^p, y)$ is determined up to equivalence by its associated $\mathbb{R}$ algebra (IV, theorem A). We also gave necessary and sufficient criteria for an $\mathbb{R}$ algebra to be associated to a stable map-germ $f : (N^n, x) \dashrightarrow (P^p, y)$ (IV, theorem B). We recall these criteria in the case $n \leq p$ here.

Let Q be a quotient of a formal power series algebra over $\mathbb{R}$. It is easily seen that there is an integer $\imath(Q)$ such that if we present Q as a quotient of a formal power series algebra over $\mathbb{R}$ in any way:

$$Q = \mathbb{R}[[x_1, \ldots, x_a]]/I$$

and let b be the minimum number of elements in a set of generators of I then $\imath(Q) = a-b$.

Clearly, there exists a C^∞ map-germ $f : (N^n, x) \dashrightarrow (P^p, y)$ such that $\hat{Q}(f) \approx Q$ if and only if $p-n \geq -\imath(Q)$.

Let $\delta(Q) = \dim_{\mathbb{R}} Q$. It is well known that $\delta(Q) < \infty$ if and only if the Krull dimension of Q is 0. It is also well known that

$$\imath(Q) \leq \text{Krull dim}\ (Q),$$

so that if $\delta(Q) < \infty$, then $\imath(Q) \leq 0$. It is easily seen that when $\delta(Q) < \infty$, there is a unique integer $\gamma(Q)$ such that

$$\dim_{\mathbb{R}} \frac{\theta(f)_x}{tf[\theta(N)_x]+f^*\mathfrak{m}_y\theta(f)_x} = \delta(Q)(p-n) + \gamma(Q)$$

for any C^∞ map-germ $f : (N^n, x) \dashrightarrow (P^p, y)$ such that $\hat{Q}(f) \approx Q$.

In IV, we showed that when $n \leq p$, there is a <u>stable</u> C^∞ map-germ $f : (N^n, x) \dashrightarrow (P^p, y)$ such that $\hat{Q}(f) \approx Q$ if and only if the following conditions are satisfied:

$$p-n \geq - \tau(Q)$$
$$\delta(Q) < \infty$$
$$\delta(Q)(p-n) + \gamma(Q) \leq p.$$

The classification of stable map-germs of type Σ_1, for $n \leq p$, follows immediately from these remarks. For if $f : (N, x) \dashrightarrow (P, y)$ is a map-germ of type Σ_1, then $\hat{Q}(f)$ is a quotient of a formal power series ring on <u>one</u> variable, and if $\delta(\hat{Q}(f)) < \infty$ (in particular, if f is stable), we have $\hat{Q}(f) \approx A_\ell$ for some positive integer ℓ, where we set $A_\ell = \mathbb{R}[[x]]/(x^{\ell+1})$. It is easily seen that

$$\delta(A_\ell) = \ell+1$$
$$\gamma(A_\ell) = \ell.$$

Hence, there is a stable map-germ $f : (N^n, x) \dashrightarrow (P^p, y)$ such that $\hat{Q}(f) \approx A_\ell$ if and only if

$$(\ell+1)(p-n) + \ell \leq p$$

i.e., if and only if

$$\ell \leq \frac{n}{p-n+1} .$$

This provides the complete classification of stable map-germs of type Σ_1 for $n \leq p$.

The results of this section were first obtained by B. Morin [4]. He also obtained "normal forms" for stable map-germs of type Σ_1, where $n \leq p$. His normal forms can be obtained by means of the procedure described in IV, lemma 5.9 and proposition 5.10 for obtaining normal forms for any stable map-germ. Another method of obtaining his normal forms in the case $n=p$ was given by Arnold [1].

§7 Classification of Stable Germs of Type $\Sigma_{2,0}$ for $n \leq p$.

In this section, we extend the arguments of the previous section to classify some more stable map-germs. The results of this section will be used later in the calculation of $\sigma_2(n, p)$.

The expression "map-germ of type $\Sigma_{2,0}$" is due to Boardman [2]. Boardman's definition is a modification of an earlier definition due to Thom [5]. Boardman defines a submanifold $\Sigma_{2,0}$ of the jet bundle $J^2(N, P)$. A germ $f : (N, x) \dashrightarrow (P, y)$ of a mapping is said to be of type $\Sigma_{2,0}$ if $j^2 f(x) \in \Sigma_{2,0}$.

Any germ of a mapping of type $\Sigma_{2,0}$ is of type Σ_2 (i.e. its kernel rank is 2) by definition. The following characterization (due to Boardman [2]) of map-germs of type $\Sigma_{2,0}$ will be all we need to know about $\Sigma_{2,0}$ for what we are going to do. Let $f : (N, x) \dashrightarrow (P, y)$ be a map-germ of type Σ_2 and let $K = \ker(df_x : TN_x \to TP_y)$ and $C = \operatorname{cok}(df_x : TN_x \to TP_y)$. Then f is of type $\Sigma_{2,0}$ if and only if the adjoint mapping $K \to \operatorname{Hom}(K, C)$ of $d^2 f_x : K \times K \to C$ is injective.

It is easily seen that if $\phi : K \times K \to C$ is a symmetric bilinear mapping, then its adjoint $K \to \operatorname{Hom}(K, C)$ fails to be injective if and only if

$$q(\phi) \approx \frac{\mathbb{R}[[x,y]]}{(x^2) + \mathfrak{m}^3}$$

or

$$q(\phi) \approx \frac{\mathbb{R}[[x,y]]}{\mathfrak{m}^3}$$

where $q(\phi)$ is the $\mathbb{R}$ algebra defined in §2 and $\mathfrak{m}$ denotes the unique maximal ideal of the formal power series ring $\mathbb{R}[[x,y]]$.

In view of the isomorphism

$$q(d^2 f(x)) \approx Q(j^2 f(x)) \approx Q_2(f_x)$$

it follows that if f is a map-germ of type Σ_2 then f fails to be of type $\Sigma_{2,0}$ if and only if

$$Q_2(f_x) \approx \frac{\mathbb{R}[[x,y]]}{(x^2) + \mathfrak{m}^3} \tag{1}$$

or

(2) $$Q_2(f_x) \approx \frac{\mathbb{R}[[x,y]]}{\mathfrak{m}^3} .$$

Now we introduce some notation. If $x_1, \ldots, x_r$ are indeterminates, $f_1, \ldots, f_q$ are formal power series in $x_1, \ldots, x_r$, and k is a positive integer or ∞ we let $V_k(x_1, \ldots, x_r; f_1, \ldots, f_q)_{n,p}$ be the set of $z \in J^\infty(n, p)$ such that

$$Q_k(z) \approx \frac{\mathbb{R}[[x_1,\ldots,x_r]]}{(f_1,\ldots,f_q)+\mathfrak{m}^{k+1}}$$

Clearly $V_k(x_1, \ldots, x_r; f_1, \ldots, f_q)$ is the lifting to $J^\infty(n, p)$ of a $\mathcal{K}^k$ orbit in $J^k(n, p)$. In general, the values of n and p will be clear from the context and we will drop them from the notation. In this section, there will always be two variables x, y, and we will drop them from the notation also so that we will write $V_k(f_1, \ldots, f_q)$ for $V_k(x, y; f_1, \ldots, f_q)_{n,p}$. Note that in this notation $V_1(-) = \Sigma_2^\infty$.

Now the arguments of §§2 and 3 give a complete determination of all $\mathcal{K}^2$ orbits in Σ_2^2. For, by proposition 2.4, these are in one-one correspondence with the set of all points of S/G, and (in this case) $S = S(0) \cup S(1) \cup S(2) \cup S(3)$. Clearly $S(0)/G$ and $S(3)/G$ each have only one point, the points of $S(1)/G$ are in one-one correspondence with equivalence classes of (nonzero) quadratic forms in 2 variables, and by the duality principle, so are the points of $S(2)/G$. Hence $\Sigma_2^2/\mathcal{K}^2$ has a total of 6 points.

In terms of the notation we have just introduced, this decomposition of Σ_2^2 into $\mathcal{K}^2$ orbits can be expressed by the following formula

(3) $$\begin{aligned}\Sigma_2^\infty &= V_2(x^2, xy, y^2) \cup V_2(x^2, y^2) \cup V_2(x^2 - y^2, xy) \\ &\cup V_2(x^2, xy) \cup V_2(xy) \cup V_2(x^2 + y^2) \cup V_2(x^2) \cup V_2(-).\end{aligned}$$

In view of what we have just proved

$$\Sigma_2^\infty - \Sigma_{2,0}^\infty = V_2(x^2) \cup V_2(-).$$

(More precisely, in terms of Boardman's notation, we have $\Sigma_{2,1}^\infty = V_2(x^2)$

and $\Sigma^\infty_{2,2} = V_2(-)$

Now we want to analyze how the various V_2's break up into $\mathcal{K}^\infty$ orbits. First of all, $V_2(x^2, xy, y^2)$, $V_2(x^2, y^2)$, and $V_2(x^2 - y^2, xy)$ are each $\mathcal{K}^\infty$ orbits. For, it is easy to see that if z is in any of these sets, then $Q(z) \approx Q_2(z)$, and this clearly implies that each of these sets is a $\mathcal{K}^\infty$ orbit.

We will write $V(f, g, \ldots)$ for $V_\infty(f, g, \ldots)$ in what follows. In terms of this notation, what we have just proved is that $V_2(x^2, xy, y^2) = V(x^2, xy, y^2)$, $V_2(x^2, y^2) = V(x^2, y^2)$, and $V_2(x^2-y^2, xy) = V(x^2-y^2, xy)$. More generally, we will prove that

$$\begin{aligned} & V_2(x^2, xy, y^2) \cup V_2(x^2, y^2) \cup V_2(x^2-y^2, xy) \\ & \cup V_2(x^2, xy) \cup V_2(xy) \\ & = \bigcup_{k \geq \ell \geq 2} V(x^k, xy, y^\ell) \cup \bigcup_{k \geq \ell \geq 2} V(x^k+y^\ell, xy) \\ & \cup \bigcup_{\substack{k \geq \ell \geq 2 \\ k,\ell \text{ even}}} V(x^k-y^\ell, xy) \cup \bigcup_{k \geq 2} V(xy, x^k) \\ & \cup V(xy) \end{aligned} \tag{4}$$

The first remark to be made is that if z is in the left side of the above equation, then

$$Q(z) \approx \mathbb{R}[[x,y]]/I$$

where I is an ideal which contains an element u whose leading term is xy. For by substituting $x_1 = x + y$, $y_1 = x-y$ we see that $V_2(x^2, y^2) = V_2(x^2+y^2, xy)$. It follows from the formal version of the Morse lemma that by change of coordinates we may assume that $u = xy$. Now any element of $\mathbb{R}[[x, y]]$ is congruent (mod xy) to an element of the form $f(x) + g(y)$, where $f(x)$ is a formal power series in x alone, and $g(y)$ is a formal power series in y alone. We may assume that there exists at least one non-zero element of I having the form $f(x) + g(y)$, for, otherwise, z is obviously in $V(xy)$. Take such an element of least order and let k denote its order. Without loss of generality we may assume that k is the order of $f(x)$. Then we have $f(x) = ax'^k$ for some formal power series

$x' = x +$ higher terms, with $a \neq 0$. Either $g(y) = 0$, or we can treat g similarly. Since xy and $x'y'$ generate the same ideal, we can suppose (substituting x', y' for x, y) that

$$f(x) + g(y) = ax^k + by^\ell \qquad \text{for some } b, \ell,$$

where $\ell \geq k$ by the choice of k.

Suppose $b \neq 0$. Then any power series P is congruent mod xy and $ax^k + by^\ell$ to a polynomial $f(x) + g(y)$ with $\deg f < k$ and $\deg g \leq \ell$. If $P \in I$, this must have order $\geq k$, so $f = 0$. If now r is the least integer with $y^r \in I$ (we know $y^{\ell+1} \in I$), it follows that I is generated by xy, $ax^k + by^\ell$ and y^r:
if $r \leq \ell$, we have $V(xy, x^k, y^r)$
if $r > \ell$ (hence $= \ell+1$) we have $V(xy, ax^k + by^\ell)$.

Similarly if $b = 0$ it is easily seen that $z \in V(x^k, xy)$ or $z \in V(x^k, xy, y^\ell)$ for some ℓ.

This proves (4).

Note that all the sets which appear on the right side of (4) are disjoint. The Hilbert-Samuel function and the $\wr$ invariant introduced above distinguish the algebras associated to all of these sets but the pair $V(x^k - y^\ell, xy)$ and $V(x^k + y^\ell, xy)$. Consider the algebras

$$Q_{\pm} = \mathbb{R}[[x, y]]/(x^k \pm y^\ell, xy)$$

which are associated to these sets.

Let $\mathfrak{m}_{\pm}$ denote the maximal ideal associated to $Q_{\pm}$. In the case when $k < \ell$ and k and ℓ are even, there exists an element in $\mathfrak{m}_- - \mathfrak{m}_-^2$ (namely x) whose k^{th} power is an ℓth power, but there exists no such element in $\mathfrak{m}_+ - \mathfrak{m}_+^2$. In the case $k = \ell$ and k and ℓ are even, there exists an element (namely $x+y$) in $\mathfrak{m}_+ - \mathfrak{m}_+^2$ whose k^{th} power vanishes, but no such element in $\mathfrak{m}_- - \mathfrak{m}_-^2$. Hence Q_+ is not isomorphic to Q_- in either of these cases.

Now we finish the description of the decomposition of $\Sigma_{2,0}^\infty$ into $\mathcal{K}^\infty$ orbits by showing

$$V_2(x^2+y^2) = \bigcup_{k\geq 3} V(x^2+y^2, x^k, x^{k-1}y) \cup \bigcup_{k\geq 3} V(x^2+y^2, x^k) \cup V(x^2+y^2). \tag{5}$$

Consider $z \in V_2(x^2+y^2)$. We have

$$Q(z) \sim \mathbb{R}[[x,y]]/I$$

where I is an ideal which contains an element whose leading term in $x^2 + y^2$. By the formal Morse lemma, we may assume that this element is $x^2 + y^2$. Now any element of $\mathbb{R}[[x, y]]$ is congruent $(\mathrm{mod}\ x^2 + y^2)$ to an element of the form $f(x) + yg(x)$. Moreover by a coordinate change we may arrange that the leading term of $f(x) + yg(x)$ is of the form αx^k. For, setting

$$x = (\cos\theta)x' + (\sin\theta)y'$$

$$y = -(\sin\theta)x' + (\cos\theta)y'$$

we find that

$$\alpha x^k + \beta y x^{k-1} \equiv (\alpha\cos k\theta - \beta\sin k\theta)x'^k + (\alpha\sin k\theta + \beta\cos k\theta)y'x'^{k-1} \quad (\mathrm{mod}(x^2+y^2) = (x'^2 + y'^2)),$$

and we may obviously choose θ so that the coefficient of $y'x'^{k-1}$ vanishes.

Now if the leading term of $f(x) + yg(x)$ is x^k, the ideal $(x^2 + y^2, f(x) + yg(x))$ contains the $(k+1)^{st}$ power of the maximal ideal. Hence if $I \neq (x^2 + y^2)$ we have $I = (x^2 + y^2, x^k)$ or $I = (x^2 + y^2, x^k, yx^{k-1})$, which proves (5).

Taking into account the fact that if f is a stable germ and $n \leq p$ then $\delta(\hat{Q}(f)) < \infty$, we get the following result.

<u>Proposition.</u> <u>If</u> $f : (N^n, x) \dashrightarrow (P^p, y)$ <u>is a stable germ of type</u> $\Sigma_{2,0}$ <u>then</u> $\hat{Q}(f)$ <u>is isomorphic to one of the following.</u>

$I_{a,b}$: $\mathbb{R}[[x, y]]/(xy, x^a + y^b)$, $\quad b \geq a \geq 2$

$II_{a,b}$: $\mathbb{R}[[x, y]]/(xy, x^a - y^b)$, $\quad b \geq a \geq 2$ even

$III_{a,b}$: $\mathbb{R}[[x, y]]/(x^a, y^b, xy)$, $b \geq a \geq 2$

IV_a : $\mathbb{R}[[x, y]]/(x^2 + y^2, x^a)$, $a \geq 3$

V_a : $\mathbb{R}[[x, y]]/(x^2 + y^2, x^a, yx^{a-1})$, $a \geq 3$

Furthermore these algebras are all distinct.

The fact that one could classify stable germs of type $\Sigma_{2,0}$ in the case $n \leq p$ was announced by the author in the fall of 1965 in a letter to B. Morin. However, the classification given was incorrect; the error was found and corrected by B. Morin. The result given above was described by Arnold in [1].

The numbers $\delta(Q)$ and $\gamma(Q)$ associated to these algebras are easily computed:

$$\delta(I_{a,b}) = \delta(II_{a,b}) = a + b$$

$$\delta(III_{a,b}) = a + b - 1$$

$$\delta(IV_a) = 2a$$

$$\delta(V_a) = 2a - 1$$

and

$$\gamma(I_{a,b}) = \gamma(II_{a,b}) = a + b$$

$$\gamma(III_{a,b}) = a + b$$

$$\gamma(IV_a) = 2a$$

$$\gamma(V_a) = 2a$$

Also $\iota(I_{a,b}) = \iota(II_{a,b}) = \iota(IV_a) = 0$ and $\iota(III_{a,b}) = \iota(V_a) = 1.$

From these facts one can determine, by using the criterion stated in the previous section, the dimensions for which these algebras occur as algebras associated to stable germs.

One can also determine the codimension of the singularities associated to these algebras, according to the formula

$$\operatorname{cod}(\Sigma_Q) = (p-n)(\delta(Q) - 1) + \gamma(Q) , \ n \leq p,$$

where Σ_Q denotes the contact class corresponding to Q.

§8 Partial Classification of Stable Germs of Type $\Sigma_{2,1}$ for $n \leq p$.

In the previous section, we have seen that $\Sigma^{\infty}_{2,1} = V_2(x^2)$; i.e., an infinity jet z is of type $\Sigma_{2,1}$ if and only if $Q_2(z) \approx \mathbb{R}[[x,y]]/(x^2)$. In this section we will partially analyze how $\Sigma_{2,1}$ splits up into $\mathcal{K}^{\infty}$ orbits. We will not give the proofs in this section; only the results. The proofs are similar to those of the previous section. The results are contained in the following formulas and propositions:

1) (Decomposition of $V_2(x^2)$ into $\mathcal{K}^3$ orbits)

$$V_2(x^2) = V_3(x^2, xy^2, y^3) \cup V_3(x^2, y^3) \cup V_3(x^2+y^3, xy^2)$$
$$\cup V_3(x^2, xy^2) \cup V_3(x^2+y^3) \cup V_3(x^2).$$

2) $V_3(x^2, xy^2, y^3)$ and $V_3(x^2, y^3)$ are $\mathcal{K}^{\infty}$ orbits.

3) $V_3(x^2+y^3, xy^2) = V(x^2+y^3, xy^2, y^4) \cup V(x^2+y^3, xy^2)$

4) $V_3(x^2, xy^2) = \bigcup_{k\geq 4} V(x^2, xy^2, y^k)$

$$\cup \bigcup_{k\geq 4} V(x^2+y^k, xy^2, y^{k+1}) \cup \bigcup_{k\geq 4} V(x^2+y^k, xy^2)$$
$$\cup \bigcup_{\substack{k\geq 4 \\ k \text{ even}}} V(x^2-y^k, xy^2, y^{k+1}) \cup \bigcup_{\substack{k\geq 4 \\ k \text{ even}}} V(x^2-y^k, xy^2)$$
$$\cup V(x^2, xy^2).$$

5) $V_3(x^2+y^3) = \bigcup_{k\geq 4} V(x^2+y^3, y^k, xy^{k-1})$

$$\cup \bigcup_{k\geq 4} V(x^2+y^3, y^k) \cup \bigcup_{k\geq 4} V(x^2+y^3, xy^{k-1}, y^{k+1}) \cup$$
$$\bigcup_{k\geq 4} V(x^2+y^3, xy^{k-1} + y^{k+1}) \cup \bigcup_{k\geq 4} V(x^2+y^3, xy^{k-1}) \cup V(x^2+y^3).$$

6) $V_3(x^2) = V_4(x^2, xy^3, y^4) \cup V_4(x^2, y^4) \cup$

$$V_4(x^2+y^4, xy^3) \cup V_4(x^2-y^4, xy^3) \cup V_4(x^2, xy^3)$$
$$\cup V_4(x^2+y^4) \cup V_4(x^2-y^4) \cup V_4(x^2).$$

7) $V_4(x^2, xy^3, y^4)$ and $V_4(x^2, y^4)$ are $\mathcal{K}^\infty$ orbits.

8) $\operatorname{cod} V_4(x^2+y^4, xy^3) = \operatorname{cod} V_4(x^2-y^4, xy^3) = 7(p-n) + 11$ and $\operatorname{cod} V_4(x^2, xy^3)$, $\operatorname{cod} V_4(x^2+y^4)$, $\operatorname{cod} V_4(x^2-y^4)$, and $\operatorname{cod} V_4(x^2)$ are all greater than this.

9) $\{V_5(x^2+y^4, xy^3 + \alpha y^5) : \alpha \in \mathbb{R}\}$ is uncountable.

§9 Calculation of $\sigma_2(n, p)$ for $n \leq p$.

Now we consider the splitting of $\Sigma^\infty_{2,2} = V_2(-)$ into $\mathcal{K}^\infty$ orbits. Let $W_i = \{z \in V_2(-) : \dim \mathfrak{m}(z)^3/\mathfrak{m}(z)^4 = i\}$. Clearly $V_2(-) = W_0 \cup W_1 \cup W_2 \cup W_3 \cup W_4$ and $W_j \subseteq \overline{W_i}$ for $j \geq i$. Clearly W_0 is a $\mathcal{K}^\infty$ orbit. Clearly the $\mathcal{K}^3$ orbits of W_1 are in one-one correspondence with equivalence classes of cubic forms in two variables; there are only a finite number of these. It follows from proposition 4.1, that all but one of these $\mathcal{K}^3$ orbits is a $\mathcal{K}^\infty$ orbit. The exceptional $\mathcal{K}^3$ orbit is the one corresponding to the quadratic form y^3. It is $V_3(x^3, x^2y, xy^2)$. It is easily seen that

$$V_3(x^3, x^2y, xy^2) = \bigcup_{k\geq 4} V(x^3, x^2y, xy^2, y^k)$$

$$\cup \bigcup_{k\geq 4} V(x^3+y^k, x^2y, xy^2) \cup \bigcup_{k\geq 4} V(x^3, x^2y+y^k, xy^2)$$

$$\cup\, V(x^3, x^2y, xy^2).$$

Next, it is easily seen that the $\mathcal{K}^3$ orbits in W_2 are in one-one correspondence to the set of orbits in the Grassmannian of two planes in $[x^3, x^2y, xy^2, y^3]$ under the natural action of the projective linear group $PGL(2)$ of substitutions in the variables x and y. Since this Grassmannian has dimension 4 and the group has dim 3, it follows that any non-empty Zariski open set of W_2 contains uncountably many $\mathcal{K}^3$ orbits.

Combining these remarks with the results of the previous two sections we see that

$$\Pi_2(n, p) = \overline{W_2} \cup \overline{V_4}\,(x^2+y^4,\ xy^3),$$

where closure is taken with respect to the Zariski topology on $\Sigma_2(n, p)$. Then

$$\begin{aligned}\sigma_2(n, p) &= \min(\mathrm{cod}\ \overline{W_2},\ \mathrm{cod}\ \ \overline{V_4}(x^2+y^4,\ xy^3)) \\ &= \min\ (7(p-n) + 10,\ 7(p-n) + 11) \\ &= 7(p-n) + 10.\end{aligned}$$

§10 Calculation of $\sigma_3(n, p)$ for $n \leqslant p$.

In this section, we will finish the calculation of $\sigma_r(n, p)$ for $n \leqslant p$ by showing that $\sigma_3 = \sigma_3^2 = 6(p-n) + 9$. The proof is similar to the calculation of σ_2 of the previous section and the calculation of $\sigma_r(n, p)$ for $r \geqslant 4$ of §6. We will show that there exists a subset W of $\Sigma_3^3(n, p)$ of codimension $6(p-n) + 10$, relative to $J^3(n, p)$, such that for all $k \geqslant 3$,

$$\Pi_{k2}^{-1}\ (\Pi_3^2(n, p)) \subseteq \Pi_3^k(n, p) \subseteq \Pi_{k2}^{-1}\ (\Pi_3^2(n, p)) \cup \Pi_{k3}^{-1}\ W.$$

The first inclusion here is obvious. To prove the second inclusion, it is enough to prove that $\Pi_{\infty 2}^{-1}\ (\Pi_3^2(n, p)) \cup \Pi_{\infty 3}^{-1}\ W$ contains only countably many $\mathcal{K}^\infty$ orbits.

By our calculations in §2, if $z \in \Sigma_3^\infty - \Pi_{\infty 2}^{-1}\ (\Pi_3^2)$ then $\dim \mathfrak{m}(z)^2/\mathfrak{m}(z)^3 \leqslant 2$. The argument of §5 shows the set of all $z \in \Sigma_3^\infty$ for which $\dim \mathfrak{m}(z)^2/\mathfrak{m}(z)^3 \leqslant 1$ meets only countably many $\mathcal{K}^\infty$ orbits. Hence we have only to consider the case $\dim\ \mathfrak{m}(z)^2/\mathfrak{m}(z)^3 = 2$.

The $\mathcal{K}^2$ orbits in this case are the points of the space $S(2)/G \approx Gr(2, 3)/\ G\ell(3) = X$ which we have studied in §3. What we must do now is see how each of these points splits up into $\mathcal{K}^\infty$ orbits.

The three real forms of the point $(i\alpha)$ of $X_{\mathbb{C}}$ have representatives z_1, z_2, z_3 such that

$$\begin{aligned}\ker d^2z_1 &= [x^{*2}-y^{*2},\ x^{*2}-z^{*2}] \\ \ker d^2z_2 &= [2x^*y^*+z^{*2},\ z^*(x^*-y^*)] \\ \ker d^2z_3 &= [x^{*2}-y^{*2},\ x^{*2}+z^{*2}].\end{aligned}$$

The corresponding V_2's are

$$V_2(x^2+y^2+z^2,\ xy,\ xz,\ yz)$$

$$V_2(x^2,\ y^2,\ z^2-xy,\ xz+yz)$$

$$V_2(x^2+y^2-z^2,\ xy,\ xz,\ yz)$$

Each of these is a $\mathcal{K}^\infty$ orbit.

The two real forms of $(i\beta)$ have representatives z_1 and z_2 such that

$$\ker d^2z_1 = [x^*(y^*-z^*),\ y^*z^*]$$

$$\ker d^2z_2 = [x^*(y^*-z^*),\ y^{*2} + z^{*2}]$$

The corresponding V_2's are

$$V_2(x^2,\ y^2,\ z^2,\ x(y+z))$$

$$V_2(x^2,\ y^2-z^2,\ x(y+z),\ yz)$$

Both of these are $\mathcal{K}^\infty$ orbits.

The two real forms of $(i\gamma)$ have representatives z_1 and z_2 such that

$$\ker d^2z_1 = [xy,\ z^2]$$

$$\ker d^2z_2 = [x^2+y^2,\ z^2]$$

The corresponding V_2's are

$$V_2(x^2,\ y^2,\ xz,\ yz)$$

$$V_2(x^2-y^2,\ xy,\ xz,\ yz)$$

We have

$$V_2(x^2,\ y^2,\ xz,\ yz) = \bigcup_{k\geq 3} V(x^2,\ y^2,\ xz,\ yz,\ z^k)$$

$$\cup \bigcup_{k\geq 3} V(x^2+z^k,\ y^2+z^k,\ xz,\ yz) \cup \bigcup_{k\geq 3} V(x^2+z^k,\ y^2-z^k,\ xz,\ yz)$$

$$\cup \bigcup_{\substack{k\geq 4\\ k \text{ even}}} V(x^2-z^k,\ y^2-z^k,\ xz,\ yz) \cup \bigcup_{k\geq 3} V(x^2+z^k,\ y^2,\ xz,\ yz)$$

$$\cup \bigcup_{\substack{k\geq 4\\ k \text{ even}}} V(x^2-z^k,\ y^2,\ xz,\ yz) \cup V(x^2,\ y^2,\ xz,\ yz)$$

and

$$V_2(x^2-y^2, xy, xz, yz) = \bigcup_{k\geq 3} (x^2-y^2, xy, xz, yz, z^k)$$

$$\cup \bigcup_{k\geq 3} V(x^2-y^2+z^k, xy, xz, yz) \cup V(x^2-y^2, xy, xz, yz)$$

The one real form of $(i\delta)$ has a representative z such that

$$\ker d^2 z = [x^*y^*, 2y^*z^* - x^{*2}]$$

The corresponding V_2 is

$$V_2(x^2+yz, xz, y^2, z^2).$$

It is a $\mathcal{K}^\infty$ orbit.

The one real form of $(i\varepsilon 1)$ has a representative z such that

$$\ker d^2 z = [y^{*2}, 2y^*z^*-x^{*2}].$$

The corresponding V_2 is

$$V_2(x^2+yz, xy, xz, z^2)$$

and we have

$$V_2(x^2+yz, xy, xz, z^2) = \bigcup_{k\geq 3} V(x^2+yz, xy, xz, z^2, y^k)$$

$$\cup \bigcup_{k\geq 3} V(x^2+yz, xy, xz, x^2+y^k) \cup \bigcup_{\substack{k\geq 4 \\ k \text{ even}}} V(x^2+yz, xy, xz, z^2-y^k)$$

$$\cup \bigcup_{k\geq 3} V(x^2+yz, xy, xz+y^k, z^2) \cup V(x^2+yz, xy, xz, z^2)$$

The two real forms of $(i\varepsilon 2)$ have representatives z_1 and z_2 such that

$$\ker d^2 z_1 = [x^{*2}, y^{*2}]$$

$$\ker d^2 z_2 = [x^{*2}-y^{*2}, x^*y^*].$$

The corresponding V_2's are

$$V_2(xy, xz, yz, z^2)$$

$$V_2(x^2+y^2, xz, yz, z^2)$$

We have

$$V_2(xy,\ xz,\ yz,\ z^2) = \bigcup_{\ell\geq m\geq 3} V(xy,\ xz,\ z^2,\ x^m,\ y^\ell)$$

$$\cup\ V(xy,\ xz,\ yz, z^2+y^\ell, x^m+y^\ell) \cup \bigcup_{\ell \text{ or } m \text{ even}} V(xy,\ xz,\ yz,\ z^2-y^\ell,\ x^m+y^\ell)$$

$$\cup \bigcup_{\ell, m \text{ even}} V(xy,\ xz,\ yz,\ z^2+y^\ell,\ x^m-y^\ell) \cup V(xy, xz, yz, z^2-y^\ell,\ x^m-y^\ell))$$

$$\cup \bigcup_{\ell\geq m\geq 3} V(xy,\ xz,\ yz,\ z^2,\ x^m+y^\ell) \cup \bigcup_{\substack{\ell\geq m\geq 4\\ \ell, m \text{ even}}} V(xy,\ xz,\ yz,\ z^2,\ x^m-y^\ell)$$

$$\cup \bigcup_{\ell, m\geq 3} V(xy,\ xz,\ yz,\ z^2+y^\ell,\ x^m) \cup \bigcup_{\substack{\ell, m\geq 3\\ \ell \text{ even}}} V(xy,\ xz,\ yz,\ z^2-y^\ell,\ x^m)$$

$$\cup \bigcup_{\ell, m\geq 3} V(xy,\ xz+y^\ell,\ yz,\ z^2,\ x^m) \cup \bigcup_{\ell\geq m\geq 3} V(xy,\ xz,\ yz,\ z^2+y^\ell+x^m)$$

$$\cup \bigcup_{\substack{\ell, m\geq 3\\ \ell \text{ even}}} V(xy,\ xz,\ yz,\ z^2-y^\ell+x^m) \cup \bigcup_{\substack{\ell\geq m\geq 4\\ \ell, m \text{ even}}} V(xy,\ xz,\ yz,\ z^2-y^\ell-x^m)$$

$$\cup \bigcup_{\ell, m\geq 3} V(xy,\ xz+y^\ell,\ yz,\ z^2+x^m) \cup \bigcup_{\substack{\ell, m\geq 3\\ m \text{ even}}} V(xy,\ xz+y^\ell,\ yz,\ z^2-x^m)$$

$$\cup \bigcup_{\ell, m\geq 3} V(xy,\ xz+y^\ell,\ yz+x^m,\ z^2) \cup W,$$

where W is of infinite codimension. Also

$$V_2(x^2+y^2,\ xz,\ yz,\ z^2) = \bigcup_{\ell\geq 3} V(x^2+y^2,\ xz,\ yz,\ z^2,\ x^\ell)$$

$$\cup \bigcup_{\ell\geq 3} V(x^2+y^2,\ xz,\ yz,\ z^2+x^\ell) \cup \bigcup_{\substack{\ell\geq 4\\ \ell \text{ even}}} V(x^2+y^2,\ xz,\ yz,\ z^2-x^\ell)$$

$$\cup \bigcup_{\ell\geq 3} V(x^2+y^2, xz+x^\ell, yz, z^2) \cup V(x^2+y^2,\ xz,\ yz, z^2).$$

The one real form of $(ii\alpha)$ has a representative z such that

$$\ker d^2 z = (x^* y^*,\ x^* z^*)$$

The corresponding V_2 is

$$V_2(x^2,\ y^2,\ yz,\ z^2).$$

It is a $\mathcal{K}^\infty$ orbit.

The one real form of $(ii\beta)$ has a representative z such that

$$\ker d^2 z = (x^{*2},\ x^* y^*)$$

The corresponding V_2 is

$$V_2(y^2, xz, yz, z^2)$$

We have

$$V_2(y^2, xz, yz, z^2) = V(y^2, xz, yz, z^2, x^3, x^2y) \cup W$$

where cod $W = 6(p-n) + 10$.

This completes the proof that $\sigma_3 = \sigma_3(n, p) = 6(p-n) + 9$.

§11 Singularities of Type Σ_{n-p+1} for $n > p$.

If f is a map-germ of type Σ_{n-p+1} for $n > p$ then $\hat{Q}(f) \approx \mathbb{R}[[x_1, \ldots, x_{n-p+1}]]/I$, where I is a principal ideal. In terms of the V notation introduced in §7, any $\mathcal{K}^k$ orbit in Σ_{n-p+1} has the form $V_k(x_1, \ldots, x_{n-p+1}, f)$. In what follows we will generally drop the mention of the variables $x_1, \ldots, x_{n-p+1}$, and just write $V_k(f)$.

Now let f be a formal power series in variables $x_1, \ldots, x_{n-p+1}$. It is easily seen that after a change of variables we can write f in the form $f(x_1, \ldots, x_{n-p+1}) = q(x_1, \ldots, x_r) + h(x_{r+1}, \ldots, x_{n-p+1})$, where q is a non-degenerate quadratic form and the order of h is ≥ 3. Furthermore we may arrange for q to be of the form $q(x_1, \ldots, x_r) = x_1^2 + \ldots + x_i^2 - x_{i+1}^2 - \ldots - x_r^2$ by a linear change of coordinates.

We call r the rank of f and i the index of f. The rank of f is an invariant of $V_k(f)$ for $k \geq 2$. The index is not an invariant of $V_k(f)$; however $i'=\min(i, r-i)$ is such an invariant, we will call i' the semi-index of f.

Proposition 11.1. *$V_k(x_{r+1}, \ldots, x_{n-p+1}; h)$ is an invariant of $V_k(x_1, \ldots, x_{n-p+1}; f)$. If there is a system of generators $y_{r+1}, \ldots, y_{n-p+1}$ of the maximal ideal of $\mathbb{R}[[x_{r+1}, \ldots, x_{n-p+1}]]$ and a formal power series $u(x_{r+1}, \ldots, x_{n-p+1})$ with positive constant term such that*

$$h(y_{r+1}, \ldots, y_{n-p+1}) = -u(x_{r+1}, \ldots, x_{n-p+1})h(x_{r+1}, \ldots, x_{n-p+1})$$

or $r = 2i'$ then the triple $(r, i', V_k(h))$ is a complete system of invariants *for* $V_k(f)$. *Otherwise the two contact classes* $V_k(q+h)$ *and* $V_k(q-h)$ *are disjoint, and these are the only contact classes whose associated system of invariants is* $(r, i', V_k(h))$.

Proof Embed $J^k(n-p+1-r, 1)$ in $J^k(n-p+1, 1)$ by $u(x_{r+1}, \ldots, x_n) \longmapsto u(0, \ldots, 0, x_{r+1}, \ldots, x_n)$. There is a subvariety $V_k(f)$ of $J^k(n-p+1, 1)$ (our original $V_k(f)$ was in $J^k(n, p)$) and the assertion for this $V_k(f)$ is equivalent to the assertion for our original $V_k(f)$. Then our assertion says $V_k(h) = V_k(f) \cap J^k(n-p+r-1, 1)$ in the case u exists or $i' = 2r$ and that $V_k(h) = (V_k(q+h) \cup V_k(q-h)) \cap J^k(n-p+r-1, 1)$ in the other case. Each side of either of these equations is a submanifold; one sees easily that the two submanifolds have the same tangent space. Since both submanifolds are orbits of group actions it follows that $V_k(h)_0 = (V_k(f) \cap J^k(n-p+1-r, 1))_0$ where the subscript 0, denotes "the connected component containing 0". Now the rest of the proof is a matter of checking components, which we leave to the reader.

The above result may be stated in more elementary form, as follows:

Corollary 11.1. Let $q_i(x_1, \ldots, x_r)$ $(i = 1, 2)$ *be non-degenerate quadratic forms of the same index* i. *Let* $h_i(x_{r+1}, \ldots, x_n)$ $(i = 1, 2)$ *be formal power series of order* ≥ 3. *Then the following are equivalent:*

(a) $\mathbb{R}[[x_{r+1}, \ldots, x_n]]/(h_1) \approx \mathbb{R}[[x_{r+1}, \ldots, x_n]]/(h_2)$.

(b) $\mathbb{R}[[x_1, \ldots, x_n]]/(q_1+h_1)$ *is isomorphic to* $\mathbb{R}[[x_1, \ldots, x_n]]/(q_2+h_2)$ *or to* $\mathbb{R}[[x_1, \ldots, x_n]]/(q_2-h_2)$.

In view of this result, the problem of classifying singularities of type Σ_{n-p+1} reduces to classifying the sets $V_k(f)$ where f is a power series of order ≥ 3. In the case of power series in 1 variable this is trivial. In the case of power series in two variables, we have the following result:

$$V_2(-) = V(x^3-xy^2) \cup V(x^3+xy^2) \cup \bigcup_{k\geq 4} V(x^3-x^2y+y^k)$$
$$\cup \bigcup_{\substack{k\geq 4 \\ k \text{ odd}}} V(x^3-x^2y-y^k) \cup V(x^3-x^2y) \cup \bigcup_{k\geq 4} V(x^3+y^k)$$
$$\cup \bigcup_{\ell>k\geq 4} V(x^3+xy^{k-1}+y^{\ell}) \cup \bigcup_{k\geq 4} V(x^3+xy^{k-1}) \cup V(x^3)$$
$$\cup V_3(-).$$

Furthermore cod $V_3(-) = 9$ and $V_3(-)$ contains no proper closed invariant subset whose complement contains only finitely many $\mathcal{K}^3$ orbits, by the usual dimension argument.

In the case of power series in 3 or more variables, $V_2(-)$ contains no proper closed invariant subset whose complement contains only finitely many $\mathcal{K}^2$ orbits.

In summary, we get

$$\Pi^{\infty}_{n-p+1} = \overline{V_3(x_1^2 + \dots + x_{n-p-1}^2)} \cup \overline{V_2(x_1^2 + \dots + x_{n-p-2}^2)}$$

We have that

$$\text{cod } V_3(x_1^2 + \dots + x_{n-p-1}^2) = n-p + 8$$

$$\text{cod } V_2(x_1^2 + \dots + x_{n-p-2}^2) = n-p + 7$$

Hence

$$\sigma_{n-p+1}(n, p) = n-p + 7 \qquad \text{if } n-p \geq 2.$$

Combining this with the observation that Σ_r has codimension $\geq n-p+7$ if $n-p \geq 3$ we get

$$\sigma(n, p) = n-p + 7 \qquad \text{if } n-p \geq 3,$$

and since $\Sigma_4 \subseteq \Pi^2(n, p)$ if $n-p=2$, we get

$$\sigma(n, p) = \text{cod } \Sigma_4 = 8 \qquad \text{if } n-p=2.$$

In the case $n-p = 1$, the variety $V_2(x_1^2 + \dots + x_{n-p-2}^2)$ is empty, so we get

$$\sigma_{n-p+1}(n, p) = \sigma_2 = 9.$$

It is clear that $\sigma_r(n, p) > 9$ when $n-p = 1$ and $r \geq 4$, so that all that remains to do to complete the computation of $\sigma(n, p)$ is show that $\sigma_3(n, p) \geq 9$ when $n - p = 1$.

§12 End of the calculation of σ_r.

The only $\sigma_r(n, p)$ which we have not yet calculated is $\sigma_3(n, p)$ in the case $n-p = 1$. To study this case we must study the decomposition of $\Sigma_3(n, p)$ into $\mathcal{K}^\infty$ orbits when $n-p = 1$. The $\mathcal{K}^\infty$ orbits, in the V notation, are of the type $V(x, y, z; f, g)$; as usual we will abbreviate and write just $V(f, g)$.

Now consider the 2-jet decomposition of $V_1(-) = \Sigma_3$. For $z \in \Sigma_3$, we have $\dim \mathfrak{m}(z)^2/\mathfrak{m}(z)^3 = 4, 5,$ or 6. In the case $\dim \mathfrak{m}(z)^2/\mathfrak{m}(z)^3 = 4$, the $\mathcal{K}^2$ orbits are in one-one correspondence with the points of the space X introduced in §3. (Compare the calculation of $\sigma_r^2(n, p)$ for $r = 3$ and $n-p = 1$ at the end of §3.) In the case $\dim \mathfrak{m}(z)^2/\mathfrak{m}(z)^3 = 5$ the orbits are in one-one correspondence with equivalence classes of quadratic forms in three variables, and in the case $\dim \mathfrak{m}(z)^2/\mathfrak{m}(z)^3 = 6$, there is only one orbit. In this way, we obtain the following formula for the decomposition of $\Sigma_3 = V_1(-)$ into $\mathcal{K}^2$ orbits:

$$\begin{aligned} V_1(-) = {} & V_2(x^2-y^2, x^2-z^2) \cup V_2(2xy + z^2, z(x-y)) \\ \cup {} & V_2(x^2-y^2, x^2+z^2) \cup V_2(x(y-z), yz) \cup V_2(x(y-z), y^2+z^2) \\ \cup {} & V_2(xy, z^2) \cup V_2(x^2+y^2, z^2) \cup V_2(xy, 2yz-x^2) \\ \cup {} & V_2(y^2, 2yz-x^2) \cup V_2(x^2, y^2) \cup V_2(x^2-y^2, xy) \\ \cup {} & V_2(xy, xz) \cup V_2(x^2, xy) \cup V_2(x^2+y^2+z^2) \\ \cup {} & V_2(x^2+y^2-z^2) \cup V_2(x^2+y^2) \cup V_2(x^2-y^2) \cup V_2(x^2) \\ \cup {} & V_2(-). \end{aligned}$$

Now $V_2(x^2-y^2, x^2-z^2)$, $V_2(2xy+z^2, x(y-z))$ and $V_2(x^2-y^2, x^2+z^2)$ are $\mathcal{K}^\infty$ orbits.

Furthermore, we have

$$V_2(x(y-z), yz) = \bigcup_{k\geq 3} V(x(y-z), yz+x^k) \cup \bigcup_{\substack{k\geq 4 \\ k \text{ even}}} V(x(y-z), yz-x^k) \cup V(x(y-z), yz).$$

$$V_2(x(y-z), y^2+z^2) = \bigcup_{k\geq 3} V(x(y-z), y^2+z^2+x^k) \cup \bigcup_{\substack{k\geq 4 \\ k \text{ even}}} V(x(y-z), y^2+z^2-x^k) \cup V(x(y-z), y^2+z^2).$$

$$V_2(xy,\ z^2) = \bigcup_{\ell \geq k \geq 3} V(xy,\ z^2+x^k+y^\ell) \cup \bigcup_{\substack{\ell, k \geq 3 \\ \ell \text{ even}}} V(xy,\ z^2+x^k-y^\ell)$$

$$\cup \bigcup_{\substack{\ell \geq k \geq 4 \\ \ell, k \text{ even}}} V(xy,\ z^2-x^k-y^\ell) \cup \{\text{infinite codimension}\}.$$

$$V_2(x^2+y^2,\ z^2) = \bigcup_{k \geq 3} V(x^2+y^2,\ z^2+x^k) \cup V(x^2+y^2,\ z^2).$$

$$V_2(xy,\ 2yz-x^2) = \bigcup_{m \geq \ell \geq 3} V(xy+xz^{\ell-1}+z^m,\ 2yz-x^2)$$

$$\cup \bigcup_{\substack{m \geq \ell \geq 4 \\ \ell \text{ even}}} V(xy-xz^{\ell-1}+z^m,\ 2yz-x^2) \cup \bigcup_{\substack{\ell \geq 4 \\ \ell \text{ even}}} V(xy-xy^{\ell-1},\ 2yz-x^2)$$

$$\cup\ V(xy,\ 2yz-x^2)$$

All the other $\mathcal{K}^2$ orbits in Σ_3 are of codimension ≥ 9. Hence $\sigma_3(n, p) \geq 9$. Combining this with the result of the previous section, we get that $\sigma(n, p) = 9$ for $n-p = 1$.

References

[1] Arnol'd, V. I. Singularities of Smooth Mappings. AMS Translations

[2] Boardman, J. M. Singularities of Differentiable Maps. Publ. Math. I.H.E.S., 33, 21-57 (1967)

[3] Mather, J. N. Stability of C^∞ Mappings:
I. The division theorem, Ann. of Math. 87, 89-104 (1968)
II. Infinitesimal stability implies stability, Ann. of Math. 89, 254-291 (1969)
III. Finitely determined map germs, Publ. Math. I.H.E.S. 35, 127-156 (1968)
IV. Classification of stable germs by IR-algebras, Publ. Math. I.H.E.S. 37, 223- (1969)
V. Transversality, Advances in Math. (1970) to appear.

[4] Morin, B. Formes canoniques des singularités d'une application différentiable. Comptes Rendus Acad. Sci. Paris 260, 5662-5665, 6503-6506 (1965)

[5] Thom, R. Les Singularitiés des applications différentiables, Ann. Inst. Fourier 6, 43-87 (1956)

A CONVERSE OF THE KUIPER-KUO THEOREM

J. Bochnak and S. Łojasiewicz

The aim of this note is to prove a converse of the following theorem due to N.H. Kuiper [1] and T.C. Kuo [2] :

An r-jet, represented by a polynomial $w : \mathbb{R}^n \to \mathbb{R}$ (of degree $\leq r$), is C^0-sufficient in C^r if $|\text{grad } w(x)| \geq c|x|^{r-1}$ holds in a neighbourhood of 0 with some $c > 0$.

The r-jet of a C^r function in a neighbourhood of 0 is identified with its r-th Taylor polynomial at 0 if $r < \infty$ or with its Taylor series at 0 if $r = \infty$; then the function is called a realization of the jet. An ω-jet is an ∞-jet which is a convergent power series.

We say that an r-jet is C^p-sufficient (resp. V-sufficient) in C^k, $(r, p, k = 0, \ldots, \infty, \omega;\ r, p \leq k)$ if for any two of its C^k realizations ϕ, ψ there is a C^p-automorphism germ g of $(\mathbb{R}^n, 0)$ such that $\phi \circ g = \psi$ in a neighbourhood of 0 (resp. the germs of $\phi^{-1}(0)$ and $\psi^{-1}(0)$ at 0 are homeomorphic). Let $v = \Sigma a_p x^p$ be an ∞-jet ; put $v_s = \sum_{|p| \leq s} a_p x^p$; we say that v is finitely C^p-(resp.V-) sufficient in C^k, $(k \geq \infty)$, iff v_s is C^p-(resp. V-) sufficient in C^k for some $s < \infty$.

Then the result is the following

Theorem. Let v be an r-jet, w a C^r realization.

A. If $r < \infty$ then the following conditions are equivalent :

(1) v is V-sufficient in C^r,

(2) v is C^0-sufficient in C^r,

(3) $|\text{grad } w(x)| \geq c|x|^{r-1}$ in a neighbourhood of 0, with some $c > 0$.

B. If $r = \infty$ then the following conditions are equivalent :

(1) v is V-sufficient in C^∞,

(2) v is C^0-sufficient in C^∞,

(3) $|\text{grad } w(x)| \geq c|x|^N$ in a neighbourhood of 0 with some $c, N > 0$ *

(4) v is finitely V-sufficient in C^∞,

(5) v is finitely C^0-sufficient in C^∞.

C. If $r = \omega$ then the following conditions are equivalent :

(1) v is finitely V-sufficient in C^ω,

(2) v is finitely C^0-sufficient in C^ω,

(3) 0 is an isolated critical point of w (or it is not a critical one).

For the proof observe first that in cases A, C the implication (2) => (1) is trivial, as well as the implications (5) => (4) => (1), (5) => (2) => (1) in case B. Next, (3) => (2) in cases A, C and (3) => (5) in case B are consequences of the Kuiper-Kuo theorem. Thus it is sufficient to prove (1) => (3) in all three cases.

Observe now that in each case the condition (1) implies that for any C^r realization $\tilde{w}$ of v (resp. of v_s in case C) the set $\tilde{w}^{-1}(0)$ must be a topological manifold of codimension 1 in a neighbourhood of 0 except 0. This follows from the fact that for some c_i the value 0 is a regular value for the function $x \to \tilde{w}(x) + \Sigma c_i \lambda(x_i)$ restricted to a punctured neighbourhood of 0; where we take $\lambda : \mathbb{R} \to \mathbb{R}$ of class C^∞, flat at 0, with no other zeros, for cases A, B; and we take $\lambda(t) = t^N$ with $N > s$ for case C. This is by an argument of Thom ** which can be formulated as follows.

Let $F : G \times H \to \mathbb{R}^m$ (G, H open in $\mathbb{R}^k, \mathbb{R}^\ell$) be a C^∞ map. If c is a regular value of F, then it is also of the map $G \ni x \to F(x, y) \in \mathbb{R}^m$ for almost every

---

* This is equivalent to the ellipticity of the ideal $\left(\frac{\partial w}{\partial x_1}, \ldots, \frac{\partial w}{\partial x_n}\right)$.

** Communicated by J. Mather.

$y \in H$. (This is an easy consequence of Sard's theorem applied to the projection of $F^{-1}(c)$ on H.)

The proof of the theorem will follow by contradiction and so the following lemmas will be used respectively :

Lemma 1. Let $n > 1$. If f is C^2 in a neighbourhood of a and such that $f(a) = 0$, $d_a f = 0$ and $d_a^2 f$ is non-degenerate, then $f^{-1}(0)$ is not a topological manifold of codimension 1 in any neighbourhood of a.

Lemma 1*. Let $n > 2$. If f is C^2 in a neighbourhood of a, $\{(x_2, \ldots, x_n) = \eta(x_1)\}$ is a (continuous) arc through a such that for $g_t : y \to f(t, y)$ we have $g_t(\eta(t)) = 0$, $d_{\eta(t)} g_t = 0$ and $d^2_{\eta(t)} g_t$ non-degenerate for any t, then $f^{-1}(0)$ is not a topological manifold of codimension 1 in any neighbourhood of a.

Both follow essentially by the Morse lemma (in a version with parameter - for lemma 1*) from the fact that the statement is true when $a = 0$ and f is a quadratic form, non-degenerate (resp. of corank 1) as for example, assuming

$$f(x) = \sum_1^{\ell} x_i^2 - \sum_1^{k} x_{\ell+j}^2 \quad \text{with } \ell + k = n \ (\text{or } n-1) \text{ and } \ell \leq k, \text{ and omitting}$$

the trivial cases of $\ell = 0$ and of $\ell = k = 1$, $n = 2$, we have $\ell - 1 < n - 1$ and for any neighbourhood W of 0 in $f^{-1}(0)$ the homotopy group $\pi_{\ell-1}(W \setminus \{0\})$ is non-trivial.

Case A. One can assume $n > 1$, $r \geq 2$, the remaining case being trivial. We need the following.

Lemma 2. If f is analytic in a neighbourhood of 0, $f(0) = 0$ and $0 < \theta < 1$ then $|x| \, |\text{grad } f(x)| \geq \theta \, |f(x)|$ in a neighbourhood of 0.

To prove it suppose, to the contrary, that 0 is in the closure of the set $\{|x| \, |\text{grad } f(x)| < \theta |f(x)|\}$; this set is semi-analytic, hence (by Bruhat-Cartan-Wallace see [3, 19, prop.2]) it contains a C^1-arc : $\{x = \gamma(t) : 0 < t \leq \epsilon\}$ where γ is C^1 in $[0, \epsilon]$, $\gamma(0) = 0$ and $|\gamma'(t)| = 1$. This implies, putting $\phi = f \circ \gamma$, that $|t| \, |\phi'(t)| \leq \rho |\phi(t)|$ in $[0, \delta]$ with some $\theta < \rho < 1$ and $0 < \delta < \epsilon$; this gives easily $|\phi(t)| \geq (t/\delta)^{\rho}|\phi(\delta)|$ which is impossible.

We are now going to prove (1) => (3) by contradiction. Suppose that $|\text{grad } w(a_v)| = o(|a_v|^{r-1})$ for a sequence $a_v \to 0$, $a_v \neq 0$; then, by lemma 2, we get $|w(a_v)| = o(|a_v|^r)$; we can assume that $|a_{v+1}| \leq \frac{1}{2}|a_v|$. Take now a C^∞ function α such that $\alpha = 0$ for $|x| \geq \frac{1}{4}$ and $\alpha = 1$ in a neighbourhood of 0 ; then

$$\phi(x) = \sum_v \alpha\left(\frac{x - a_v}{|a_v|}\right)\left(w(a_v) + (d_{a_v} w)(x - a_v) + \tfrac{1}{2}\epsilon_v|x - a_v|^2\right)$$

is C^r and $o(|x|^r)$ provided that $\epsilon_v = o(|a_v|^{r-2})$. Since $\phi(a_v) = w(a_v)$, $d_{a_v}\phi = d_{a_v} w$ and $(d^2_{a_v}\phi)(u, u) = \epsilon_v|u|^2$ we can choose ϵ_v so that $d^2_{a_v}(w - \phi)$ is non-degenerate. Therefore, by lemma 1, $(w - \phi)^{-1}(0)$ is not a topological manifold of codimension 1 in any punctured neighbourhood of 0 . But $w - \phi$ is a C^r realization of v, and we have a contradiction.

<u>Case B.</u> The only difference in the proof is that having $|\text{grad } w(a_v)| = o(|a_v|^N)$ for every N and making ϵ_v satisfy $\epsilon_v = O(|a_v|^N)$ for each N we get a function ϕ of class C^∞ and ∞-flat at 0 .

<u>Case C.</u> The proof of (1) => (3) follows by contradiction. Suppose (by Bruhat-Cartan-Wallace) that $\{\text{grad } w = 0\}$ contains a semi-analytic arc λ ending at 0; then $w = 0$ on λ .

Let first $n = 2$. Then $w = g^2h$ in a neighbourhood of 0 with g, h analytic and such that g vanishes on λ (in a neighbourhood of 0); this follows from unique factorisation of w because (using the preparation theorem) any irreducible factor has no critical zeros in a neighbourhood of 0 except 0 (argument with discriminant) and two different irreducible factors have no common zero in a neighbourhood of 0 except 0 (argument with resultant). Now (using the preparation theorem) the set $\{g(x) + \epsilon|x|^{2N} = 0\}$ contains a semi-analytic arc at 0 for N large enough and some $\epsilon = \pm 1$, and the germ at 0 of $\{h = 0\}$ does not contain that of $\{g + \epsilon|x|^{2N} = 0\}$ for N sufficiently large (otherwise the germ of $\{h = 0\}$ would contain an infinity of disjoint germs of semi-analytic arcs at 0); hence the germs of $\{w = 0\}$ and $\{g(g + \epsilon|x|^{2N})h = 0\}$ are not homeomorphic, as the number of germs of semi-analytic arcs at 0 contained in the first one is smaller than that of the second one. This contradicts the finite V-sufficiency of v .

Let now $n > 2$. After a linear substitution we can have $\lambda = \{(x_2, \dots, x_n) = \eta(x_1),\ 0 < x_1 < \delta\},\ \delta < 1$, with η analytic in $(0, \delta)$, $\lim_{t\to 0} \eta = 0$, and $H_i(t, \eta_i(t)) = 0$ where $H_i(t, s)$ are distinguished pseudo-polynomials with discriminants $\not\equiv 0$, $i = 2, \dots, n$, (see [3, 11, prop. 2 and 16, th.1]); it follows that $\frac{\partial H_i}{\partial s}(t, \eta_i(t)) \neq 0$. Take now $\phi(x) = \frac{1}{2}x_1^N \sum_2^n H_i(x_1, x_i)^2$.

Put $\psi_t(y) = \phi(t, y)$ (where $y = (x_2, \dots, x_m)$), so $d_{\eta(t)}\psi_t = 0$ and $(d^2_{\eta(t)}\psi_t)(u, u) = t^N \sum_2^n \left(\frac{\partial H_i}{\partial s}(t, \eta_i(t))\right)^2 u_i^2$. Put $g_t(y) = w(t, y)$; then $d_{\eta(t)}g_t = 0$ and the characteristic equation of the quadratic form $d_u^2 g_t$ is $\Delta(t, u, \sigma) = 0$ where Δ is analytic. The set $\{\Delta(t, \eta(t), \sigma) = 0,\ 0 < t < \delta\}$ is semi-analytic of dimension 1 (as a projection of $\{\Delta(t, u, \sigma) = 0,\ u = \eta(t),\ 0 < t < \delta\}$ of dimension 1 (see [3, 23, th.1])) Hence, by regular separation properties (see [3, 18]), $\Delta(t, \eta(t), \sigma) = 0,\ \sigma \neq 0$ implies $t^{N'} < |\sigma|$ provided N' is large enough. This gives $t^N\left(\frac{\partial H_i}{\partial s}(t, \eta(t))\right)^2 < |\sigma|$ in $0 < t < \delta,\ i = 2, \dots, n$, for any characteristic root σ of $d^2_{\eta(t)}g_t$, provided that N is sufficiently large; but then $d^2_{\eta(t)}(g_t + \psi_t)$ is non-degenerate (and $d_{\eta(t)}(g_t + \psi_t) = 0$) so, by lemma 1*, $(w + \phi)^{-1}(0)$ is not a topological manifold of codimension 1 in any punctured neighbourhood of 0 . But $w + \phi$ is a C^ω realization of v_s , provided $N > s$, and we have a contradiction.

Remark By an additional argument which permits us to replace the H_i by polynomials, one can get the statement C for a polynomial w with the class of polynomials instead of C^ω .

An application

Let J^k denote the space of k-jets (of functions $\mathbb{R}^n \to \mathbb{R}$) ; let $\pi : J^{r+1} \to J^r$ be the natural projection and put $N = \dim \ker \pi$. We will give a proof of the following

<u>Theorem</u>. <u>For every</u> $v \in J^r$ <u>the set of</u> $(r+1)$-<u>jets of</u> $\pi^{-1}(v)$ <u>which are not</u> C^0-<u>(or, equivalently, not V-) sufficient in</u> C^{r+1} <u>is semi-algebraic of dimension</u> $< N$ <u>and therefore contained in some proper algebraic subset of</u> $\pi^{-1}(v)$.

In the general case of $\mathbb{R}^n \to \mathbb{R}^p$ (and for V-sufficiency in C^∞) this theorem was announced by R. Thom [4, theorem 3] and recently proved by T.C. Kuo [5]. It is however of some interest to give a proof in the case $p = 1$ as an application of the previous theorem of this note.

<u>Proof</u>. Put $H_c(x) = \sum_{|\alpha|=r+1} c_\alpha x^\alpha$ for $c = \{c_\alpha\} \in \mathbb{R}^N$. Let Ω be the set of all c such that 0 is an isolated critical point of $v + H_c$ (or is not critical). Then (by the above argument of Thom based on Sard's theorem) the set $\mathbb{R}^N \setminus \Omega$ is of measure 0. Let B be any open ball in $\mathbb{R}^N$. Since $\Omega \cap B$ is the union of the sets $\{c \in B : \operatorname{grad}(v + H_c)(x) \neq 0 \text{ for } 0 < |x| < \frac{1}{n}\}$, $n = 1, 2, \ldots,$ which are (by Seidenberg) semi-algebraic, one of them must be of dimension N (otherwise, by a theorem of Lelong (see [3, 18, prop.2]), $\Omega \cap B$ would be of measure 0) and therefore contain an interior point. This means that there is some $\epsilon > 0$ and an open ball $B_0 \subset B$ such that

$$(*) \qquad \operatorname{grad}(v + H_c)(x) \neq 0 \quad \text{for} \quad 0 < |x| < \epsilon \quad \text{and} \quad c \in B_0 .$$

It is sufficient to prove that if $c \in B_0$ then for some $\alpha, \gamma > 0$

$$(**) \qquad |\operatorname{grad}(v + H_c)(x)| \geq \gamma |x|^r \quad \text{for} \quad |x| < \alpha .$$

For this implies that the set of c satisfying $(**)$ is dense in $\mathbb{R}^N$; this set being (by Seidenberg) semi-algebraic, its complement must be of dimension $< N$.

Let $c \in B_0$ and suppose, to the contrary, that $\delta_\nu = \operatorname{grad}(v + H_c)(a_\nu) = o(|a_\nu|^r)$ for a sequence $a_\nu \to 0$, $a_\nu \neq 0$. We get a contradiction with $(*)$ by finding $b_\nu \in \mathbb{R}^N$ such that

$$(***) \qquad \operatorname{grad} H_{b_\nu}(a_\nu) = \delta_\nu \quad \text{and} \quad b_\nu \to 0 ,$$

since then $\operatorname{grad}(v + H_{c-b_\nu})(a_\nu) = 0$ and $c - b_\nu \in B_0$, $|a_\nu| < \epsilon$ for ν large enough.

Let $L_\nu : \mathbb{R}^n \to \mathbb{R}^n$ be orthogonal and such that $L_\nu(a_\nu/|a_\nu|) = (1, 0, \ldots, 0)$; let $L_\nu(\delta_\nu) = (\theta_{\nu 1}, \ldots, \theta_{\nu n})$, so $\theta_{\nu i} = o(|a_\nu|^r)$, $i = 1, \ldots, n$. Take $H_{b_\nu} = |a_\nu|^{-r} G_\nu \circ L_\nu$ where $G_\nu(x) = \frac{\theta_{\nu 1}}{r+1} x_1^{r+1} + \sum_2^n \theta_{\nu i} x_1^r x_i$, so $\operatorname{grad} G_\nu(1, 0, \ldots, 0) = L_\nu(\delta_\nu)$. Then (***) holds.

Addenda

The following theorem characterizes the C^0- and V-sufficiencies of r-jets in C^{r+1} :

<u>Theorem</u>. <u>If</u> w <u>is an r-jet, the following conditions are equivalent</u> :

(1) w <u>is C^0-sufficient in</u> C^{r+1}

(2) w <u>is V-sufficient in</u> C^{r+1}

(3) $|\operatorname{grad} w(x)| \geq c|x|^{r-\delta}$ <u>in a neighbourhood of</u> 0 <u>with some</u> $c > 0$, $\delta > 0$.

<u>Proof</u>. The implication (3) => (1) is a theorem of T.C. Kuo ([6], th. 0). The implication (1) => (2) being trivial, we only need to prove (2) => (3).

Let Q be the set of all homogeneous polynomials of degree $r + 1$. The condition (2) implies obviously that for each $H \in Q$ the $(r + 1)$-jet $w + H$ is V-sufficient in C^{r+1} . Therefore it is sufficient to prove that the condition

(*) $H \in Q \Rightarrow \{|\operatorname{grad}(w + H)(x)| \geq c|x|^r$ in a neighbourhood of 0 with some $c > 0\}$,

implies (3).

For this observe that $\min_{|x|=s} |\operatorname{grad} w(x)| = |(\operatorname{grad} w)(x(s))|$ for some semi-algebraic function $x(s)$ defined say on $[0,\epsilon]$, C^1 and such that $|x(s)| = s$; (one applies Bruhat-Cartan-Wallace to the semi-algebraic set $\{x: |\operatorname{grad} w(x)| = \min_{|u| = |x|} |\operatorname{grad} w(u)|\}$). Supposing, to the contrary, that (3) fails, and using the semi-algebraicity of the functions $\gamma : s \to (\operatorname{grad} w)(x(s))$ and $s \to s^{-r}\gamma(s)$ we deduce that the limit $a = \lim_{s \to 0} s^{-r}\gamma(s)'$ exists. Put $b = x'(0)$ and choose $H_1, \ldots, H_n \in Q$ such that $\operatorname{grad} H_i(b)$, $i = 1, \ldots, n$, form a base of $\mathbb{R}^n$. Then $a + \Sigma\lambda_i \operatorname{grad} H_i(b) = 0$ for some λ_i and we have

$\operatorname{grad}(w + \Sigma\lambda_i H_i)(x(s)) = \gamma(s) - s^r a + s^r \Sigma\lambda_i((\operatorname{grad} H_i)(\frac{x(s)}{s}) - \operatorname{grad} H_i(b)) = o(s^r)$,
which contradicts (*) .

Remark. The C^0-(as well as V-) sufficiencies in C^r and C^{r+1} for an r-jet, are, in general, not equivalent. Namely, it may happen for an r-jet w that
$c \leq s^{-r+\sigma} \min_{|x|=s} |\operatorname{grad} w(x)| \leq C$ in a neighbourhood of 0 with some $c, C > 0$ and $0 < \sigma < 1$, which implies that w is sufficient in C^{r+1} but not in C^r, as e.g. for the jet $w(x, y) = x^{r-1} - y^{r-1}x$, $r > 3$, considered by T.C. Kuo in [6].

(To see the second inequality put $x = \left(\frac{y^{r-1}}{r-1}\right)^{\frac{1}{r-2}}$; for the first one consider it separately in the sets $A = \{|(r-1)x^{r-2} - y^{r-1}| \leq |x|^{r-2}\}$, $(\mathbb{R}^2 \setminus A) \cap B$ and $\mathbb{R}^2 \setminus A \setminus B$ where $B = \{(r-1)|x|^{r-2} \geq \frac{1}{2}|y|^{r-1}\}$.)

REFERENCES

[1] N.H. Kuiper, C^1-equivalence of functions near isolated critical points. Proc. Symp. in Infinite Dimensional Topology. (Baton Rouge, 1967).

[2] T.C. Kuo, On C^0 sufficiency of jets of potential functions, Topology 8 (1969), 167-171.

[3] S. Łojasiewicz, Ensembles semi-analytiques. Lecture notes, Inst. Hautes Etudes Sci., Bures-sur-Yvette, 1965.

[4] R. Thom, Local topological properties of differentiable mappings, Differential Analysis, Bombay Colloq. 1964.

[5] T.C. Kuo, Criteria for v-sufficiency of jets. Preprint, Manchester, 1970.

[6] T.C. Kuo, A Complete Determination of C^0-Sufficiency in $J^r(2, 1)$, Inv. Math. 8 (1969), 226-235.

REMARKS ON FINITELY DETERMINED ANALYTIC GERMS

Jacek Bochnak and Stanislaw Łojasiewicz

§1. The results.

Denote by K a complete field of characteristic zero with non-trivial valuation and let A be the ring of germs at $0 \in K^n$ of analytic functions of n variables (we say that a K-valued function defined in an open subset of K^n is analytic if it is locally developable as a power series with coefficients in K which has a non-zero radius of convergence with respect to the given valuation).

We say that a polynomial of degree $\leq r$ (i.e. r-jet) $w: K^n \to K$ is analytically sufficient if for any analytic function $f = \sum_{|\alpha|=0}^{\infty} a_\alpha x^\alpha$, defined in a neighbourhood of $0 \in K^n (\alpha = (\alpha_1, \ldots, \alpha_n), \quad x = (x_1, \ldots, x_n))$ such that the r-jet $j^{(r)}(f) = \sum_{|\alpha| \leq r} a_\alpha x^\alpha = w$, there exists a local analytic isomorphism h, with $h(0) = 0$, such that $f \circ h = w$ in a neighbourhood of 0.

An analytic function f defined in a neighbourhood of $0 \in K^n$ (or a germ $f \in A$) is called analytically r-determined if its r-jet $j^{(r)}(f)$ is analytically sufficient; f is analytically finitely determined if f is analytically r-determined for some $r \in \mathbb{N}$.

As an immediate consequence of this definition, we get that an analytically finitely determined germ $f \in A$ is analytically equivalent to a germ of a polynomial.

The main theorems of this note are Theorem 2 and 3. The first of these, which has also been proved (in the case $K = \mathbb{R}$ or $\mathbb{C}$) by different methods by J. Mather [3] and J.-C. Tougeron [7], gives a characterization of finitely determined functions. The second is closely related with Thom's Bombay Theorem 4 [6].

For $f \in \mathcal{F}$, denote by $\mathcal{J}(f) = \left(\frac{\partial f}{\partial x_1}, \ldots, \frac{\partial f}{\partial x_n}\right)$ the ideal of the ring $\mathcal{F} = K[[x]]$ of power series in $x = (x_1, \ldots, x_n)$ generated by the formal power series $\frac{\partial f}{\partial x_1}, \ldots, \frac{\partial f}{\partial x_n}$ of partial derivatives of f. Let $\mathfrak{m}$ be the maximal ideal of $\mathcal{J}$.

Theorem 1. Let $f \in \mathcal{A}$; assume $\mathfrak{m}^s \subset \mathcal{P}(f)$ (where $s \in \mathbb{N}$). Then the 2s-jet of f is analytically sufficient.

The proof of Theorem 1 will be given in §2. The method of proof is purely algebraic. First, given $g \in f + \mathfrak{m}^{2s+1}$, we construct by the standard method (see e.g. [5]) a "formal isomorphism" $\tilde{h}$, such that $f \circ \tilde{h} = g$. Next, by a theorem of M. Artin [2], we can replace the formal solution by an analytic one. This argument will give somewhat more than we have stated. In fact we can choose an analytic solution h such that in a neighbourhood of $0 \in K^n$ $h(x) = x + \hat{h}(x)$, where $\hat{h}_i \in \mathfrak{m}^{s+1}$ $(i = 1, \ldots, n)$.

Let $\Omega(s) = \{\alpha = (\alpha_1, \ldots, \alpha_n) : 0 \leq |\alpha| \leq s\}$ and for $f \in \mathcal{J}$ form the matrix

$$A(f,s) = (u^{\beta}_{(i,\alpha)}) \qquad (i,\alpha)\in\{1,\ldots,n\}\times\Omega(s),\ \beta\in\Omega(s),$$

where $u^{\beta}_{(i,\alpha)}$ denotes the coefficient of the monomial x^{α} in the series $x^{\beta} \frac{\partial f}{\partial x_i}$.

The proof of the following proposition will be given in §3; we shall use some arguments which are essentially due to Tougeron [7].

Proposition 1. (α) $\dim \mathcal{J}/\mathcal{P}(f) \leq s \Rightarrow \mathfrak{m}^s \subset \mathcal{P}(f)$

(β) $\mathfrak{m}^{s+1} \subset \mathcal{P}(f) \Rightarrow \dim \mathcal{J}/\mathcal{P}(f) \leq \dim \mathcal{J}/\mathfrak{m}^{s+1} = \binom{n+s}{n}$

(γ) $\operatorname{rank} A(f,s) < \binom{n+s}{n} - s \Leftrightarrow \dim \mathcal{J}/\mathcal{P}(f) > s$.

Corollary 1. If, for $f \in \mathcal{A}$ and $s \in \mathbb{N}$, $\operatorname{rank} A(f,s) \geq \binom{n+s}{n} - s$ then the 2s-jet of f is analytically sufficient.

Example 1. Let $f(x_1,x_2) = \Sigma^{\infty}_{i,j=0}\, a_{ij}\, x_1^i\, x_2^j$ be an analytic function of two variables $(n = 2)$. If a_{01} or a_{10} is nonzero, then $\mathcal{P}(f) = \mathcal{A}$ and f is 1-determined (the results do not apply for $s = 0$).

If $a_{01} = a_{10} = 0$, $A(f,1)$ has rank $2 \geq 3 - 1$ if and only if $a_{11}^2 \neq 4a_{20}a_{02}$: when this holds, f is 2-determined. If it does not, consider the matrix $A(f,2)$:

$$
A(f,2) = \begin{pmatrix}
a_{10} & a_{11} & 2a_{20} & a_{12} & 2a_{21} & 3a_{30} \\
a_{01} & 2a_{02} & a_{11} & 3a_{03} & 2a_{12} & a_{21} \\
0 & a_{10} & 0 & a_{11} & 2a_{20} & 0 \\
0 & a_{01} & 0 & 2a_{02} & a_{11} & 0 \\
0 & 0 & a_{10} & 0 & a_{11} & 2a_{20} \\
0 & 0 & a_{01} & 0 & 2a_{02} & a_{11} \\
 & & & a_{10} & 0 & 0 \\
 & & & a_{01} & 0 & 0 \\
 & 0 & & 0 & a_{10} & 0 \\
 & & & 0 & a_{01} & 0 \\
 & & & 0 & 0 & a_{10} \\
 & & & 0 & 0 & a_{01}
\end{pmatrix}
$$

We can write $a_{20} = \lambda^2$, $a_{11} = 2\lambda\mu$, $a_{02} = \mu^2$. Then rank $A(f,2) = 4$ if and only if

$$\lambda^3 a_{03} - \lambda^2\mu\, a_{12} + \lambda\mu^2\, a_{21} - \mu^3 a_{30} \neq 0.$$

In this case, the theorem shows that f is analytically 4-determined. (In fact, it is 3-determined, and equivalent to $x_1^2 + x_2^3$ if K is algebraically closed.) In the sequel we will say that $0 \in K^n$ is an isolated critical point of $f \in A$ if and only if in some neighbourhood of 0, $\operatorname{grad} f(x) = \left(\frac{\partial f}{\partial x_1}(x), \dots, \frac{\partial f}{\partial x_n}(x)\right)$ vanishes only at $x = 0$.

Proposition 2. Assume that K is algebraically closed. Then for any $f \in A$, 0 is an isolated critical point of f if and only if $\mathfrak{m}^s \subset \mathfrak{I}(f)$, for some positive integer s.

Proof. If $\mathfrak{m}^s \subset \mathfrak{I}(f)$, every monomial x_j^s $(j = 1, \dots, n)$ can be expressed as a linear combination $x_j^s = \sum_{i=1}^{n} a_{ij} \frac{\partial f}{\partial x_j}$, with coefficients $a_{ij} \in A$, which implies that 0 is an isolated critical point.

Conversely, suppose that 0 is an isolated critical point of f. Then by the Hilbert-Rückert Nullstellensatz [1] for each j there exists $q \in \mathbb{N}$ such that $x_j^q \in \mathfrak{I}(f)$, so $\mathfrak{m}^s \subset \mathfrak{I}(f)$ for s large enough.

Remark 1. Proposition 2 is not true for a field which is not algebraically closed. But clearly the condition $\mathfrak{m}^s \subset \mathcal{P}(f)$ (for some $s \in \mathbb{N}$) implies always that 0 is an isolated critical point of f (or is not critical).

We say that K is a π-field if the following condition holds: For every polynomial $f \in K[x]$ with $\frac{\partial f}{\partial x_1}(0) = \dots = \frac{\partial f}{\partial x_n}(0) = 0$, there exists a polynomial $g \in K[x]$, of order greater than the degree of f, such that 0 is an isolated critical point of $f + g$.

Remark 2. The most important examples of π-fields are the fields $\mathbb{R}$ and $\mathbb{C}$. The fact that $\mathbb{R}$ and $\mathbb{C}$ are π-fields can be proved by Sard's theorem, using arguments of Thom [3]. Moreover, if $f \in \mathbb{C}[x]$ has real coefficients, then g can be chosen with real coefficients such that 0 is a complex isolated critical point of $f + g$.

For $f \in A$ consider the following conditions:

(a) f is analytically finitely determined,

(b) $\mathfrak{m}^s \subset \mathcal{P}(f)$ for some $s \in \mathbb{N}$,

(c) $\dim \mathcal{J}/\mathcal{P}(f) < \infty$,

(d) $\operatorname{rank} A(f,s) \geq \binom{n+s}{n} - s$ for some $s \in \mathbb{N}$,

(e) 0 is an isolated critical point of f (or is not critical).

Theorem 2. Assume that K is a π-field and $f \in A$.

(i) If K is algebraically closed then (a)<=>(b)<=>(c)<=>(d)<=>(e).

(ii) If $K = \mathbb{R}$ then (a)<=>(b)<=>(c)<=>(d) => (e).

Proof. We can assume that 0 is a critical point of f.

(i) The equivalence of the conditions (b), (c) and (d) follows from Proposition 1. Proposition 2 implies that (b)<=>(e), and Theorem 1 implies (b) =>(a). It is sufficient to prove that (a) =>(e). Suppose that the r-jet $j^{(r)}(f) = w$ of f is analytically sufficient. By the definition of π-field we can choose a polynomial g such that $j^{(r)}(g) = 0$ and 0 is an isolated critical point of $w + g$. Sufficiency of w implies the existence of a local analytic isomorphism h, $h(0)=0$ such that $f \circ h = w + g$; so 0 must be an isolated critical point of f.

(ii) It is sufficient to prove (a) =>(d). Suppose that r-jet $j^{(r)}(f) = w$ is analytically sufficient and consider the complexification $\tilde{f}$ of the germ f. By

Remark 2 we can choose a polynomial g with real coefficients such that $j^{(r)}(g)=0$ and 0 is a complex isolated critical point of $\tilde{w} + g$. The fact that w is analytically sufficient for real analytic functions implies easily the existence of a complex local analytic isomorphism h, $h(0) = 0$, such that $\tilde{f} \circ h = \tilde{w} + g$, which implies, by (i), that rank $A(\tilde{f},s) \geq \binom{n+s}{n}-s$ for some $s \in \mathbb{N}$. The trivial observation that $A(\tilde{f},s) = A(f,s)$ completes the proof.

Example 2. In the real case condition (e) does not imply in general condition (a). Indeed, any real analytic function f which satisfies (e) and such that $0 \in \mathbb{C}^n$ is not an isolated critical point of $\tilde{f}$ fails to be analytically finitely determined. This follows immediately from Theorem 2. (Compare also Case C in [3]).

From Remark 2 and the proof of Theorem 2 follows:

Lemma 1. Assume that K is an algebraically closed π-field or $K = \mathbb{R}$. Then for any polynomial w over K (in n variables) there exists a polynomial g, of order greater than the degree of w, such that rank $A(w+g, s) \geq \binom{n+s}{n}-s$, for some $s \in \mathbb{N}$.

Denote by J^q the vector space of polynomials (over K) of degree $\leq q$ (space of q-jets), and for fixed $r \in \mathbb{N}$ and $s \geq r$ let $\pi_s : J^s \to J^r$ be the natural projection.

Theorem 3. Suppose that K is an algebraically closed π-field or $K = \mathbb{R}$. Then for any $w \in J^r$ there exists $s \geq r$ and a proper algebraic subset Σ of $\pi_s^{-1}(w)$ such that every s-jet $\phi \in \pi_s^{-1}(w)-\Sigma$ is analytically sufficient.*

Proof. Choose a polynomial g such that $j^{(r)}(g) = 0$ and the rank $A(w+g, p) \geq \binom{n+p}{p}-p$, for some $p \in \mathbb{N}$ (Lemma 1). Put $s = \max\{2p, \deg g\}$ and

$$\Sigma = \pi_s^{-1}(w) \cap \{\phi \in J^s : \text{rank } A(\phi,p) < \binom{n+p}{p}-p\}.$$

It is clear that Σ is an algebraic subset of $\pi_s^{-1}(w)$. It is proper, since $w+g \in \pi^{-1}(w)-\Sigma$, and by Corollary 1 every s-jet $\phi \in \pi_s^{-1}(w)-\Sigma$ is analytically sufficient.

---

* One can prove that s depends only on r and n.

§2. Proof of Theorem 1.

Suppose that for an analytic function g, $g-f \in \mathfrak{m}^{2s+1}$. Assume that we have the infinite sequence of systems of n series

$$a^1 = (a_1^1, \dots, a_n^1),\ a^2 = (a_1^2, \dots, a_n^2), \dots, a^p = (a_1^p, \dots, a_n^p), \dots$$

such that

(1) $a_j^p \in \mathfrak{m}^{s+p}$ $\quad (j=1, \dots, n)$

and

(2) $f(x + a^1(x) + \dots + a^p(x)) - g(x) \in \mathfrak{m}^{2s+p+1}$.

Condition (1) ensures the existence of the limit

$$y_j = \lim_{p\to\infty} \sum_{r=1}^{p} a_j^r, \quad j = 1, \dots, n,$$

where $y_j \in \mathfrak{m}^{s+1}$, and by (2)

(3) $f(x + y) - g(x) = 0$.

In virtue of the theorem of Artin [2], we can replace the series y_j by convergent series $\tilde{h}_j \in \mathfrak{m}^{s+1}$, which also satisfy equation (3). It is known that $h(x) = x + \tilde{h}(x)$ is an analytic isomorphism in a neighbourhood of $0 \in K^n$. Clearly $f \circ h = g$ in a sufficiently small neighbourhood of 0.

Now we shall find by induction the sequence $a^1, a^2, \dots$ satisfying (1) and (2). First observe that if $\mathfrak{m}^s \subset \mathfrak{I}(f)$, then every element $\phi \in \mathfrak{m}^p (p > s)$ can be expressed in the form $\phi = \sum_{j=1}^{n} \alpha_j \frac{\partial f}{\partial x_j}$, where $\alpha_j \in \mathfrak{m}^{p-s}$.

As $a^1 = (a_1^1, \dots, a_n^1)$ we can get the coefficients in the expression

$$g - f = \sum_{j=1}^{n} a_j^1 \frac{\partial f}{\partial x_j}, \qquad a_j^1 \in \mathfrak{m}^{s+1},$$

because by the Taylor formula

$$f(x + a^1(x)) = f(x) + \sum_{j=1}^{n} a_j^1(x) \frac{\partial f}{\partial x_j} + \tfrac{1}{2} \sum_{i,j=1}^{n} a_i^1(x) a_j^1(x) \frac{\partial^2 f}{\partial x_i \partial x_j} + \dots =$$

$$= g(x) + \tfrac{1}{2} \sum_{i,j=1}^{n} a_i^1(x) a_j^1(x) \frac{\partial^2 f}{\partial x_i \partial x_j} + \dots$$

so $f(x + a^1(x)) - g(x) \in \mathfrak{m}^{2s+2}$, i.e. the condition (2) is satisfied.

Suppose now that we have already chosen $a^1, \dots, a^p$, satisfying (1) and (2).

We shall construct a^{p+1}. Let $\tilde{f}(x) = f(x + a^1(x) + \dots + a^p(x))$ and observe that $\mathfrak{m}^s \subset \mathfrak{J}(f)$ implies $\mathfrak{m}^s \subset \mathfrak{J}(\tilde{f})$. So the inductive hypothesis and the remark above imply that

$$g(x) - \tilde{f}(x) = \sum_{j=1}^{n} b_j(x) \frac{\partial \tilde{f}}{\partial x_j}, \quad b_j \in \mathfrak{m}^{s+p+1}.$$

Put $a^{p+1}(x) = b(x) + \sum_{\nu=1}^{n} (a^\nu(x + b(x)) - a^\nu(x))$. It is easy to verify that $a^{p+1} \in \mathfrak{m}^{s+p+1}$ and

$$f(x + a^1(x) + \dots + a^{p+1}(x)) = f(x + b(x) + a^1(x + b(x)) + \dots + a^p(x + b(x)) =$$

$$\tilde{f}(x + b(x)) = \tilde{f}(x) + \sum_{j=1}^{n} b_j(x)\frac{\partial \tilde{f}}{\partial x_j} + \frac{1}{2}\sum_{i,j=1}^{n} b_i(x) b_j(x) \frac{\partial^2 \tilde{f}}{\partial x_i \partial x_j} + \dots =$$

$$g(x) + \frac{1}{2}\sum_{i,j=1}^{n} b_i(x) b_j(x) \frac{\partial^2 \tilde{f}}{\partial x_i \partial x_j} + \dots$$

so $f(x + a^1(x) + \dots + a^{p+1}(x)) - g(x) \in \mathfrak{m}^{2s+2p+2} \subset \mathfrak{m}^{2s+p+2}$, which ends the proof of Theorem 1.

§3. Proof of Proposition 1.

(α) Assume that $\dim \mathcal{F}/\mathcal{J}(f) \leq s$ and suppose that $\mathfrak{m}^s$ is not contained in $\mathcal{J}(f)$. Then the sequence of vector spaces

$$\{0\} \subset \mathfrak{m}^s + \mathcal{J}(f)/\mathcal{J}(f) \subset \mathfrak{m}^{s-1} + \mathcal{J}(f)/\mathcal{J}(f) \subset \ldots \subset \mathfrak{m} + \mathcal{J}(f)/\mathcal{J}(f) \subset \mathcal{F}/\mathcal{J}(f)$$

is, by Nakayama's lemma, strictly increasing, which implies $\dim \mathcal{F}/\mathcal{J}(f) > s$.

Implication (β) is trivial.

(γ) Observe first that

$$(*)\ \dim \mathcal{F}/\mathcal{J}(f)+\mathfrak{m}^{s+1} + \dim \mathcal{J}(f)+\mathfrak{m}^{s+1}/\mathfrak{m}^{s+1} = \dim \mathcal{F}/\mathfrak{m}^{s+1} = \binom{n+s}{n}$$

and $(**)$ $\operatorname{rank} A(f,s) = \dim \mathcal{J}(f)+\mathfrak{m}^{s+1}/\mathfrak{m}^{s+1}$.

($\Rightarrow$) The condition : $\operatorname{rank} A(f,s) < \binom{n+s}{n}-s$: is by $(**)$ and $(*)$ equivalent to the condition : $\dim \mathcal{F}/\mathcal{J}(f) + \mathfrak{m}^{s+1} > s$; which, by surjectivity of the homomorphism $\mathcal{F}/\mathcal{J}(f) \to \mathcal{F}/\mathcal{J}(f)+\mathfrak{m}^{s+1}$, implies $\dim \mathcal{F}/\mathcal{J}(f) > s$.

($\Leftarrow$) Suppose now that $\dim \mathcal{F}/\mathcal{J}(f) > s$; then $\dim \mathcal{F}/\mathcal{J}(f)+\mathfrak{m}^{s+1} > s$. Indeed, if this is not true, then by (α) we have $\mathfrak{m}^s \subset \mathcal{J}(f) + \mathfrak{m}^{s+1}$, which implies (Nakayama) that $\mathfrak{m}^s \subset \mathcal{J}(f)$. One gets a contradiction, because then $\mathcal{J}(f) + \mathfrak{m}^{s+1} = \mathcal{J}(f)$ so $\dim \mathcal{F}/\mathcal{J}(f) \leq s$. Condition $\dim \mathcal{F}/\mathcal{J}(f)+\mathfrak{m}^{s+1} > s$ implies by $(**)$ and $(*)$ that $\operatorname{rank} A(f,s) < \binom{n+s}{n}-s$.

References.

[1] S. S. Abyankar, Local analytic geometry, Acad. Press, New York and London, (1964)

[2] M. Artin, On the solution of analytic equations, Invent. Math. 5 277-291 (1968)

[3] J. Bochnak and S. Łojasiewicz, A converse of the Kuiper-Kuo theorem, this volume.

[4] J. Mather, Stability of C^∞ mappings III, Publ. Math. I.H.E.S. 35, 127-156 (1969)

[5] J. Milnor, Singular points of complex hypersurfaces, Princeton Univ. Press, (1968)

[6] R. Thom, Local topological properties of differentiable mappings, Bombay Colloquium on differential analysis, Oxford Univ. Press (1964)

[7] J-C. Tougeron, Idéaux de fonctions differentiables I, Ann. Inst. Fourier 8, 177-240 (1968)

NONDEGENERATE CRITICAL POINTS OF DEFICIENCY ONE

Samir Khabbaz and Everett Pitcher

0. Introduction

We shall consider differentiable functions $f : M^m \to R^n$, $m \geqq n$, from an m-dimensional manifold to euclidean n-space. A component S of the set K of critical points of f will be subject to two conditions. First, there is an integer $r>0$ such that at each point of S the rank of the functional matrix of f is r. The points of S will be called points of <u>deficiency</u> $d=n-r$,* while points of $M^n \setminus K$ are called <u>ordinary</u>. Second, each critical point P is nondegenerate in the sense that if $f = (f^1, \ldots, f^n)$ then there are multipliers $(\lambda_1, \ldots, \lambda_n)$, with $\lambda_i\lambda_i = 1$, defining a projection $\Lambda : R^n \to R$ such that $F = \Lambda\circ f = \lambda_i f^i$ has a nondegenerate critical point in the classical sense.

We shall show that S is an r-dimensional manifold. We shall determine the topological structure of S in the case $d = 1$ and shall relate the topological structure of K to that of M in case M is compact and every component of K is of deficiency 1. The theorems generalize classical theorems on critical point theory to higher dimensions.

---

* In the Thom-Boardman notation, these present singularities of type Σ^{m-r}.

1. The Principal Results

Let

$$f : M^m \to R^n$$

be a smooth function from an m-dimensional manifold, not necessarily compact and possibly with boundary, to Euclidean n-space R^n, with $2 \leqq n \leqq m$. Let K denote the set of critical points of f and let S denote a component of K. We shall say that f is nondegenerate with a single deficiency d over S if it satisfies the following conditions

(1) Each point of S is a critical point of deficiency d.

(2) Corresponding to each point P_0 of S is a projection $\Lambda : R^n \to R$ such that $\Lambda\circ f : M^m \to R$ has a nondegenerate critical point at P_0.

We emphasize that if n were 1 this concept would reduce directly to that of nondegenerate critical point in classical critical point theory.

The definitions agree with those of [K-Pl], though for expository purposes we do not here require M^m to be compact. In the context of this paper the definitions will be restated in local coordinates with the proofs. A principal theorem is the following.

Theorem 1. *Suppose that* f *is nondegenerate with a single deficiency* d *over* S. *Then* S *is a submanifold of* M^m *of dimension* $r=n-d$. *Further, near each point of* S, *the multipliers* $(\lambda_1, \ldots, \lambda_n)$ *at points of* S *may be so chosen that an appropriate set of* r *of them are admissible local coordinates on* S.

We shall continue with the case that f is nondegenerate over S with the single deficiency 1.

Associated with a critical point P_0 there are two sets of multipliers, (λ^0) and $(-\lambda^0)$. Suppose the notation so chosen that $F = \lambda_i^0 f^i$ has a critical point of index $k \leqq m/2$. This index is constant over a component S of K and is by definition the index of S.

Theorem 2. *Suppose* S *is a compact component of the critical set of* f, *that* f *is nondegenerate with the single deficiency* 1 *over* S, *and that the index of* S *is* k. *Then* S *is the standard (n-1)-sphere with one exception: If* m *is even and* $k=m/2$, *then* S *is either the standard (n-1)-sphere or the standard projective* (n-1)-space. If $m=n\geq 9$, S is also a sphere.

We shall show later by example that the exceptional case of the projective space can occur.

In the course of the proof, we shall observe with $n\geqq 3$ that the case of the sphere corresponds to the possibility of choosing one sense of the multipliers (λ) over all of S. The corresponding composite function has two critical points with indices k and $n-k$ on S. The case of the projective space corresponds to the identification of multipliers (λ) and $(-\lambda)$ over S and to the fact that the composite function has just one critical point, with index $k=m/2$.

With this in mind, consider the case that M^m is compact, so that K is compact, and that f is nondegenerate with a single deficiency 1 over each component of K. Assume $n\geqq 3$, since the case $n=2$ differs slightly and will be discussed separately. For a fixed line R through the origin of R^n, the projection $\Lambda : R^n \to R$ defines a composite function $F = \Lambda \circ f$ whose critical points can be counted through the Morse inequalities and their extensions admitting torsion (see [P]). The result is a count of the components of K, as follows.

For $0\leqq k<m/2$, let $N(k)$ denote the number of components (necessarily spheres) of K of index k. If $m=2h$, let N^S and N^P denote the number of components of K of index h that are spheres and projective spaces respectively and let $N(h) = 2N^S+N^P$. Let

$$\mathfrak{N}(k) = N(k) - N(k-1) + \dots + (-1)^k N(0).$$

Let R_k denote the rank of the homology group $H_k(M)$ with a field or integers as coefficients and let η_k denote the number of torsion coefficients in $H_k(M)$ in the case of integer coefficients. Let

$$\mathfrak{R}(k) = R_k - R_{k-1} + \dots + (-1)^k R_0+\eta_k.$$

The inequalities take the following form

$$N(k) \geqq R_k+\eta_k+\eta_{k-1} \quad k\leqq m/2$$

$$N(k) \geqq R_{m-k}+\eta_{m-k}+\eta_{m-k-1} \quad k>m/2$$

and

$$\mathfrak{N}(k) \geqq \mathfrak{R}(k) \qquad k \leqq m/2$$

$$\mathfrak{N}(k) \geqq \mathfrak{R}(m-k-1)+(-1)^{m-k}\chi(M) \quad k>m/2,$$

where χ is the Euler-Poincaré characteristic. If Poincaré duality holds for M, the second inequality of each pair is the same as the first.

The equality takes the form

$$0 = \chi(M)$$

if m is odd, a relation that is transparent if Poincaré duality holds but not in general. If $m=2h$ is even, then

$$2\mathfrak{N}(h-1)+(-1)^h N(h) = \chi(M).$$

Since $N(h)= 2N^S+N^P$, $\chi(M)$ and N^P have the same parity and, if Poincaré duality holds for M, then N^P and R_h have the same parity.

2. Examples

Suppose $2 \leqq n \leqq m$. Let S^m denote the unit sphere in R^{m+1}, with $0 \in R^n \subset R^{m+1}$. The function f, defined as the projection of R^{m+1} onto R^n cut down to S^m has $S^{n-1} = S^m \cap R^n$ as its critical set. The function is nondegenerate with deficiency 1 over S^{n-1} and the index of f over S^{n-1} is 0.

The map $z \mapsto z^2$ of the complex plane, regarded as a function $f : (x, y) \mapsto (x^2-y^2, 2xy)$ of R^2 into R^2 has the origin as its critical set, with deficiency 2. Every projection Λ from R^2 to a line R through the origin induces a function $\Lambda \circ f$ with a nondegenerate critical point at the origin. That is, the maximum deficiency can occur, and with arbitrary multipliers. Compare $(x, y) \mapsto (x^4-y^4, 2xy)$, where the origin is the critical set and the composite function is nondegenerate for some multipliers but not for others.

The map $f : (x, y) \mapsto (x^2+y^2, 2xy)$ has the lines $x=\pm y$ as its critical set. The origin is of deficiency 2. Corresponding exactly to the multipliers (λ, μ) with $\lambda \neq \pm\mu$ the composite function has a nondegenerate critical point at the origin.

Suppose $n \geq 3$ and suppose M^{2n-2} is the tangent bundle to P^{n-1} conveniently constructed as follows. Regard P^{n-1} as the space of $(n-1)$-planes through the origin of R^n and let M^{2n-2} consist of the points (x, π) of $R^n \times P^{n-1}$ such that x lies on π. Identify P^{n-1} with $\{0\} \times P^{n-1}$. The function $f: M^{2n-2} \to R^n$ is defined by $f(x, \pi) = x$. The critical set of f consists of P^{n-1}, over which f is nondegenerate with deficiency one. The calculations are readily performed in terms of obvious local coordinates. The example illustrates the exceptional case of Theorem 2.

3. Definitions and proofs

The point P_0 on M with local coordinates (x_0) in the system (x) is ordinary if the matrix

$$(f^i_{x^j}\ (x_0)) \qquad i=1, \ldots, n; j=1, \ldots, m$$

has rank n and is critical with deficiency d if it has rank $n-d=r<n$. Thus it is critical if and only if there is a projection Λ^0, defined by multipliers $(\lambda_1^0, \ldots, \lambda_n^0)$ with $\lambda_i^0 \lambda_i^0 = 1$, such that $F=\lambda_i^0 f^i = \Lambda^0 f$ has a critical point at (x_0) in the more elementary sense. The nondegeneracy of f refers to the existence of a set of multipliers Λ^0 for which the critical point of F is nondegenerate in that

$$(F_{x^j x^k}(x_0))$$

is nonsingular.

Proof of Theorem 1. The theorem is local. It will be proved first in case* $d=1$. One wishes to solve $m+1$ equations

$$\lambda_i f^i_{x^j}\ (x) = 0 \qquad\qquad j=1, \ldots m$$

$$\lambda_i \lambda_i = 1$$

* This part was stated in [K-P 1]. In the summary of proof the indices were scrambled. Instead of trying to state the correction we have chosen to present the proof in slightly different form.

in $m+n$ variables $(x; \lambda)$ for $m+1$ of them in terms of the rest near an initial solution $(x_0; \lambda^0)$. Suppose the notation such that $\lambda_n^0 \neq 0$ and seek $(x^1, \ldots, x^m; \lambda_n)$ in terms of $(\lambda_1, \ldots, \lambda_{n-1})$. The implicit function theorem applies provided

$$\begin{vmatrix} \lambda_i^0 f^i_{x^j x^k}(x_0) & f^n_{x^j}(x_0) \\ 0 & \lambda_n^0 \end{vmatrix} \neq 0.$$

This is the case because

$$\left| \lambda_i^0 f^i_{x^j x^k}(x_0) \right| = \left| F_{x^j x^k}(x_0) \right| \neq 0.$$

The solution is of the form

$$x^i = \phi^i(\hat{\lambda})$$

$$\lambda_n = \psi(\hat{\lambda})$$

where $(\hat{\lambda}) = (\lambda_1, \ldots, \lambda_{n-1})$.

One must then show that

$$(\phi^i_{\lambda_\alpha}(\hat{\lambda}_0)) \qquad i=1, \ldots, m; \alpha = 1, \ldots, n-1.$$

has rank $n-1$. Suppose that

$$c_\alpha \phi^i_{\lambda_\alpha}(\lambda^0) = 0.$$

We shall show $(c) = (0)$. To that end, observe that

$$\lambda_\alpha f^\alpha_{x^j}(\phi(\hat{\lambda})) + \psi(\hat{\lambda}) f^n_{x^j}(\phi(\hat{\lambda})) = 0$$

Differentiate these equations with respect to λ_β at $(\hat{\lambda}^0)$, multiply by c_β, and sum on $\beta=1, \ldots, n-1$, to find that

$$c_\beta f^\beta_{x^i}(x_0) + c_\beta \psi_{\lambda_\beta}(\hat{\lambda}^0) f^n_{x^j}(x_0) = 0.$$

thus

$$(c_1, \ldots, c_{n-1}, c_\beta \psi_{\lambda_\beta}(\hat{\lambda}^0)) = (t\lambda^0),$$

because the multipliers (λ^0) are unique except for sign. Recall that

$$\lambda_\alpha \lambda_\alpha + \psi^2(\hat{\lambda}) = 1.$$

Differentiate with respect to λ_β at $(\hat{\lambda}^0)$ to learn that

$$\lambda_\beta{}^0 + \lambda_n{}^0 \psi_{\lambda_\beta}(\hat{\lambda}^0) = 0.$$

Multiply by c_β and sum on β, introducing $(t\lambda^0)$ when it appears, to learn that $t=0$. The statement on rank is proved.

The proof that $\dot{S}$ is a manifold of dimension r in the general case is proved by reducing locally to the case $d=1$. To that end, suppose $P_0 \in S$. There are coordinates in R^n such that $f = (f^1, \ldots, f^n)$ has the following two properties. First, f has a nondegenerate critical point at P_0. Second, the function $g = (f^1, f^{n-r+1}, \ldots, f^n)$ has a nondegenerate critical point of deficiency 1 with multipliers $(1, 0, \ldots, 0)$ at P_0. The sets of critical points of f and of g near P_0 are coextensive. On one hand each critical point of g is a critical point of f; on the other hand a critical point of f is of deficiency d and is therefore a critical point of g, which is defined from f by dropping only $d-1$ components. The critical points of g near P_0 form an r-manifold by virtue of the case $d=1$ already proved, so that the critical points of f form an r-manifold also.

Proof of Theorem 2. The set S is clearly S^1 when $n=2$. Hence the proof continues with $n>2$. It is clear from Theorem 1 that a component S of K is a compact $(n-1)$-dimensional maniold on which a set $(\hat{\lambda})$ of $n-1$ of the multipliers (λ) at (x) on S are admissible local coordinates. That is, there is a map

$$\alpha : S \to P^{n-1}$$

associating with each point (x_0) of S the one dimensional vector space of multipliers at (x_0). The map is locally 1-1. Let

$$p : S^{n-1} \to P^{n-1}$$

be the usual universal covering map. If z and $-z$ are diametrically opposite points of S^{n-1}, we have $p(z) = p(-z)$. One can factor p through α by a map β

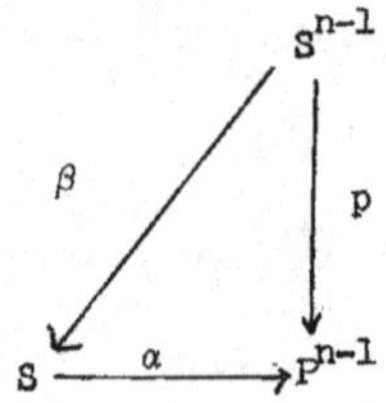

There are two cases

I α is 2-1 and β is 1-1

II α is 1-1 and β is 2-1.

The proof lies in showing when Case II cannot hold. In Case II, corresponding to $(\lambda) \in S^{n-1}$ let $\chi = \phi(\lambda) = \alpha^{-1}p(\lambda)$ and recall that $p(\lambda) = p(-\lambda)$ so that $\phi(\lambda) = \phi(-\lambda)$. Look at $F = \lambda_i{}^0 f^i$ near $x_0 = \phi(\lambda^0)$. The function F has a critical point of some index k that does not change as (λ^0) changes on S^{r-1}. Since $(-\lambda^0)$ lies on S^{r-1} with (λ^0), the critical point of $\lambda_i{}^0 f^i$ at $\phi(\lambda^0)$ and of $(-\lambda_i{}^0)\, f^i$ at $\phi(-\lambda^0)$ have the same index k. Since $\phi(-\lambda^0) = \phi(\lambda^0)$ it also follows that the second critical point has index $m-k$. Accordingly in Case II, m is even and k-$m/2$. The proof of the first part is complete.

The following argument for $m=n$ can obviously be refined. Let $m=n$, then using the differential df, one obtains the equation

$$\tau^{m-1} \oplus L \oplus \nu = \tau(M)|S \oplus L = 0^m \oplus \alpha,$$

where τ^{m-1} is the tangent bundle of S, ν is its normal bundle in M, 0^m is the trivial m-dimensional bundle, L is the line bundle defined as the subset of $S \times \tau(R^m)$ consisting of all pairs (s, ν) where $s \in S$ and ν is a vector in the tangent bundle $0^m = \tau(R^m)$ of R^m based at $f(s)$ and normal to $df\ (\tau(M)|S)$, and α is defined by the equation. An equation such as $\tau^{m-1} \oplus L \oplus \nu = 0^m + \alpha$ does not hold for P^{m-1}, $m \geq 9$ as may be verified directly or with the aid of results from immersion theory.

4. The case n=2.

The facts in the case $n=2$ are readily summarized. In Theorem 2 the exceptional case does not arise. However, a multiplicity Δ may be assigned to the critical set S, namely the degree Δ of the covering $\Lambda : S \to S^1$ that assigns to each critical point x the multipliers $\Lambda(x)$. This map is locally one-to-ne. The function $\Lambda \circ f$ has 2Δ critical points on the component S, nondegenerate and with multiplicities $k \leq [m/2]$ and $m-k$. Then $N(k)$ is defined as before, but with each component counted according to its multiplicity. (Recall that the exceptional case does not arise.) Also $\mathfrak{N}(k)$ is defined from $N(k)$ as before. Then the inequalities hold when $n=2$.

The results for $n=2$ overlap with work of Levine (see [L]) where, for instance, one may find in his Theorem 1 the equality for $m=2h$ and $n=2$.

5. Immersion of the critical set.

An added assumption, namely in the situation of Theorem 2 and its sequellae, that $f|S$ is an immersion, allows both a geometric interpretation and a simplification. First, the multipliers (λ) at $P \in S$ are direction numbers of the normal to fS at P; that is, fS admits the direction numbers of the normal as local coordinates. Second, S is always the standard $(n-1)$-sphere; the exceptional case that S is P^{n-1} does not appear. We prove a somewhat more general statement, as follows.

Lemma Let W^m $(m \geq 2)$ be a connected compact manifold without boundary immersed in $\mathbb{R}^{m+1}$ with the property that the direction of the normal provides a local coordinate system. Then W is diffeomorphic to S^m, and is embedded as the boundary of a convex set.

Proof Write P^m for the projective space whose points are the lines through the origin of $\mathbb{R}^{m+1}$. Then we have a map $\gamma : W \to P^m$ which assigns to each point $x \in W$ the direction of the normal to W at x.

The hypothesis of the lemma tells us that γ is a local diffeomorphism, and hence a covering map. Since $\pi_1(P^m)$ has order 2, γ is a diffeomorphism or is two to one; in the latter case it provides a diffeomorphism from W^m to the double covering space S^m of P^m.

Now if γ is a diffeomorphism, for each direction ν there is only one point P of W at which the normal is parallel to ν. Hence orthogonal projection of W on ν has only one critical point: at P. But this is absurd, since there must be at least two critical points: a maximum and a minimum (W is compact).

The final statement follows from a well-known theorem of differential geometry (see, for example, the paper by Kuiper in Part II of these Proceedings).

The count of spheres in the critical set on a compact manifold without boundary in which f is nondegenerate with a deficiency 1 on the critical set K is then also simplified by the assumption that $f|K$ is an immersion. With $f : M^m \to R^n$ and $m = 2h$, one has $N(h) = 2N^S$ where N^S is now the number of components of K of index h.

The assumption that $f|K$ is an immersion also permits one to observe that the index of any component S of K is at most $m-n+1$, so that one has the following.

<u>Theorem 3.</u> <u>If</u> M^m <u>is compact without boundary,</u> f <u>is of deficiency 1 over the critical set</u> k, $f|K$ <u>is an immersion, and</u> $n > [(m+1)/2] + 1$, <u>then</u>

$$N(k) = 0 \qquad m-n+1<k\leq [m/2]$$

<u>and so</u>

$$R_k = 0 \qquad m-n+1<k<n-1$$

$$\eta_k = 0 \qquad m-n+1\leq k<n-1$$

<u>and</u>

$$\mathfrak{N}(m-n+1) = \mathfrak{R}(m-n+1).$$

This theorem may be read somewhat like a non-immersion theorem; if $R_k \neq 0$ for some k with $m-n+1<k<n-1$ then there is no map f admitting only a simple type of singularity.

Bibliography

[K-P1] Khabbaz, S. and Pitcher, E. Critical submanifolds of differentiable mappings, Bull. Amer. Math. Soc., 73, 164-168 (1967).

[K-P2] Khabbaz, S. and Pitcher, E. Critical submanifolds of differentiable mappings II, Bull. Amer. Math. Soc., 73, 310-314 (1967).

[L] Levine, Harold I. Mappings of manifolds into the plane, Amer. J. Math., 88, 357-365 (1966).

[P] Pitcher, E. Inequalities of critical point theory, Bull. Amer. Math. Soc., 64, 1-30 (1958).

MAPS WITH O-DIMENSIONAL CRITICAL SET

J.G. Timourian

Our object is to describe, topologically, the local structure of a class of differentiable maps $f : M^n \to N^p$. Such a map f is said to be locally topologically equivalent at $x \in M^n$ to a map $\rho : \mathbb{R}^n \to \mathbb{R}^p$ if there exist neighbourhoods U of x, V of $f(x)$, and homeomorphisms α, β such that the following diagram commutes :

$$\begin{array}{ccc} U & \xrightarrow{\ \alpha\ } & R^n \\ \Big\downarrow {\scriptstyle f|U} & & \Big\downarrow {\scriptstyle \rho} \\ V & \xrightarrow{\ \beta\ } & R^p \end{array}$$

The critical set $X \subset M^n$ is the set of points at which the rank of f is less than maximal. According to the Rank Theorem, at each $x \in M^n$ for which the rank of f is p, f is locally topologically (in fact, diffeomorphically) equivalent to

(1) $\rho : \mathbb{R}^n \to \mathbb{R}^p$ defined by $\rho(x_1, x_2, \ldots, x_n) = (x_1, x_2, \ldots, x_p)$.

Some examples in which the critical set X contains the origin $\{0\}$ are :

(2) $\rho : \mathbb{R}^2 \to \mathbb{R}^2$ defined by $\rho(z) = z^m$, $m = 2, 3, \ldots$ (consider $\mathbb{R}^2$ to be the complex numbers). In this case ρ is an m - to - 1 covering map on the complement of $X = \{0\} = \rho^{-1}(0)$.

(3) $\rho : \mathbb{R}^2 \times \mathbb{R}^2 \to \mathbb{R}^3$ defined by $\rho(z, w) = (2z \cdot \bar{w}, |w|^2 - |z|^2)$.
In this case ρ is topologically equivalent to the natural extension of the Hopf fibration $\psi : S^3 \to S^2$ to the cone map $c(\psi)$ mapping the cone $\mathbb{R}^4$ of S^3 onto the cone $\mathbb{R}^3$ of S^2 . The map p is a fibration on the complement of $X = \{0\} = \rho^{-1}(0)$ with fiber S^1 .

(4) $\rho : \mathbb{R}^2 \times \mathbb{R}^2 \to \mathbb{R}^2$ a map which is locally topologically equivalent to ρ in example (1) at each point in $\mathbb{R}^4$ except $\{0\}$, with $\rho^{-1}(\rho(0)) \neq \{0\}$. Some polynomial examples are given by $\rho(z, w) = z^i + w^j$, $i > j \geq 2$; more information on the local structure of these functions can be found in [D].

__Theorem__. __Let__ $f : M^n \to N^p$ __be__ C^n __with__ $n - p = 0$, 1 or 2, $p > 1$, __and__ $\dim X = \dim f(X) \leq 0$. __Then at each__ $x \in M^n$, f __is locally topologically equivalent to a map__ ρ __from the examples__ (1) - (4) .

The proof is the result of joint work with P.T. Church; it is rather complicated, so we shall sketch a coarse outline of part of it in order to exhibit some of the techniques involved. References to previous work cited in the lecture are omitted in this summary, but can be found in the papers [A], [B] and [C]. The details of the part of the proof discussed here are in [C]. The theorem in case $n - p = 1$ is a weaker version of the main theorem of [A] (for example, the assumption that $\dim f(X) \leq 0$ can be omitted). The preliminary edition of [B] contains most of the details sufficient for the theorem when $n - p = 2$, and a slightly stronger result can be proved in this case also.

__Proof__. Let us restrict ourselves to the case when $n \neq p$, f is proper, and $\dim f^{-1}(f(B_f)) \leq 0$, where $B_f \subset X \subset M^n$ is the set of points at which f fails to have the local structure described in example (1).

We split the proof into two parts. First suppose that (*) if $p \geq 3$, then for each $y \in f(B_f)$ there are arbitrarily small euclidean neighbourhoods W of y such that $\pi_1(W - f(B_f)) = 0$. If a suitable such W is chosen there is a component L of $f^{-1}(W)$ so that $f|L$ can be factored in the manner indicated by the diagram

$$\begin{array}{ccc} & & L \\ & {}^{g}\swarrow & \downarrow{\scriptstyle f|L} \\ K & \xrightarrow{\ h\ } & W \end{array}$$

where g is the quotient map which collapses components of $f^{-1}(z) \cap L$ to points (this is known as the monotone-light factorization of $f|L$). In this particular factorization $B_g = L \cap B_f$, g is monotone, K is a manifold, and $h|(K - g(B_g))$ is a finite-to-one covering map (and thus for $p \geq 3$ is a homeomorphism). Using the Leray spectral sequence (with possibly twisted coefficie-nts if $p = 2$) for the restriction of g to $\bar{g} : L - B_g \to K - g(B_g)$ (which turns out to be a bundle map since $B_{\bar{g}} = \emptyset$ and g is proper) we prove that B_g is discrete. Some standard techniques in bundle theory and results on fibrations of spheres enable us to prove that at points in B_g the map g (and hence f) has the topological structure of example (3).

Now suppose that the embedding of $f(B_f)$ in N^p does not have the simple connectivity property (*) described in the previous paragraph. It is possible to find an embedding $\lambda : S^m \times \mathbb{R}^{p-m} \to N^p$, $m \neq p$, such that $\lambda(S^m \times \{t\}) \cap f(B_f) \neq \emptyset$ for each t and f is transverse regular on each layer $\lambda(S^m \times \{t\})$. Then there are neighbourhoods V of $\lambda(S^m \times \{0\})$, U of $f^{-1}(\lambda(S^m \times \{0\}))$, and diffeomorphisms α, β such that the diagram

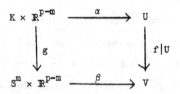

$$\begin{array}{ccc} K \times \mathbb{R}^{p-m} & \xrightarrow{\ \alpha\ } & U \\ \downarrow{\scriptstyle g} & & \downarrow{\scriptstyle f|U} \\ S^m \times \mathbb{R}^{p-m} & \xrightarrow{\ \beta\ } & V \end{array}$$

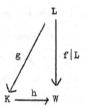

commutes, where K is an $(n - p + m)$-manifold, $\alpha(B_g) = B_f \cap \alpha(K \times \mathbb{R}^{p-m}) \neq \emptyset$, $g(K \times \{t\}) \subset S^m \times \{t\}$, and if $g_t = g|K \times \{t\}$, then $B(g_t) \neq \emptyset$ for each t. In addition, $\bigcup_t B(g_t) = B_g$. If $m \geq 3$ and the 0-dimensional set $g_t(B(g_t))$ does not have the property (*) for each $t \in R^{p-m}$, then it is possible to re-layer the map g and eventually reduce ourselves to the situation in which $g_t(B(g_t))$ does have the required property for each t. Now applying the previous paragraph to g_t we discover that each $B(g_t)$ is discrete. In fact, the cardinality of $B(g_t)$ is related to the Euler characteristic of K, and using this we can show that if x is a point of B_g in layer t, then nearby layers must also have a point of B_g contained in a small neighbourhood of x. In this manner we show $\dim B_g > 0$, which is a contradiction, and hence f has the desired structure.

[A] P.T. Church and J.G. Timourian, Differentiable maps with 0-dimensional critical set I (to appear).

[B] P.T. Church and J.G. Timourian, Differentiable maps with 0-dimensional critical set II (to appear).

[C] P.T. Church and J.G. Timourian, Fiber bundles with singularities. Indiana J. of Math. and Mech. 18 (1968), 71-90.

[D] J. Milnor, Singular Points of Complex Hypersurfaces. Annals. of Math Studies No.61, Princeton University Press, Princeton, N.J., 1968.

SIMPLE SINGULARITIES OF MAPS

I.R. Porteous

Introduction

This paper was first circulated in mimeographed form from Columbia University in 1962. It is here published basically unaltered. The only essential additions are further details in the proof of Proposition 1.3, a short note at the end of §1, more explicit references to intrinsic differentials in §2 and an extra example with comments at the end of §2. The singularities discussed in this paper, and the intrinsic differentials in particular, have also been discussed by J.M. Boardman [14]. The problem of computing Thom polynomials for higher order singularities raised in §2 has been tackled recently by F. Ronga [17] and by B. Bellew.

In the paper we are concerned with a category $\mathcal{M}$ of "manifolds", C^∞ in the sense that a tangent bundle functor T is defined from $\mathcal{M}$ to an appropriate category of vector bundles $\mathcal{E}$, the projection maps of the bundles being themselves morphisms in $\mathcal{M}$. We call morphisms in $\mathcal{M}$ <u>maps</u> and morphisms in $\mathcal{E}$ <u>homomorphisms.</u> Proper maps are defined in $\mathcal{M}$ by a compactness condition. In particular, the inclusion of a closed submanifold is proper, and the projection map of a fibre bundle is proper if the fibres are closed. A theory of Chern classes is defined for the pair $(\mathcal{M}, \mathcal{E})$ in terms of a value functor A from $\mathcal{M}$ to the category $\mathcal{G}$ of augmented commutative graded rings with unit. (The augmentation ring $A^0 = A^0(X)$, for all $X \in \mathcal{M}$. Also $A^k(X) = 0$, for each negative integer k.)

Some simple types of singularity for maps in $\mathcal{M}$ are defined and described in terms of this structure. In making these definitions we adopt a heuristic or opportunist approach, imposing transversality conditions on maps whenever these conditions guarantee further progress. The work complements earlier results of Whitney [13], Thom [11], and Levine [9] by providing more detailed information on specific examples. For example, the polynomial expressions derived in Proposition 1.3 and the last paragraph of §2 are evidently Thom polynomials. [4]. As a byproduct we recover many of the standard formulas of classical algebraic geometry including the formulas relating the Todd-Eger canonical classes to Jacobians of linear systems [12]. We postpone details of this last application to a second paper [16] .

The paper is divided into three sections. §0 contains basic notations and some elementary lemmas. The first order singularities are discussed in §1 and some higher order singularities in §2. Note that we use symbols such as Z_a , Z_{ab}, etc., where Whitney and Thom use S_i, S_jS_i, etc., to denote various singularity types. Our conventions regarding the use of such symbols differ in several respects from theirs.

§0. Notations.

We do not give a formal axiomatisation of the categories $\mathcal{M}$, $\mathcal{E}$ or of the functor A introduced above. However, any properties we impose at any stage will hold in each of the following typical examples :

a) $\mathcal{M}$ the category of (connected, paracompact) C^∞-manifolds and maps, $\mathcal{E}$ the category of C^∞ vector bundles, $A = H^*(\quad , A^0)$, with $A^0 = \underline{\underline{Z}}_2$.

b) $\mathcal{M}$ the category of C^ω-manifolds and maps, $\mathcal{E}$ the category of C^ω vector bundles, $A = H^*(\quad , A^0)$, with $A^0 = \underline{\underline{Z}}_2$.

c) $\mathcal{M}$ the category of complex analytic manifolds and maps, $\mathcal{E}$ the category of complex analytic vector bundles
$A = H^{2*}(\quad , A^0)$, with $A^0 = \underline{\underline{Z}}$.

d) $\mathcal{M}$ the category of irreducible non-singular quasi-projective algebraic manifolds over a field, k , of characteristic zero, $\mathcal{E}$ the category of algebraic k-vector bundles,. A the rational equivalence (Chow ring)functor, with $A^0 = \underline{\underline{Z}}$. For the theory of Chern classes for such categories (including Stiefel-Whitney, Chern and Todd-Eger) we follow [8] .

In the study of a particular map $f : X \to Y$ the only auxiliary maps introduced will all belong to a diagram, $R(f)$, of maps terminating with f, commutative whenever alternative routes are defined. In such a situation each map is uniquely determined by its source and target (domain and range). We take advantage of this to make certain notational conventions. For example, if $g : X_2 \to X_1 \in R(f)$ and $E \in \mathcal{E}(X_1)$, the induced bundle on X_2 is commonly denoted by g^*E. Instead we write "$E \in \mathcal{E}(X_2)$" or "$E \searrow X_2$" or "E over X_2". Also we take advantage of the universal property of the induced bundle to make the convention that the source and target bundles of any morphism have the same base. Any commutative exact diagram (simply diagram) of bundles will then be over a single manifold, X_1 say. This is indicated by the symbol $\searrow X_1$ placed to the right of the diagram. It is often convenient to regard a vector space V as a vector bundle over a point. Then $V \searrow X_1$ denotes the trivial bundle over X_1 with fibre V.

Let $f : X \to Y \in \mathcal{M}$. The map f induces the homomorphism $f_T : T(X) \to T(Y) \in \mathcal{E}(X)$. Further properties of T include :
If Y is a submanifold of X, with $i : Y \to X$ the inclusion map, then i_T is a monomorphism. The normal bundle, $T(X, Y)$, is defined by the exact sequence

$$0 \longrightarrow T(Y) \xrightarrow{i_T} T(X) \longrightarrow T(X, Y) \longrightarrow 0 \qquad \searrow Y .$$

If $\pi : X \to Y$ is the projection map of a fibre bundle, then π_T is an epimorphism and the connected components of the fibres are submanifolds of X. The tangent bundle along the fibres, also denoted by $T(X, Y)$, is defined by the exact sequence

$$0 \longrightarrow T(X, Y) \longrightarrow T(X) \xrightarrow{\pi_T} T(Y) \longrightarrow 0 \qquad \searrow X .$$

The correct interpretation of $T(X, Y)$ should always be clear from the context. An apparent counterexample to this remark is provided by a vector bundle $E \searrow X$ if we identify the base X with the image $z(X)$ of the zero section $z : X \to E$. In this case $T(E, X) \searrow E$ denotes the tangent bundle along the fibres, so $T(E, X) \searrow X$ denotes both the restriction of this bundle to X and the normal bundle of X in E. However, there is a canonical isomorphism between the two so the choice of interpretation does not matter. In fact, it is easy to see that if

$s : X \to E$ is any section of $E \searrow X$, then, over $s(X)$, $T(E, X) = T(E, s(X))$. ($=$ denotes canonical isomorphism). If $x \in X \in \mathcal{M}$ then $T(X, x)$ denotes the tangent space to X at x. $T(X, x)$ is canonically isomorphic to $T(X, \{x\})$.

Let $X, Y \in \mathcal{M}$. We write $\dim X = \mathrm{rk} T(X)$ and $\dim(X, Y) = \dim X - \dim Y$. If $Y \subset X$, $\dim(X, Y) = \mathrm{codim}_X Y$. If $h : E \to F$ belongs to $\mathcal{E}(X)$ we define the <u>rank</u> of h, $\mathrm{rk}\, h = \mathrm{rk}(h(E))$, and the <u>kernel rank</u> (or <u>nullity</u> or <u>corank at the source</u>) of h, $\mathrm{kr}\, h = \mathrm{rk}\, E - \mathrm{rk}\, h$. Both $\mathrm{rk}\, h$ and $\mathrm{kr}\, h$ are non-negative integral valued functions on X. If $f : X \to Y \in \mathcal{M}$ we define $\mathrm{rk}\, f = \mathrm{rk}\, f_T$ and $\mathrm{kr}\, f = \mathrm{kr}\, f_T$. In fact, it is convenient to omit the letter T whenever its absence is apparent from the context.

We recall briefly some basic "Advanced Calculus" (cf. [6]). Let V, W be (finite-dimensional!) vector spaces over $\underline{\underline{R}}$, A an open set of V and $f : A \to W$ a C^∞-function. We denote by $Df(a) : V \to W$ the <u>derivative</u> of f <u>at</u> a and by $Df : A \to \mathrm{Hom}\,(V, W) = \check{V} \otimes W$ the <u>derivative</u> of f ($\check{}$ denotes dual). The <u>second derivative</u> of f is the map :

$$D(Df) = D^2 f : A \to \check{V} \circledcirc \check{V} \otimes W .$$

This associates to each point of A a <u>symmetric</u> bilinear map from $V \times V$ to W. ($\circledcirc$ denotes the symmetric product.) In like fashion the pth derivative of f is the map

$$D(D^{p-1} f) = D^p f : A \to \circledcirc^p \check{V} \otimes W .$$

The <u>tangent space</u> to V at $a \in V$, $T(V, a)$ ($= V$), can be defined as the pair (a, V), and the <u>tangent bundle</u>, $T(A)$, (A any open subset of V) as the trivial bundle $V \searrow A$. Then the action of $f_T : T(A) \to T(W)$ is explicitly given by the formula

$$f_T(a, v) = (f(a), Df(a)(v)), \quad v \in V .$$

It is well-known how the above ideas can be lifted to the category of C^∞-manifolds to give an explicit description of the functor T. The symmetry in the second and higher derivatives gives rise to symmetries under repeated application of the functor T. We assume that such symmetry arises in $\mathcal{M}$ exactly where it would arise in the category of C^∞-manifolds and maps. Frequently jets and jet-bundles are used (cf. [7], [4]) to handle higher order derivatives. We prefer to use an alternative prolongation process, obtaining expressions involving the higher order derivatives directly in coordinate-free form. It would be of interest to correlate our approach with the jet bundle approach and with the theory of partial differential equations, but we do not undertake such a correlation in this paper.

The symmetric product.

In a canonical way $V \otimes V = (V \circledcirc V) \oplus (V \wedge V)$, where $\circledcirc$ denotes the symmetric product, and $\wedge$ the exterior or antisymmetric product (V a vector space). In the sequel it will be convenient to extend the use of the symbol $\circledcirc$ to cover situations such as the following : Let $0 \to V' \to V \to V'' \to 0$ be an exact sequence of vector spaces. Then we define $V'' \circledcirc V$ by the diagram

$$\begin{array}{ccccccccc} & & & & 0 & & 0 & & \\ & & & & \vdots\downarrow & & \downarrow & & \\ 0 & \to & V'' \otimes V' & \dashrightarrow & V'' \circledcirc V & \dashrightarrow & V'' \circledcirc V'' & \to & 0 \\ & & \| & & \vdots\downarrow & & \downarrow & & \\ 0 & \to & V'' \otimes V' & \to & V'' \otimes V & \to & V'' \otimes V'' & \to & 0 \\ & & & & \downarrow & & \downarrow & & \\ & & & & V'' \wedge V'' & = & V'' \wedge V'' & & \\ & & & & \downarrow & & \downarrow & & \\ & & & & 0 & & 0 & & \end{array}$$

where the right-hand vertical sequence is the split exact sequence defined by the canonical direct sum decomposition of $V'' \otimes V''$. (In dual fashion, we could also define $V' \circledcirc V$, where V' is a subspace of V , but we shall not require this usage.) The construction of $V'' \circledcirc V$ generalises at once to give a definition of $E'' \circledcirc E$, where $E \in \mathcal{E}$ and E'' is a quotient of E .

Sections transversal to zero

Let $E \in \mathcal{E}(X)$, with $s : X \to E$ a section of E , and the inclusion map $z : X \to E$ the zero section, X being identified with $z(X)$. Let $Z = s(X) \cap X$. We say the section s is transversal to zero if $s(X)$ intersects X transversally in E, that is if at each point $x \in Z$, $T(X, x)$ and $T(s(X),x)$ together span $T(E, x)$. Then by the inverse function theorem, assumed to hold in our category, $Z \in \mathcal{M}$ and is a proper submanifold of X. Also $\dim(X, Z) = \mathrm{rk}\, E$.

Proposition 0.1. *If s is a section of* $E \in \mathcal{E}(X)$, *transversal to zero, then* $T(X, Z) = E \searrow Z$.

Proof : Apply the functor T to the commutative diagram of inclusions

$$\begin{array}{ccc} Z & \to & s(X) \\ \downarrow & & \downarrow \\ X & \to & E \end{array}$$

and to the projection $\pi : E \to X$.

We then have the diagram (of *split* exact sequences!)

$$\begin{array}{ccccccccc} & & 0 & & 0 & & 0 & & \\ & & \downarrow & & \downarrow & & \downarrow & & \\ 0 & \to & T(Z) & \to & T(s(X)) & \to & T(s(X), Z) & \to & 0 \\ & & \downarrow & & \downarrow & & \downarrow & & \\ 0 & \to & T(X) & \to & T(E) & \to & T(E, X) & \to & 0 \\ & & \downarrow & & \downarrow & & \downarrow & & \\ 0 & \to & T(X, Z) & \to & T(E, X) & \to & 0 & & \\ & & \downarrow & & \downarrow & & & & \\ & & 0 & & 0 & & & & \end{array} \searrow Z,$$

where the zero in the lower right-hand corner is a direct consequence of transversality. Proposition 0.1 follows since $T(E, X) = E \searrow X$.

Corollary 1. Let $h : E \to F \in \mathcal{E}(X)$. Then h naturally determines a section, also denoted by h, of $\check{E} \otimes F$ (= Hom (E, F)). We say the *homomorphism* h is *transversal to zero* if the induced *section* is transversal to zero. *Then*

$Z = \{x \in X;\ (\mathrm{kr}\, h)(x) = \mathrm{rk}\, E\}$ *is a submanifold of* X *and* $T(X, Z) = \check{E} \otimes F \searrow Z$.

Grassmannians

Let V be a vector space $\in \mathcal{M}$. We denote by $G_a(V)$ the *Grassmann manifold* of a-planes through 0 in V. For each direct sum decomposition $V' \oplus V''$ of V with $V' \in G_a(V)$ the map $\mathrm{Hom}(V', V'') \to G_a(V)$; $t \rightsquigarrow$ graph t is a chart on $G_a(V)$.

We refer to such a chart as a <u>standard chart</u> on $G_a(V)$. Clearly, $\dim G_a(V) = a(p-a)$, where $p = \dim V$. Over $G_a(V)$ sits the <u>canonical</u> a-<u>plane</u> <u>bundle</u> K_a which associates to each point of $G_a(V)$ the a-plane which it represents. We then have a natural exact sequence

$$0 \to K_a \to V \dashrightarrow M_a \dashrightarrow 0 \in \mathcal{E}(G_a(V)) ,$$

defining M_a, a sequence which we call the <u>canonical sequence</u> over $G_a(V)$. Note that <u>a</u> is here a label and not a mark of rank. In fact, $\mathrm{rk}\, K_a = a$, but $\mathrm{rk}\, M_a = p - a$.

<u>Proposition 0.2</u> $T(G_a(V)) = \check{K}_a \otimes M_a \quad (= \mathrm{Hom}\,(K_a, M_a))$.

<u>Proof</u> : We show that $T(G_a(V) \times G_a(V), \Delta G_a(V)) = \check{K}_a \otimes M_a \searrow \Delta G_a(V)$, where $\Delta : G_a(V) \to G_a(V) \times G_a(V)$ is the diagonal map.

Let π_1 and $\pi_2 : G_a(V) \times G_a(V) \to G_a(V)$ be the first and second projections. Over $\Delta G_a(V)$, $\pi_1 = \pi_2$. Then $\Delta G_a(V)$ is the set of zeros of the homomorphism $\pi_1^* K_a \to V \to \pi_2^* M_a$ on $G_a(V) \times G_a(V)$. It is readily checked, using standard charts on $G_a(V)$, that the induced section of the bundle $\pi_1^* \check{K}_a \otimes \pi_2^* M_a \searrow G_a(V) \times G_a(V)$ is transversal to zero everywhere on $\Delta G_a(V)$. The proposition follows at once by Proposition 0.1.

<u>Corollary 1</u> <u>For any</u> $q \in G_a(V)$, <u>there is a natural isomorphism</u>

$$T(G_a(V), q) = \check{K}_a \otimes M_a \searrow \{q\} .$$

The operation G_a on vector spaces extends naturally to vector bundles. If $E \in \mathcal{E}(X)$ and if $0 \to K_a \to E \to M_a \to 0 \searrow G_a(E)$ is the canonical exact sequence, then we have

<u>Corollary 2</u> $T(G_a(E), X) = \check{K}_a \otimes M_a$.

<u>Chern classes.</u>

We assume that the categories $\mathcal{M}$ and $\mathcal{E}$ and the functor $A : \mathcal{M} \to \mathcal{G}$ satisfy the axioms of Grothendieck [8] for a theory of Chern classes. For any $R \in \mathcal{G}$, let R_1 denote the multiplicative subgroup of R formed by the elements of R of augmentation 1. We recall that c, the Chern class function, is a function from $\mathcal{E}(X)$ to $(A(X))_1$ for each $X \in \mathcal{M}$.

The functor A is contravariant. A map $f : X \to Y \in \mathcal{M}$ induces a ring homomorphism $f^* : A(Y) \to A(X)$, preserving degree. The map f also induces $f^* : \mathcal{E}(Y) \to \mathcal{E}(X)$, and $f^*c = cf^*$. (Under our convention the symbol f^* on the right hand side of this equation will often be omitted.) The induced functor from the subcategory of $\mathcal{M}$ consisting of manifolds in $\mathcal{M}$ and proper maps to the category of abelian groups is covariant. To be precise, a proper map $f : X \to Y$ induces group homomorphisms $f_* : A^k(X) \to A^\ell(Y)$ where $\ell = k - \dim(X, Y)$. Therefore, f_* lowers degree if $\dim X > \dim Y$ and raises degree if $\dim X < \dim Y$. The ring structure is not preserved. If $f : X \to Y$ is proper and if $a \in A(X)$, $b \in A(Y)$, then $f_*(f^*b \,.\, a) = b \,.\, f_*a$. If the diagram of maps

$$\begin{array}{ccc} Y' & \xrightarrow{j} & X' \\ \downarrow g & & \downarrow f \\ Y & \xrightarrow{i} & X \end{array}$$

is commutative, with f, g proper and i, j inclusions, then $i^*f_* = g_*j^*$.

Let the map $f : X \to Y$ be proper and such that there exists an open submanifold U of X such that X is the closure of $U, f|U$ is an embedding, and $f(U) \cap f(X - U) = \emptyset$. Then $f_*1 \in A(Y)$ can be regarded in some appropriate sense, depending on the category we are in, as the "cohomology class" of the "singular submanifold" $f(X)$ in Y . In the applications we consider f will be locally analytic, even in case a). For a direct definition of the fundamental cycle of a singular real or complex analytic manifold see [2] and [3]. If $i : Y \to X$ is a proper inclusion we define $\alpha(X, Y) = i_*1$. If $E \in \mathcal{E}(X)$ and Z is the submanifold of zeros of a section $s : X \to E$ transversal to zero then

$$\alpha(X, Z) = c_p(E) \in A^p(X) , \quad p = \operatorname{rk} E .$$

In such a situation Z is always a proper submanifold of X. Further properties of c are :

$$c_i(E) = 0, \quad i > \operatorname{rk} E ,$$
$$c_i(\check{E}) = (-1)^i c_i(E) ,$$

while if $0 \to E' \to E \to E'' \to 0$ is exact, $c(E) = c(E') \,.\, c(E'')$, $E', E, E'' \in \mathcal{E}$. There are also universal formulas for $c(E \otimes F)$, $c(E \circledS E)$, $c(E \wedge E)$, $c(E'' \circledcirc E)$

(E'' a quotient of E), etc. We state such formulas if and when we require them. Finally, we define $\bar{c}(E) = (c(E))^{-1}$ and $c(F - E) = c(F) . (c(E))^{-1}$.

Proposition 0.3 *Let* $E \in \mathcal{E}(X)$, $\operatorname{rk} E = p$, *and let* M_a *be the canonical quotient bundle of* E *over* $G_a(E)$. *Let* $\pi_a : G_a(E) \to X$ *be the bundle projection map and let* $m_i = c_i(M_a)$. *Then* $\pi_{a*}((m_{p-a})^a) = 1$ *and the image under* π_{a*} *of any other a-fold product of the* m_i *(with* $m_o = 1$ *a possible factor) is zero.*

Proof : By Proposition 0.2, Corollary 1 and 2, $\dim (G_a(E), X) = a(p - a)$. So $\pi_{a*}((m_{p-a})^a) \in A^o(X) = A^o$. To evaluate this element it is sufficient to evaluate it at a point of X . Now we have the diagram of maps

$$\begin{array}{ccc} G_a(V) & \xrightarrow{j} & G_a(E) \\ \downarrow \theta & & \downarrow \pi_a \\ pt & \xrightarrow{i} & X \end{array}$$

with i and j inclusions and $i^*(\pi_a)_* = \theta_* j^*$, θ the trivial map. Here V denotes the fibre of E at "pt". Since $\operatorname{rk} M = p - a$, $j^* m_{p-a}$ is represented by the zeros of any section of M_a over $G_a(V)$ that is transversal to zero. Such a section is induced by any non-zero vector in V by composing its inclusion in V over $G_a(V)$ with the canonical homomorphism $V \to M_a \searrow G_a(V)$. Since an a-plane in V is uniquely determined by $\underline{a}$ linearly independent vectors, $(j^* m_{p-a})^a = j^*((m_{p-a})^a)$ is represented on $G_a(V)$ by any point. Therefore

$$i^*(\pi_{a*}(m_{p-a})^a) = \theta_* j^*((m_{p-a})^a) = 1 .$$

So $\pi_{a*}(m_{p-a})^a = 1$.

The second part of the proposition is an immediate consequence of the relations $m_i = 0$, $i > p - a$.

In applications of Proposition 0.3, lemmas 0.4 - 0.7 are of use. We state them without proof.

Let $A \in \mathcal{G}$ with $A^k = 0$, $k < 0$, and let A_1 be the multiplicative subgroup of A of elements of augmentation 1. For any $c \in A_1$, let $\bar{c} = c^{-1}$ and let

$\check{c} \in A_1$ be defined by $\check{c}_i = (-1)^i c_i$.

<u>Lemma 0.4</u> <u>Let</u> $c \in A_1$, $c_k = 0$, $k > a$. <u>Then the set of a-fold products of the</u> $\bar{c}_i$ <u>(with</u> $\bar{c}_0 = 1$ <u>a possible factor) spans the</u> A^o<u>-module</u> $A^o[c_1, c_2, \ldots, c_a]$.

A homogeneous element of this module, of degree k say, is called an <u>isobaric polynomial</u> on A_1 of degree k. Also if $|b_{ij}|_a$ denotes the determinant of the $a \times a$ matrix (b_{ij}) with i, j th term b_{ij}, then a function $A_1 \to A^k$ of the form $c \to |c_{\lambda_i + \mu_j - i + j}|_a$, $\lambda_i, \mu_j \in \underline{\underline{Z}}$, $1 \leq i \leq a$, $1 \leq j \leq a$, $k = \sum_{i=1}^{a} (\lambda_i + \mu_i)$, is called an <u>isobaric determinant</u> on A_1 of degree k.

<u>Lemma 0.5</u> <u>Let</u> k <u>be a positive integer and let</u> $\lambda = \{\lambda_i\}_{1 \leq i \leq n(\lambda)}$ <u>denote a partition of</u> k <u>by the positive integers</u> λ_i. <u>Then the set of isobaric determinants on</u> A_1 <u>of the form</u> $|c_{\lambda_i - i + j}|_a$ <u>is a basis for the</u> A^o<u>-module of isobaric polynomials on</u> A_1 <u>of degree</u> k.

Aitken [1] has a duality theorem for isobaric determinants. As a particular case of this theorem we note

<u>Lemma 0.6</u> $$|c_{b - i + j}|_a = |\check{c}_{a - i + j}|_b , \quad c \in A_1 .$$

Using Aitken's theorem it is not difficult to prove

<u>Lemma 0.7</u> <u>Let</u> $c \in A_1$, $c_k = 0$, $k > a$, <u>and let</u> $\lambda_i, \mu_i \in \underline{\underline{Z}}$, $1 \leq i \leq a$. <u>Then</u>

$$|\bar{c}_{\lambda_i - i + j}|_a \, |\bar{c}_{\mu_j - i + j}|_a = |\bar{c}_{\lambda_i + \mu_j - i + j}|_a .$$

§1. First-order singularities.

Let $h : E \to F \in \mathcal{E}(X)$. In particular, we have in mind the homomorphism $f_T : T(X) \to T(Y) \in \mathcal{E}(X)$ induced by a map $f : X \to Y$ in $\mathcal{M}$. Let $Z_a(h)$ (respectively $Z_a(f)$) denote the subset of X where $\mathrm{kr}\, h$ (respectively $\mathrm{kr}\, f$) $= a$. We enquire whether, if h (or f) is suitably "nice", something nice can be said about $Z_a(h)$ (or $Z_a(f)$).

Let $x \in Z_a(h)$. At x exactly one a-plane in E_x is mapped to zero by h. Such an a-plane is represented by a point of $G_a(E)$. Let $s_a : Z_a(h) \to G_a(E)$ be the set map thus defined. Now consider

$$\begin{array}{ccccccccc} 0 & \to & K_a & \to & E & \to & M_a & \to & 0 \\ & & & \searrow^{h^a} & \downarrow h & & & \searrow & G_a(E) \\ & & & & F & & & & \end{array}$$

defining h^a, the horizontal exact sequence being the canonical sequence over $G_a(E)$. Also let $\pi_a : G_a(E) \to X$ be the canonical projection.

Proposition 1.1 i) $s_a(Z_a(h)) \subset Z_a(h^a)$,

ii) $\pi_a s_a = 1$,

iii) $\pi_a(Z_a(h^a) - s_a Z_a(h)) = \bigcup_{a'>a} Z_{a'}(h)$.

These assertions are all trivial to prove.

We say h is a-transversal to zero if h^a is transversal to zero on $s_a(Z_a(h))$; h is fully a-transversal to zero if h^a is transversal to zero on $Z_a(h^a)$. We say that $f : X \to Y$ is (fully) a-transversal to zero if f_T is (fully) a-transversal to zero. If h is a-transversal to zero, then $s_a \in \mathcal{M}$.

Suppose h is a-transversal to zero. Then $M_a = \mathrm{im}\, h \searrow s_a(Z_a(h))$ and we have a diagram

$$
\begin{array}{ccccccccc}
 & & & & & & 0 & & \\
 & & & & & & \downarrow & & \\
0 & \to & K_a & \to & E & \to & M_a & \to & 0 \\
 & & & h^a = 0 \searrow & & \searrow h & \downarrow & & \\
 & & & & & & F & & \searrow s_a(Z_a(h)) \\
 & & & & & & \downarrow & & \\
 & & & & & & Q_a & & \\
 & & & & & & \downarrow & & \\
 & & & & & & 0 & &
\end{array}
$$

defining $Q_a \searrow s_a(Z_a(h))$ (<u>not</u> $\searrow Z_a(h^a)$).

<u>Proposition 1.2</u> <u>If</u> $h : E \to F$ <u>is a-transversal to zero</u>

$$T(X, Z_a(h)) = \check{K}_a \otimes Q_a \searrow Z_a(h) .$$

<u>Proof</u> : By Proposition 0.1, Corollary 1, $T(G_a(E), Z_a(h^a)) = \check{K}_a \otimes F \searrow Z_a(h^a)$. Also by Proposition 0.2, Corollary 2, $T(G_a(E),X) = \check{K}_a \otimes M_a \searrow G_a(E)$. Now, over $s_a(Z_a(h))$, the tangent vectors along the fibre of $G_a(E) \searrow X$ will be normal in $G_a(E)$ to $Z_a(h^a)$. The resultant monomorphism $T(G_a(E),X) \to T(G_a(E), Z_a(h^a)) \searrow s_a(Z_a(h))$ is in fact that induced by the monomorphism $M_a \to F$ of the above diagram, as one can easily check, bearing in mind the proof of Proposition 0.2. Therefore, $T(G_a(E), \pi_a^{-1}(Z_a(h))) = \check{K}_a \otimes Q_a \searrow Z_a(h^a)$. It follows, via s_a, that $T(X, Z_a(h)) = \check{K}_a \otimes Q_a \searrow Z_a(h)$.

<u>Corollary 1</u> : <u>If</u> $h : E \to F$ <u>is a-transversal to zero</u>,

$$\dim(X, Z_a(h)) = a(p - a) \qquad p = \operatorname{rk} E .$$

<u>Corollary 2</u> : <u>If</u> $f : X \to Y$ <u>is a-transversal to zero</u>,

$$\dim(X, Z_a(f)) = a(n - a) \qquad n = \dim X .$$

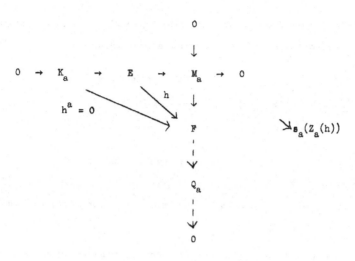

An important case of a map which is not a-transversal to zero is afforded by the dilatation projection map $f : X' \to X$ associated to the dilatation in a manifold X of a submanifold Y of codimension $(a+1)$ in X . The resultant structure can be summarised by the commutative diagram of maps :

$$\begin{array}{ccc} Y' & \xrightarrow{j} & X' \\ \downarrow g & & \downarrow f \\ Y & \xrightarrow{i} & X \end{array} \qquad \begin{array}{l} f,\ g \text{ projections} \\ i,\ j \text{ inclusions ,} \end{array}$$

where $\dim(X', Y') = 1$, Y' is a projective bundle over Y, with $\dim(Y', Y) = a$, and $f|X' - Y'$ is a bijection. In this case $Y' = Z_a(f)$ and, if $L = T(X', Y')$, then, in contrast to the result of Proposition 1.2, we have [10] the formula $K_a = \check{L} \otimes Q_a \searrow Z_a(f)$.

For our next proposition, we require the stronger transversality condition.

<u>Proposition 1.3</u>.

If $h : E \to F \in \mathcal{E}(X)$ *is fully a-transversal to zero*

$$\pi^a_* 1 = |c_{q-p+a-i+j}(F - E)|_a ,$$

where $\pi^a = \pi_a | Z_a(h^a)$, *and where* $p = \operatorname{rk} E$, $q = \operatorname{rk} F$.

<u>Proof</u> : $Z_a(h^a)$ is the set of zeros of a section of $\check{K}_a \otimes F \searrow G_a(E)$ transversal to the zero section. Therefore $Z_a(h^a)$ is a proper submanifold of $G_a(E)$ and $\alpha(G_a(E), Z_a(h^a)) = c_{aq}(\check{K}_a \otimes F)$.

Now $c_{aq}(\check{K}_a \otimes F) = |c_{q-i+j}(F - K)|_a$. For let $c(K_a) = \prod_{i=1}^{a} (1 + \kappa_i)$ and let $c(F) = \prod_{j=1}^{q} (1 + \phi_j)$ (formal factorizations). Then

$$c(\check{K}_a \otimes F) = \prod_{1 \le i \le a, 1 \le j \le q} (1 + \phi_j - \kappa_i). \text{ So } c_{aq}(\check{K}_a \otimes F) = \prod_{i,\,j} (\phi_j - \kappa_i) .$$

It should come as no surprise that this can be expressed as the determinant of a matrix. That the matrix is the one stated is proved in a manner suggested by the usual proof that

$$\begin{vmatrix} 1 & 1 & 1 \\ a & b & c \\ a^2 & b^2 & c^2 \end{vmatrix} = (b-c)(c-a)(a-b) .$$

(If $\phi_j = \kappa_i$, for some i, j, then there is an isomorphism between the jth formal line-bundle factor of K_a and the ith formal line-bundle factor of F. In that case $F - K_a = F' - K'_a$, where F' and K'_a are the formal complements of these line bundles. But then

$$|c_{q-i+j}(F - K_a)|_a = |c_{q-i+j}(F' - K'_a)|_a = 0$$

as can be seen by multiplying the columns of the matrix $(c_{q-i+j}(F' - K'_a))$ by κ'_{a-j}, $1 \leq j \leq a$, and adding. The parallel elementary symmetric function argument shows that each $(\phi_j - \kappa_i)$ is a factor of $|c_{q-i+j}(F - K_a)|_a$. That the multiplier as 1 can be seen by considering the case that K_a is a trivial bundle.) Therefore

$$\pi^a_* 1 = \pi_{a*}\alpha(G_a(E), Z_a(h^a)) = \pi_{a*}|c_{q-i+j}(F - K_a)|_a = \pi_{a*}|c_{q-i+j}(F - E + M_a)|_a .$$

Now E and F have their home on X. Therefore by Proposition 0.3 and the formula relating π_{a*} and π^*_a the result follows.

Notice that, since $\mathrm{rk}\,\check{h} = \mathrm{rk}\,h$, $Z_{q-p+a}(\check{h}) = Z_a(h)$. So we have alternative formulas

$$\pi^a_* 1 = |c_{a-i+j}(\check{E} - \check{F})|_{q-p+a} = |\check{\bar{c}}_{a-i+j}(F - E)|_{q-p+a} .$$

(cf. Lemma 0.6.)

Note also that, since π^a is locally analytic in any of the examples for $\mathcal{M}$ cited, $\pi^a_* 1$ can be regarded as the "cohomology" class in X of the closure of $Z_a(h)$.

Theorem 1.3 is essentially due to Thom. In [11] he gives a description of the singularities of a map in terms of certain Schubert cycles, and he states that polynomial expressions could be deduced using formulas of Chern [5]. However, he does not state the formulas explicitly except in one or two special cases. Our method of computing the Thom polynomials almost avoids the introduction of Schubert cycles. (The simplest are present in Proposition 0.3.) The Thom polynomials for $Z_1(f)$, dim X $\leq$ dim Y, and for $Z_1(\check{f})$, dim X $\geq$ dim Y, have immediate application in algebraic geometry in the particular case where Y is a projective space.[†] By a trick we can include the more usual situation where the map f, with target projective space, is not regular, but rational. In this way, we recover the classical formulas relating the Todd-Egar classes of a manifold X to Jacobians of linear systems on X (read 'the singular sets $(Z_1(f)$ or $Z_1(\check{f}))$ of rational maps from X to projective space.) We discuss this application in a second paper [16] (published in this volume).

§2. Higher order singularities

Let $f : X \to Y \in \mathcal{M}$. In §1 we travelled from X to $Z_a(f)$ via $G_a(X)$ $(=G_a(T(X)))$ and $Z_a(f^a)$ under the hypothesis that f is a-transversal to zero. This is the beginning of the route $R(f)$ referred to in §0. Under the stronger hypothesis that f is fully a-transversal to zero we have obtained a description of $Z_a(f)$, or rather of its closure, in the ring $A(X)$.

Suppose then that f is a-transversal to zero, so that $Z_a(f)$ is a submanifold of X with normal bundle $\check{K}_a \otimes Q_a$ and let f''_a denote both the restriction of f to $Z_a(f)$ and the induced homomorphism $(f''_a)_T$. Define $Z_{ab}(f) = Z_b(f''_a)$. Following our previous procedure we ask when something nice can be said about $Z_{ab}(f)$.

---

† Cf. Todd [18], especially 5.06. It is an essential feature of Todd's examples that $Z_2(f) = \phi$. See also the note at the end of §3.

Notice, first, that we have an exact diagram

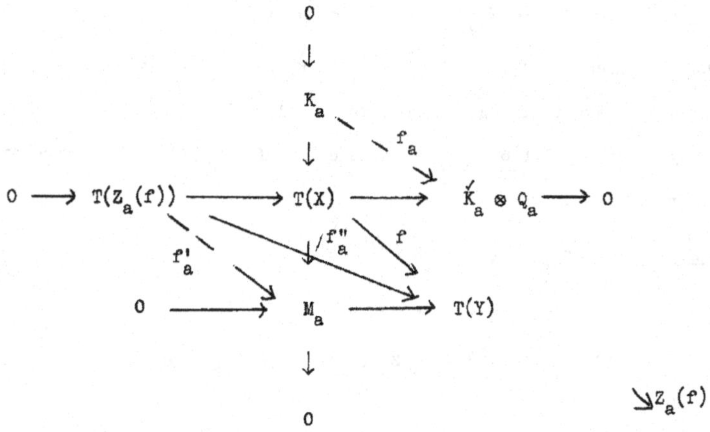

defining the morphisms f'_a and f_a over $Z_a(f)$ (or over ${}_aZ_a(f)$, but not over $Z_a(f^a)$ in general).

<u>Proposition 2.1</u> <u>For any local charts at</u> $x \in X$ <u>and at</u> $f(x) \in Y$ <u>the map</u> $f_a|(K_a)_x : (K_a)_x \to (\check{K}_a)_x \otimes (Q_a)_x$ <u>is the map</u>

$$(K_a)_x \xrightarrow{i} T(X, x) \xrightarrow{D^2 fx} T(\check{X}, x) \otimes T(Y, f(x)) \xrightarrow{p} (\check{K}_a)_x \otimes (Q_a)_x ,$$

<u>where</u> i <u>is the inclusion and</u> p <u>is the obvious projection.</u>

<u>Proof</u> : Local analysis of Proposition 0.1.

It can easily be verified that the map $K_a \to \check{K}_a \otimes Q_a$ induced by D^2f is well-defined everywhere that $\operatorname{kr} f = a$, whether or not the transversality conditions are satisfied. Since this paper was first circulated I have called this map the <u>second</u> <u>intrinsic</u> <u>derivative</u> or <u>differential</u> of f , the homomorphism $f_T : T(X) \to T(Y)$ being the first one.

<u>Proposition 2.2</u> <u>In the above diagram,</u> $\operatorname{kr} f''_a = \operatorname{kr} f'_a = \operatorname{kr} f_a$.

<u>Proof</u> : An elementary diagram chasing argument.

<u>Corollary 1.</u> $Z_{ab}(f)$ $(= Z_b(f''_a))$ $= Z_b(f_a)$ $\quad 0 \leq b \leq a$.

Now let $G_{ab}(f) = G_b(K_a \searrow Z_a(f))$, and let $\pi_{ab} : G_{ab}(f) \to Z_a(f)$ be the bundle projection map. Let

$$0 \to K_{ab} \to K_a \to M_{ab} \to 0 \qquad \searrow G_{ab}(f)$$

be the canonical sequence. Also let $f_a^b : K_{ab} \to \check{K}_a \otimes Q_a \searrow G_{ab}(f)$ be defined by composition, analogously to the definition of f^a in §1. Let s_{ab} be the set map associating to each point of $Z_{ab}(f)$ the point of $G_{ab}(f)$ which represents the kernel there of f_a.

<u>Proposition 2.3.</u> i) $s_{ab}(Z_{ab}(f)) \subset Z_b(f_a^b)$,

ii) $\pi_{ab} s_{ab} = 1$,

iii) $\pi_{ab}(Z_b(f_a^b) - s_{ab} Z_{ab}(f)) = \bigcup_{b' > b} Z_{ab'}(f)$.

Straightforward proof. Cf. Proposition 1.1.

Suppose now that f_a is b-transversal to zero. Then $s_{ab} \in \mathcal{M}$ and f_a^b is represented canonically by a section of the bundle $\check{K}_{ab} \circledcirc \check{K}_a \otimes Q_a \searrow G_{ab}(f)$, zero exactly on $Z_b(f_a^b)$. (The symmetric tensor product arises here from the symmetry of the second derivative, which appears in Proposition 2.1) To make further progress we assume that the section of $\check{K}_{ab} \circledcirc \check{K}_a \otimes Q_a$ representing f_a^b is transversal to zero.

<u>Proposition 2.4</u> <u>Let</u> $f : X \to Y \in \mathcal{M}$, <u>let</u> f^a <u>be transversal to zero and let</u> f_a^b , <u>regarded as a section of</u> $\check{K}_{ab} \circledcirc \check{K}_a \otimes Q_a$, <u>also be transversal to zero</u>. <u>Then over</u> $s_{ab}(Z_{ab}(f))$ <u>we have an exact sequence</u> $0 \to M_{ab} \to \check{K}_a \otimes Q_a \dashrightarrow Q_{ab} \dashrightarrow 0$ <u>defining</u> Q_{ab}, <u>and</u> $T(Z_a(f), Z_{ab}(f)) = \check{K}_{ab} \circledcirc Q_{ab} \searrow Z_{ab}(f)$, where $\check{K}_{ab} \circledcirc Q_{ab}$ is <u>defined by the exact sequence</u>

$$0 \to \check{K}_{ab} \otimes M_{ab} \to \check{K}_{ab} \circledcirc \check{K}_a \otimes Q_a \dashrightarrow \check{K}_{ab} \circledcirc Q_{ab} \dashrightarrow 0 \qquad \searrow Z_{ab}(f)$$

<u>induced in the obvious way</u>.

<u>Proof</u> : A straightforward analogue of the proof of Proposition 1.2.

<u>Corollary 1</u>. (Levine [9]) <u>Under the above transversality conditions</u>

$$\dim(X, Z_{ab}(f)) = \dim(X, Z_a(f)) + \dim(Z_a(f), Z_{ab}(f))$$
$$= \mathrm{rk}(\check{K}_a \otimes Q_a) + \mathrm{rk}(\check{K}_{ab} \circledcirc Q_{ab})$$

$$= a(m-n+a) + (b(a-b) + \tfrac{1}{2}b(b+1))(m-n+a) - b(a-b),$$

where $n = \dim X$, $m = \dim Y$.

To put these propositions into a familiar context consider the following example :

Let X be an n-dimensional C^{∞}-manifold, let $Y = \mathbb{R}$, and let $f : X \to Y$ be (fully!) n-transversal to zero. Then $K_n = T(X)$ and $Q_n = T(Y) = I$, the trivial line bundle, all these bundles being defined over $Z_n(f)$. Also $\dim Z_n(f) = 0$. The homomorphism $f^n : \check{T}(X) \to T(X) \otimes I \searrow Z_n(f)$ is represented in this instance by a section over $Z_n(f)$ of $\check{T}(x) \circledS \check{T}(X)$. Explicitly, in terms of local coordinates, $f^n\left(\frac{\partial}{\partial x^i}\right) = \frac{\partial^2 y}{\partial x^i \partial x^j} dx^j \left(\otimes \frac{\partial}{\partial y}\right)$. This exhibits the usual quadratic forms associated one to each critical point.

There is no difficulty in extending our definitions and constructions to obtain descriptions of nice higher order singularities. The appropriate transversality conditions are clearly defined at each stage as is the latent symmetry and also the generic codimensions of the singular manifolds. Since this paper was first written J.M. Boardman has integrated this approach with his own more powerful one [14]. From my point of view candidates for higher-order intrinsic differentials are clear. The third one is a map

$$f_{ab} : K_{ab} \to \check{K}_{ab} \circledS Q_{ab} ,$$

where $K_{ab} = \ker f_a$ over $Z_{ab}(f)$ and $Q_{ab} = \operatorname{coker} f_a$ over $Z_{ab}(f)$, and $\circledS$ denotes the appropriate symmetric tensor product; and so on. Their definition depends, however, on the successive transversality conditions being satisfied. What one would like to do is to extend f_{ab} so that it is defined directly everywhere that f_a has kernel rank b , independently of any transversality assumption, just as we were able to do for f_a ; and so on, for every order. This I have been unable to do satisfactorily except in some special cases. The difficulty is that the higher order intrinsic differentials may depend on the lower ones in a fairly complex way. See, for instance, the example given at the end of [15].

We have as yet only a partial solution of the problem of computing the Thom polynomials for higher order singularity types. The problem is to find a non-singular model of the closure of the singular set presented as the full set of zeros of a vectorbundle homomorphism.

If $f : X \to Y \in \mathcal{M}$ and f^a and f^b_a are transversal to zero, as in Proposition 2.4, and if $Z_{a'}(f) = \emptyset$, $a' > a$, this can be done, and a computation analogous to the computation of Proposition 1.3. can be performed. We append a sketch of this computation. If $Z_{a'}(f) \neq \emptyset$, the problem is still open. (But see Ronga [17].)

We work along the route :

$$\begin{array}{ccccccc}
Z_b(f^a) & \xrightarrow{j'} & G_{ab}(f) & & Z_a(f^a) & \xrightarrow{j} & G_a(T(X)) \\
\uparrow s' & & \searrow \pi' & \nearrow s & & & \downarrow \pi \\
Z_{ab}(f) & \xrightarrow{i'} & Z_a(f) & & \xrightarrow{i} & & X \xrightarrow{f} Y ,
\end{array}$$

abbreviating notations for the various canonical imbeddings and projections as shown. Since for $a' > a$ $Z_{a'}(f) = \emptyset$, $Z_b(f^a)$ covers the closure of $Z_{ab}(f)$ in X. Also i and therefore s is proper. By analogy with Proposition 1.3 we have to compute

$$i_* \pi'_* j'_* 1 = \pi_* j_* s_* \pi'_* j'_* 1 .$$

Now $j'_* 1$ is the top Chern class of $\check{K}_{ab} \circledcirc \check{K}_a \otimes Q_a$. This can be expressed by means of universal formulas as a polynomial in the Chern classes of K_{ab}, K_a and Q_a. Proposition 0.3. applied to M_{ab}, taken in conjunction with Lemma 0.4 enables us to evaluate the image of this class under π'_*. This image is in the image of $s^* j^*$, $s_* = 1$, and we know $j_* 1$ from the proof of Proposition 1.3; so we get as far as $G_a(X)$. Finally, using Lemmas 0.5 and 0.7 to compute π_*, we obtain the answer expressed as the value of a polynomial on $(A(X))_1$, depending only on a, b and $\dim(Y, X)$, with argument $c(T(Y) - T(X))$. (The degree of this polynomial is already known by Proposition 2.4, Corollary 1.)

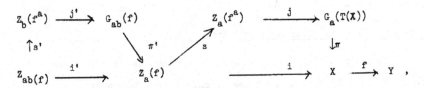

Example : $m = n$, Z_{11} . In this case,

$$\begin{aligned} j'_{*}1 &= c_1(K_1^{\vee 2} \otimes Q_1) \\ &= c_1(Q_1 - 2K_1) \\ &= c_1(T(Y) - T(X) - K_1) \\ &= c_1 - \kappa_1 , \\ j_* j'_* 1 &= (c_1 - \kappa_1) j_* 1 \\ &= (c_1 - \kappa_1) c_n(T(Y) - K_1) \\ &= c_1 c_n(T(Y) - K_1) + c_{n+1}(T(Y) - K_1) \end{aligned}$$

and $\pi_* j_* j'_* 1 = c_1^2 + c_2$,

that is $= (c_1(T(Y) - T(X)))^2 + c_2(T(Y) - T(X))$.

For maps f with $Z_2 f = \emptyset$, the Thom polynomials P_{1^k} for Z_{1^k} satisfy the recurrence relation

$$P_{1^k} = \sum_{i=1}^{k} \frac{(k-1)!}{(k-i)!} c_i P_{1^{k-i}}$$

. This formula, due to Todd in a special case (see the reference given in the footnote at the end of §2), was proved in my thesis - Cambridge 1960 (unpublished). This, my first computation of the P_{1^k} , used the Grothendieck Riemann - Rooh theorem. Cf. [10] .

REFERENCES

[1] Aitken, A.C. Note on dual symmetric functions. Proc. Edinburgh Math. Soc. (2) 2 (1931) , 164-167.

[2] Atiyah, M.F. Analytic cycles on complex manifolds. Topology 1 (1962).

[3] Borel, A. et Haefliger, A. La classe d'homologie fondamentale d'un espace analytique. Bull. Soc. Math. France 89 (1961) 461-513.

[4] Cartan, H. Quelque questions de topologie. Seminaire E.N.S. (1956-57). [Exposé 7 : Haefliger, A. Les singularités des applications différentiables. Exposé 8 : Haefliger, A. et Kosinski, A. Un théorème de Thom sur les singularites des applications différentiables.]

[5] Chern, S. On the multiplication in the characteristic ring of a sphere bundle. Ann. Math. (2) 49 (1948), 362-372.

[6] Dieudonné, J. Foundations of Modern Analysis. Academic Press New York 1960.

[7] Ehresmann, C. Les prolongements d'une variété différentiable: I. Calcul des jets, prolongement principal. II. L'espace des jets d'ordre r de V_n dans V_m. C.r. Acad. Sci. (Paris) 233 (1951), 598-600 et 777-779.

[8] Grothendieck, A. La théorie des classes de Chern. Bull. Soc. Math. France 86 (1958), 137-154.

[9] Levine, H.I. Singularities of differentiable mappings. This volume., pp.1-89.

[10] Porteous, I.R. Blowing up Chern classes. Proc. Camb. Phil. Soc. 56 (1960), 118-124.

[11] Thom, R. Les ensembles singuliers d'une application différentiable et leurs propriétés homologiques. Séminaire de Topologie de Strasbourg, Dec, 1957.

[12] Todd, J.A. Canonical systems on algebraic varieties. Bol. Soc. Mat. Mexicana (2) 2 (1957), 26-44.

[13] Whitney, H. Singularities of mappings of Euclidean spaces. Symposium Internacional de Topologica Algebrica, Mexico (1958), 285-301.

[14] Boardman, J.M. Singularities of differentiable maps. Publ. Math. I.H.E.S. 33 (1967), 21-57.

[15] Porteous, I.R. Normal singularities of submanifolds. To appear in J. Diff. Geom.

[16] Porteous, I.R. Todd's canonical classes. This volume.,pp.308-312.

[17] Ronga, F. La calcul des classes duales aux singularités de Boardman d'ordre 2. A paraître dans les Commentarii Mathematici Helvetici.

[18] Todd, J.A. Invariant and covariant systems on an algebraic variety. Proc. Lond. Math. Soc. (2) 46 (1940), 199-230.

TODD'S CANONICAL CLASSES

I.R. Porteous

The aim of this note is to recover Todd's original definition (1937) of his canonical classes in terms of singularities of maps.

For simplicity we put ourselves in the category of complex analytic manifolds and maps. Characteristic classes of vector bundles will be Chern classes. The singular set of a map $f : X \to Y$, the set of points where the kernel rank of its differential is less than $\inf\{\dim X, \dim Y\}$, will be denoted by Σf.

Let X be a complex manifold of dimension n, let L be a complex line bundle over X and let $h : L \to \underset{\sim}{C}_X^{r+1}$ be a bundle map over X ($\underset{\sim}{C}_X^{r+1}$ denoting the trivial bundle over X with fibre $\underset{\sim}{C}^{r+1}$). The dual map $h^\vee : \underset{\sim}{C}^{r+1} \to L^\vee$ induces sections of $L^\vee$, one induced by each vector in $\underset{\sim}{C}^{r+1}$. Suppose that S is the manifold of zeros of some section of $L^\vee$, transversal to the zero section. Then the dual cohomology class carried by S is $c_1(L^\vee)$. We denote this class by s. So $c(L^\vee) = 1 + s$, $c(L) = 1 - s$.

Consider $h' : L \to \underset{\sim}{C}^{r+1}$, the composite of h with the projection map $\underset{\sim}{C}_X^{r+1} \to \underset{\sim}{C}^{r+1}$. Then the differential of h' maps each tangent vector to the manifold L to a vector of $\underset{\sim}{C}^{r+1}$. Now $\underset{\sim}{C}$ acts on L and therefore on TL by multiplication along the fibre. Let QL denote the vector bundle of tangent vector fields on L invariant under this action (cf. [1]). This bundle has rank $n + 1$, for there is an exact sequence

$$\{0\} \to \underset{\sim}{C}_X \xrightarrow{i} QL \xrightarrow{p} TX \to \{0\},$$

the image of i being the invariant tangent fields along the fibres of L and p being induced by the differential of the projection map $L \to X$. Moreover, the differential of h' induces a bundle map

$$\chi : QL \to \mathrm{Hom}(L, \underset{\sim}{C}_X^{r+1}),$$

in the obvious way. The section of $\text{Hom}(L, \underset{\sim}{C}_X^{r+1})$ induced by the composite $\chi i : \underset{\sim}{C}_X \to \text{Hom}(L, \underset{\sim}{C}_X^{r+1})$ and $1 \in \underset{\sim}{C}$ is just that induced by h .

Now suppose that $\Sigma h = \emptyset$. Then, since the image by h' of each fibre of L is a line through 0 in $\underset{\sim}{C}^{r+1}$, we obtain a map $f : X \to \underset{\sim}{C}P^{r+1}$. Let K denote the Hopf bundle on $\underset{\sim}{C}P^{r+1}$ associating to each point of $\underset{\sim}{C}P^{r+1}$ itself as a line in $\underset{\sim}{C}^{r+1}$. Then $L = f^*K$. Moreover there is a natural identification $QK = \text{Hom}(K, \underset{\sim}{C}^{r+1})$ and a commutative diagram

$$
\begin{array}{ccccccccc}
\{0\} & \to & \underset{\sim}{C}_X & \to & QL & \to & TX & \to & \{0\} \\
& & \| & & \downarrow \chi & & \downarrow Tf & & \\
\{0\} & \to & \underset{\sim}{C}_X & \to & \text{Hom}(L,\underset{\sim}{C}^{r+1}) & \to & f^*T(\underset{\sim}{C}P^{r+1}) & \to & \{0\} \\
& & & \diagup\diagup & & & \| & & \\
& & f^*\underset{\sim}{C}_{CP^{r+1}} & & f^*\text{Hom}(K,\underset{\sim}{C}^{r+1}) & & & &
\end{array}
$$

with exact rows inducing a natural isomorphism between $\ker \chi$ and $\ker Tf$. As we have seen, χ is defined even where $\Sigma h \neq \emptyset$. We define $J(h) = \Sigma\chi$ to be the __Jacobian__ of the __linear system__ h . This coincides with the singular set of f together with some, possibly, of the points of Σh , the __base locus__ of h , where f is not defined.

Now I have shown elsewhere [9] that, for any bundle map $h : E \to F$ over a manifold X with $\text{rk}\,E \leqslant \text{rk}\,F$, the dual cohomology class of Σh is $c_{\text{rk}\,F - \text{rk}\,E + 1}(F - E)$, provided that the induced section of the bundle $\text{Hom}(E, F)$ is transversal to the zero section.

As a first example consider $h : L \to \underset{\sim}{C}_X^{r+1}$. Then, if transversality is satisfied,

$$
\begin{aligned}
D\Sigma(h) &= c_{r+1}(\underset{\sim}{C}_X^{r+1} - L) \\
&= \left(1 \Big/ c(L)\right)_{r+1} \\
&= \left(1 \Big/ {1 - s}\right)_{r+1} \\
&= s^{r+1} ,
\end{aligned}
$$

which makes sense in terms of the interpretation of Σh as the base locus of h , and of s as the dual class of the set of zeros S of a section of L (transversal to the zero section, of course).

Secondly, consider $\chi : QL \to \mathrm{Hom}(L, \underset{\sim}{C}_X^{r+1})$, or rather its dual $\chi^\vee : \mathrm{Hom}(L, \underset{\sim}{C}_X^{r+1})^\vee \to (QL)^\vee$, in the case that $r \leq n$. Then $J(h) = \Sigma\chi = \Sigma\chi^\vee$, so that, with the transversality assumption fulfilled,

$$D(J(h)) = c_{n-r+1}((QL)^\vee - \mathrm{Hom}(L, \underset{\sim}{C}_X^{r+1})^\vee)$$

$$= \left(\frac{c((TX)^\vee)}{(1-s)^{r+1}}\right)_{n-r+1}$$

$$= ((1 + \frac{s}{1-s})^{r+1} \cdot c((TX)^\vee))_{n-r+1} \quad . \qquad (*)$$

Our final task is to give a geometrical interpretation to each term in the expansion of the right hand side of this equation.

Let S be the set of zeros of a section of $L^\vee$ transversal to the zero section and let $j : S \to X$ be the inclusion map. Then $s = j_*1$. Now, from the exact sequence

$$\{0\} \to TS \to j^*TX \to NS \to \{0\} ,$$

or rather its dual, and the elementary fact that $NS = j^*L^\vee$, we find that

$$c(j^*L^\vee)\, c((TS)^\vee) = c(j^*(TX)^\vee),$$

that is

$$j^*(1 - s) \,.\, c(TS)^\vee = j^*c(TX)^\vee ,$$

from which it follows, by applying j_* to both sides, and using the relation $j_*(a \,.\, j^*b) = (j_*a) \,.\, b$, that

$$(1 - s) \,.\, j_*c(TS)^\vee = s \,.\, c(TX)^\vee ,$$

or

$$j_*c(TS)^\vee = \left(\frac{s}{1-s}\right) \,.\, c(TX)^\vee ,$$

a formula known classically as the <u>adjunction formula</u>. This is the interpretation we required. If we now suppose the existence of $r + 1$ sections of L fully transversal to each other (Todd emphasises that Geometry is an experimental science!) then formula (*) can be turned to provide an inductive definition of the Chern classes of $(TX)^\vee$ in terms of the corresponding classes of lower-dimensional manifolds (the intersections of the zero manifolds of the various sections) and the Jacobians of the linear systems involved.

This is how these classes were first defined by Todd [10], except that Todd worked in an algebraic category, the cohomology ring available in the case of complex analytic manifolds being replaced by a ring of rational equivalence [2] whose status at that time was still somewhat in doubt. He termed his classes the canonical classes of X. About the same time a similar definition was given of the classes by Eger [3], [11], who used r bundle maps $L \to \underset{\sim}{C}_X^2$ in place of the single map $L \to \underset{\sim}{C}_X^{r+1}$ that Todd used. A generalisation of both methods was given later by Monk [7]. The methods of the present paper can easily be generalised to cover these cases. The relationship between the canonical classes of Todd and Eger and the characteristic classes of Chern was proved by Hodge [5] and Nakano [8] using other formulas than those discussed here. The line bundle presentation of linear systems is due to Hirzebruch [4] (where an excellent introduction to characteristic classes is to be found). Jacobians of bundle maps $h : L \to \underset{\sim}{C}_X^{r+1}$ in the case that $r > n$ have been studied from a classical point of view more recently by Ingleton and Scott [6].

(This paper is a simplified version of a previously unpublished part of the author's thesis (Cambridge 1960). The results were announced in a short talk at the I.C.M., Stockholm, 1962.)

REFERENCES

[1] Atiyah, M.F. Complex analytic connections in fibre bundles. Trans. Amer. Math. Soc. 85 (1957), 181-207.

[2] Chevalley, C. Anneaux de Chow et applications. Séminaire E.N.S. (1958).

[3] Eger, M. Sur les systèmes canoniques d'une variété algébrique. C.R. Acad. Sci. (Paris) 204 (1937), 92-94 et 217-219.

[4] Hirzebruch, F. Topological methods in algebraic geometry. (3rd edition) Springer-Verlag New York, 1966.

[5] Hodge, W.V.D. The characteristic classes on algebraic varieties. Proc. Lond. Math. Soc. (3) 1 (1951), 138-151.

[6] Ingleton, A.D., and Scott, D.B. The tangent direction bundle of an algebraic variety and generalized Jacobians of linear systems. Annali Mat. pura ed app. (IV) 56 (1961), 359 - 374.

[7] Monk, D. Jacobians of linear systems on an algebraic variety. Proc. Camb. Phil. Soc. 52 (1956), 198-201.

[8] Nakano, S. Tangent vector bundles and Todd canonical systems of an algebraic variety. Mem. Coll. Sci. Kyoto (A) 29 (1955), 145-149.

[9] Porteous, I,R. Simple singularities of maps. Columbia University Notes, 1962. Reprinted, immediately preceding this paper, pp.286-307.

[10] Todd, J.A. The geometrical invariants of algebraic loci. Proc. Lond. Math. Soc. (2) 43 (1937), 127-138.

[11] Todd, J.A. The geometrical invariants of algebraic loci (second paper). Proc. Lond. Math. Soc. (2) 45 (1939), 410-424.

LE CALCUL DE LA CLASSE DE COHOMOLOGIE ENTIERE DUALE A $\bar{\Sigma}^k$

F. Ronga

Soient $\xi = (E \to X)$ et $\eta = (F \to X)$ des fibrés vectoriels réels différentiables (de classe C^∞) ; soit $HOM(\xi, \eta)$ le fibré associé de fibré les homomorphismes de E_x dans F_x. Posons $\Sigma^k(\xi, \eta) = \{\alpha \in HOM(\xi, \eta) | \dim(\ker(\alpha)) = k\}$; c'est une sous-variété de codimension $k.(\text{rang}(\eta) - \text{rang}(\xi) + k)$ de $HOM(\xi, \eta)$ (voir [4]), qu'on notera encore Σ^k.

Proposition

Supposons que la variété $HOM(\xi, \eta)$ soit orientable ; si $k = 2i$ et $\text{rang}(\eta) - \text{rang}(\xi) = 2r$, i et r entiers, l'adherence $\bar{\Sigma}^k$ porte une classe fondamentale entière.

En effet, $\bar{\Sigma}^{2i}$ est un espace de type VS_m (voir [1], §2), où m est la dimension de Σ^{2i}, qui est l'ouvert épais homéomorphe à une variété. On a que $\dim(\bar{\Sigma}^k - \Sigma^k) \leq m - 3$; d'après ([2], prop. 2.3), pour s'assurer l'existence d'une classe fondamentale, il suffit alors de s'assurer que Σ^{2i} est orientable. Soient $\ker = \{(\alpha, v) \in \Sigma^{2i} \times E | v \in \ker(\alpha)\}$, $Im = \{(\alpha, w) \in \Sigma^{2i} \times F | w \in Im(\alpha)\}$, $coker = \pi^*(\eta)/Im$; ce sont des fibrés vectoriels sur Σ^{2i} de fibré respectivement $\ker(\alpha)$, $Im(\alpha)$, $coker(\alpha)$. Le fibré normal à Σ^{2i} dans $HOM(\xi, \eta)$ s'identifie au fibré $HOM(\ker, coker)$(voir p. ex. [3], prop. 1.1) ; or

$w_1(HOM(\ker, coker)) = \text{rang}(\ker) . w_1(coker) + \text{rang}(coker) . w_1(\ker) = 0$, vu que rang (ker) et rang (coker) sont pairs. Puisque la variété $HOM(\xi, \eta)$ est orientable, il s'ensuit que Σ^{2i} est orientable.

La classe de la cohomologie entière de $HOM(\xi, \eta)$ obtenue en appliquant la dualité de Poincaré à l'image de la classe fondamentale de $\bar{\Sigma}^{2i}$ dans l'homologie de $HOM(\xi, \eta)$ est appelée classe duale à $\bar{\Sigma}^{2i}$. On sait (voir [2]) qu'elle s'obtient par évaluation sur les classes caractéristiques de $\pi^*(\xi)$ et $\pi^*(\eta)$ d'un polynôme universel, qu'on se propose de déterminer.

La classe duale à $\bar{\Sigma}^{2i}$ pourra être considérée comme élément de la cohomologie de X par l'isomorphisme induit en cohomologie par π : HOM $(\xi, \eta) \to X$. Soit $G_{n,N}$ la grassmannienne des n-plans dans $\mathbb{R}^{n+N}$ et γ_n le fibré canonique sur $G_{n,N}$; soit $\bar{\eta}$ un fibré sur X tel que $\eta \oplus \bar{\eta}$ soit trivial et $f : X \to G_{n,N}$ une application classifiante pour le fibré $\xi \oplus \bar{\eta}$, qu'on peut supposer différentiable. La classe duale à $\bar{\Sigma}^{2i}(\xi, \eta)$ est égale à l'image par l'application induite en cohomologie par f de la classe duale à $\bar{\Sigma}^{2i}$ $(\gamma_n ; \theta^{n+2r})$, où θ^{ℓ} designe le fibré trivial de rang ℓ (cela se démontre de la même manière que le théorème 4.2 de [3]) .

Théorème

La classe duale à $\bar{\Sigma}^{2i}(\gamma_n; \theta^{n+2r})$ en cohomologie entière est déterminée par sa réduction modulo deux et par sa réduction rationnelle. La réduction modulo deux est le déterminant de la matrice $(\bar{w}_{2(i+r)+k-\ell})_{k,\ell=1,\dots,2i}$, où $\bar{w}_k$ désigne la k-ième classe de Stiefel-Whitney duale de γ_n . La réduction rationnelle est le déterminant de la matrice $(\bar{p}_{i+r+k-\ell})_{k,\ell=1,\dots,i}$, où $\bar{p}_k$ désigne la k-ième classe de Pontrjagin rationnelle duale de γ_n.

En effet, on sait que les éléments de torsion de la cohomologie entière de $G_{n,N}$ sont d'ordre 2, ce qui entraîne la première affirmation.

L'expression de la réduction modulo deux de la classe duale à $\bar{\Sigma}^{2i}$ est bien connue; on en trouvera par exemple le calcul dans [4] .

L'expression de la réduction rationnelle de la classe duale à $\bar{\Sigma}^{2i}(\gamma_n, \theta^{n+2r})$ s'obtient par des propositions analogues à celles de ([3], IIème partie), où on remplace la cohomologie entière par la cohomologie rationnelle et les classes de Chern par les classes de Pontrjagin rationnelles.

Par exemple, déterminons la classe entière duale à $\bar{\Sigma}^{2}(\xi, \eta)$, où rang (η) - rang $(\xi) = 2r$. La réduction modulo deux est égale à $w_{2r+2}^2 + w_{2r+1} \cdot w_{2r+3}$, ou $w_k = w_k(\eta - \xi)$; la réduction rationnelle est égale à $p_{r+1}(\eta - \xi)$. Remarquons que w_{2r+2}^2 est la réduction modulo deux de la (r+1)-ième classe de Pontrjagin entière, dont la classe duale à $\bar{\Sigma}^2$ diffère par une classe entière d'ordre deux. Cette dernière est caractérisée par sa réduction modulo deux, qui est $w_{2r+1} \cdot w_{2r+3}$.

Soit $f : M^4 \to V^4$ une application différentiable entre variétés orientables de dimension 4, Σ^k-générique pour $k \geq 2$; la classe duale à $\bar{\Sigma}^2(f) = \{x \in M | \dim (\ker(d f_x)) = 2\}$ (Σ^k est vide si $k \geq 3$) est alors égale à $P_1 (f^* (T(V)) - T(M))$. Si on l'évalue sur la classe fondamentale de M, on trouve $3 . (d^o (f) . \sigma(V) - \sigma(M))$, où σ désigne la signature et $d^o(f)$ le degré de f .

La théorème précédent permet aussi de calculer la classe entière duale à $\bar{\Sigma}^{k,\ell}(\xi, \eta)$ (voir [3], §2), qui existe si $k = 2i$, $\ell = 4j$ et rang (η) - rang $(\xi) = 2r$; en effet, dans [3] on ramène le calcul de la classe duale à $\bar{\Sigma}^{k,\ell}$ à celui de la classe duale à la singularité $\bar{\Sigma}^h$ d'un certain morphisme de fibrés, pour un certain h .

BIBLIOGRAPHIE

[1] A. Borel et A. Haefliger, La classe d'homologie fondamentale d'un espace analytique. Bull. Soc. Math. France, 89 (1961) 461-513.

[2] A. Haefliger et A. Kosinski, Un théorème de Thom sur les singularités des applications différentiables. Séminaire H. Cartan, E.N.S. 1956-1957, exposé 8.

[3] F. Ronga, La calcul des classes duales aux singularités de Boardman d'ordre 2. A paraître dans les Commentarii Mathematici Helvetici.

[4] R. Thom, Ann. Inst. Fourier (Grenoble), 6, 1955-56, p.43-87.

SINGULARITIES OF VECTOR BUNDLE MAPS

R. MacPherson

We will consider situations with two smooth real or complex vector bundles E^n and F^p (superscripts denoting dimension) over a compact manifold N and a vector bundle map from E to F. For example, if $f : N^n \to P^p$ is a differentiable map of manifolds, the differential $df : TN \to f^{-1}TP$ is such a situation. Suppose first that the vector bundle map is nonsingular, i.e. everywhere of maximal rank, such as the differential of a fibration or an immersion. Then if $c\ell$ is any characteristic class we have trivially

$$c\ell(E) = c\ell(K \oplus I)$$

where K is the kernel bundle and I is the image bundle. We will generalise this equation to the case that the vector bundle map has generic singularities in the sense of the next paragraph.

The total space of the bundle Hom (E, F) has submanifolds Σ^i consisting of points representing linear transformations whose kernel has dimension i. A vector bundle map from E to F is equivalent to a section of the bundle $\mathrm{Hom}(E, F)$; the map is called <u>generic</u> if this section is transverse to all the Σ^i. By the transversality theorem generic vector bundle maps are open and dense in the C^∞ topology; so are maps of manifolds with generic differentials.

The pullback of Σ^i via the section is called the singularity Σ^i in N. In general the closure of Σ^i will not be a manifold but will have "singularities" at Σ^{i+1}, Σ^{i+2}, etc. Neither will the kernel subbundle of E and the image subbundle of F over Σ^i extend over the closure. We resolve both these difficulties at once by finding a manifold $\tilde{\Sigma}^i$ and a map $\phi_i : \tilde{\Sigma}^i \to N$ with the following properties:

1) $\tilde{\Sigma}^i$ is a compact manifold.
2) $\phi_i^{-1}(\Sigma^i)$ is dense in $\tilde{\Sigma}^i$.
3) ϕ_i restricted to $\phi_i^{-1}(\Sigma^i)$ is a diffeomorphism to Σ^i.

4) There exist subbundles $K^i \subset \phi_i^{-1} E^n$ and $I^{n-i} \subset \phi_i^{-1} F^p$ that extend the pull-backs of the kernel and image subbundles over Σ^i.

To construct such a $\tilde{\Sigma}^i$, we consider $G_i(E^n)$, the Grassmann bundle of i-planes, and $G_{n-i}(F^p)$. The fiber product $G_i(E) \times_N G_{n-i}(F)$ has a section over Σ^i given by the kernel and image bundles. $\tilde{\Sigma}^i$ is the closure of the image of this section.

Over $\tilde{\Sigma}^i$, let C^{p-n+i} (for cokernel) be $\phi_i^{-1}(F^p)/I^{n-i}$. Let π_i be the projection of the Grassmann bundle $G_i(K^i \oplus C^{p-n+i})$ onto $\tilde{\Sigma}^i$; ξ^i be the canonical bundle over $G_i(K \oplus C)$ and I^{n-i} denote also $\pi_i^{-1} I^{n-i}$. The notation for these constructions is summarised in the following diagram :

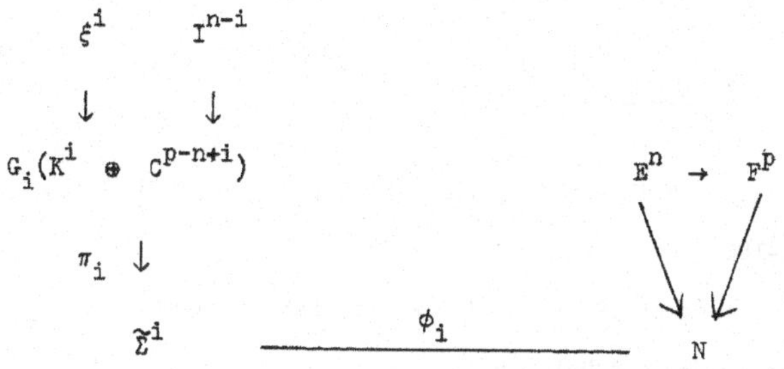

The map $\phi_i \circ \pi_i$ is orientable for cohomology with coefficients in a ring R if the bundles are complex, if $n - p$ is even, or if R has characteristic two ; we assume at least one of these conditions holds. Selecting a canonical orientation is then not difficult but rather technical [2].

<u>Theorem</u> : <u>Let</u> $c\ell$ <u>be any characteristic class in</u> $H^*(\ ; R)$. <u>Then</u>

$$c\ell(E) = \sum_{i \geq 0} (\phi_i \pi_i)_* c\ell(\xi^i \oplus I^{n-i})$$

The leading term of the summation, where $i = \max(0, n - p)$, is the analogue to $c\ell(K \oplus I)$ in the nonsingular case ; the other terms are "corrections" for the singularities. If $c\ell$ is in the mod two Whitney ring or the integral Chern ring, then ϕ and π are individually orientable and $\pi_* c\ell(\xi \oplus I)$ may be evaluated by standard techniques as a polynomial in the classes of K, C, and I. Similarly, if $c\ell$ is in the rational Pontrjagin ring then all the odd terms are zero while for i even ϕ_i and π_i are again individually orientable and the above remark applies.

As an example, let $cl = c$, the total Chern or Whitney class, and let $n = p$. In this case all these polynomials $\pi_* \, c\,(\xi^i \oplus I^{n-i})$ vanish except when $i = 0$ or 1, and the formula becomes

$$c(E) = c(F) + \phi_{1*}\, c(I^{n-1}) .$$

This formula was found by Levine [1] for the case that Σ^2 is empty. Similarly, if $cl = p$, the total rational Pontrjagin class, then

$$p(E) = p(F) + \phi_{2*}\, p(I^{n-2}) .$$

REFERENCES

[1] H. Levine, The singularities S^q , Illinois Math. Journal 8 (1964) 152-186.

[2] R. MacPherson, Singularities of Maps and Characteristic classes. Thesis, Harvard University, 1970.

AUTHOR ADDRESSES

J. Bochnak, Faculté des Sciences, Université de Paris, Batiment 425, Mathematique, 91-ORSAY, France.

G. Glaeser, Departement de Mathematique, Université de Strasbourg, Rue René Descartes, 67-Strasbourg, France.

S.A. Khabbaz, Department of Mathematics and Astronomy, College of Arts and Sciences, Lehigh University, Bethlehem, PA.18015, U. S. A.

T.-C. Kuo, Department of Mathematics, The University, Manchester.

H.I. Levine, Department of Mathematics, Brandeis University, Waltham, Ma. 02154, U. S. A.

S. Lojasiewicz, Instytut Matemetyczny U.J., Krakow, Reymonta 4, Poland.

R. MacPherson, Department of Mathematics, Brown University, Providence, Rhode Island, U. S. A.

J.N. Mather, Department of Mathematics, Harvard University, 2 Divinity Avenue, Cambridge, Mass.02138, U. S. A.

L. Nirenberg, New York University, Courant Institute of Math. Sciences, 251 Mercer Street, New York, N.Y.10012, U. S. A.

E. Pitcher, Department of Mathematics and Astronomy, College of Arts and Sciences, Lehigh University, Bethlehem, PA. 18015, U. S. A.

I.R. Porteous, Department of Pure Mathematics, The University, P.O. Box 147, Liverpool, L69 3BX, England.

F. Ronga, Institut de Mathematiques, Université de Geneve, 16 bd d'Yvoy, 1211 Geneve, Switzerland.

D. Sullivan, Department of Mathematics, Massachusetts Institute of Technology, Cambridge, Mass. 02139, U. S. A.

R. Thom, Institut des Hautes Etudes Scientifiques, Rt. de Chartres, Bures sur Yvette, France.

J.G. Timourian, Department of Mathematics, University of Alberta, Edmonton, Alberta, Canada.

C.T.C. Wall, Department of Pure Mathematics, The University, P.O. Box 147, Liverpool, L69 3BX, England.

Lecture Notes in Mathematics

Lecture Notes in Mathematics – Lecture Notes in Physics

Vol. 144: Seminar on Differential Equations and Dynamical Systems, II. Edited by J. A. Yorke. VIII, 268 pages. 1970. DM 20,– / $ 5.50

Vol. 145: E. J. Dubuc, Kan Extensions in Enriched Category Theory. XVI, 173 pages. 1970. DM 16,– / $ 4.40

Vol. 146: A. B. Altman and S. Kleiman, Introduction to Grothendieck Duality Theory. II, 192 pages. 1970. DM 18,– / $ 5.00

Vol. 147: D. E. Dobbs, Cech Cohomological Dimensions for Commutative Rings. VI, 176 pages. 1970. DM 16,– / $ 4.40

Vol. 148: R. Azencott, Espaces de Poisson des Groupes Localement Compacts. IX, 141 pages. 1970. DM 14,– / $ 3.90

Vol. 149: R. G. Swan and E. G. Evans, K-Theory of Finite Groups and Orders. IV, 237 pages. 1970. DM 20,– / $ 5.50

Vol. 150: Heyer, Dualität lokalkompakter Gruppen. XIII, 372 Seiten. 1970. DM 20,– / $ 5.50

Vol. 151: M. Demazure et A. Grothendieck, Schémas en Groupes I. (SGA 3). XV, 562 pages. 1970. DM 24,– / $ 6.60

Vol. 152: M. Demazure et A. Grothendieck, Schémas en Groupes II. (SGA 3). IX, 654 pages. 1970. DM 24,– / $ 6.60

Vol. 153: M. Demazure et A. Grothendieck, Schémas en Groupes III. (SGA 3). VIII, 529 pages. 1970. DM 24,– / $ 6.60

Vol. 154: A. Lascoux et M. Berger, Variétés Kähleriennes Compactes. VII, 83 pages. 1970. DM 8,– / $ 2.20

Vol. 155: Several Complex Variables I, Maryland 1970. Edited by J. Horváth. IV, 214 pages. 1970. DM 18,– / $ 5.00

Vol. 156: R. Hartshorne, Ample Subvarieties of Algebraic Varieties. XIV, 256 pages. 1970. DM 20,– / $ 5.50

Vol. 157: T. tom Dieck, K. H. Kamps und D. Puppe, Homotopietheorie. VI, 265 Seiten. 1970. DM 20,– / $ 5.50

Vol. 158: T. G. Ostrom, Finite Translation Planes. IV. 112 pages. 1970. DM 10,– / $ 2.80

Vol. 159: R. Ansorge und R. Hass. Konvergenz von Differenzenverfahren für lineare und nichtlineare Anfangswertaufgaben. VIII, 145 Seiten. 1970. DM 14,– / $ 3.90

Vol. 160: L. Sucheston, Constributions to Ergodic Theory and Probability. VII, 277 pages. 1970. DM 20,– / $ 5.50

Vol. 161: J. Stasheff, H-Spaces from a Homotopy Point of View. VI, 95 pages. 1970. DM 10,– / $ 2.80

Vol. 162: Harish-Chandra and van Dijk, Harmonic Analysis on Reductive p-adic Groups. IV, 125 pages. 1970. DM 12,– / $ 3.30

Vol. 163: P. Deligne, Equations Différentielles à Points Singuliers Réguliers. III, 133 pages. 1970. DM 12,– / $ 3.30

Vol. 164: J. P. Ferrier, Seminaire sur les Algebres Complètes. II, 69 pages. 1970. DM 8,– / $ 2.20

Vol. 165: J. M. Cohen, Stable Homotopy. V, 194 pages. 1970. DM 16,– / $ 4.40

Vol. 166: A. J. Silberger, PGL_2 over the p-adics: its Representations, Spherical Functions, and Fourier Analysis. VII, 202 pages. 1970. DM 18,– / $ 5.00

Vol. 167: Lavrentiev, Romanov and Vasiliev, Multidimensional Inverse Problems for Differential Equations. V, 59 pages. 1970. DM 10,– / $ 2.80

Vol. 168: F. P. Peterson, The Steenrod Algebra and its Applications: A conference to Celebrate N. E. Steenrod's Sixtieth Birthday. VII, 317 pages. 1970. DM 22,– / $ 6.10

Vol. 169: M. Raynaud, Anneaux Locaux Henséliens. V, 129 pages. 1970. DM 12,– / $ 3.30

Vol. 170: Lectures in Modern Analysis and Applications III. Edited by C. T. Taam. VI, 213 pages. 1970. DM 18,– / $ 5.00.

Vol. 171: Set-Valued Mappings, Selections and Topological Properties of 2^X. Edited by W. M. Fleischman. X, 110 pages. 1970. DM 12,– / $ 3.30

Vol. 172: Y.-T. Siu and G. Trautmann, Gap-Sheaves and Extension of Coherent Analytic Subsheaves. V, 172 pages. 1971. DM 16,– / $ 4.40

Vol. 173: J. N. Mordeson and B. Vinograde, Structure of Arbitrary Purely Inseparable Extension Fields. IV, 138 pages. 1970. DM 14,– / $ 3.90.

Vol. 174: B. Iversen, Linear Determinants with Applications to the Picard Scheme of a Family of Algebraic Curves. VI, 69 pages. 1970. DM 8,– / $ 2.20.

Vol. 175: M. Brelot, On Topologies and Boundaries in Potential Theory. VI, 176 pages. 1971. DM 18,– / $ 5.00

Vol. 176: H. Popp, Fundamentalgruppen algebraischer Mannigfaltigkeiten. IV, 154 Seiten. 1970. DM 16,– / $ 4.40

Vol. 177: J. Lambek, Torsion Theories, Additive Semantics and Rings of Quotients. VI, 94 pages. 1971. DM 12,– / $ 3.30

Vol. 178: Th. Bröcker und T. tom Dieck, Kobordismentheorie. XVI, 191 Seiten. 1970. DM 18,– / $ 5.00

Vol. 179: Seminaire Bourbaki – vol. 1968/69. Exposés 347-363. IV. 295 pages. 1971. DM 22,– / $ 6.10

Vol. 180: Séminaire Bourbaki – vol. 1969/70. Exposés 364-381. IV, 310 pages. 1971. DM 22,– / $ 6.10

Vol. 181: F. DeMeyer and E. Ingraham, Separable Algebras over Commutative Rings. V, 157 pages. 1971. DM 16,– / $ 4.40

Vol. 182: L. D. Baumert. Cyclic Difference Sets. VI, 166 pages. 1971. DM 16,– / $ 4.40

Vol. 183: Analytic Theory of Differential Equations. Edited by P. F. Hsieh and A. W. J. Stoddart. VI, 225 pages. 1971. DM 20,– / $ 5.50

Vol. 184: Symposium on Several Complex Variables, Park City, Utah, 1970. Edited by R. M. Brooks. V, 234 pages. 1971. DM 20,– / $ 5.50

Vol. 185: Several Complex Variables II, Maryland 1970. Edited by J. Horváth. III, 287 pages. 1971. DM 24,– / $ 6.60

Vol. 186: Recent Trends in Graph Theory. Edited by M. Capobianco/ J. B. Frechen/M. Krolik. VI, 219 pages. 1971. DM 18.– / $ 5.00

Vol. 187: H. S. Shapiro, Topics in Approximation Theory. VIII, 275 pages. 1971. DM 22,– / $ 6.10

Vol. 188: Symposium on Semantics of Algorithmic Languages. Edited by E. Engeler. VI, 372 pages. 1971. DM 26,– / $ 7.20

Vol. 189: A. Weil, Dirichlet Series and Automorphic Forms. V, 164 pages. 1971. DM 16,– / $ 4.40

Vol. 190: Martingales. A Report on a Meeting at Oberwolfach, May 17-23, 1970. Edited by H. Dinges. V, 75 pages. 1971. DM 12,– / $ 3.30

Vol. 191: Séminaire de Probabilités V. Edited by P. A. Meyer. IV, 372 pages. 1971. DM 26,– / $ 7.20

Vol. 192: Proceedings of Liverpool Singularities – Symposium I. Edited by C. T. C. Wall. V, 319 pages. 1971. DM 24,– / $ 6.60

Lecture Notes in Physics

Bisher erschienen/Already published

Vol. 1: J. C. Erdmann, Wärmeleitung in Kristallen, theoretische Grundlagen und fortgeschrittene experimentelle Methoden. II, 283 Seiten. 1969. DM 20,– / $ 5.50

Vol. 2: K. Hepp, Théorie de la renormalisation. III, 215 pages. 1969. DM 18,– / $ 5.00

Vol. 3: A. Martin, Scattering Theory: Unitarity, Analytic and Crossing. IV, 125 pages. 1969. DM 14,– / $ 3.90

Vol. 4: G. Ludwig, Deutung des Begriffs physikalische Theorie und axiomatische Grundlegung der Hilbertraumstruktur der Quantenmechanik durch Hauptsätze des Messens. XI, 469 Seiten. 1970. DM 28,– / $ 7.70

Vol. 5: M. Schaaf, The Reduction of the Product of Two Irreducible Unitary Representations of the Proper Orthochronous Quantummechanical Poincaré Group. IV, 120 pages. 1970. DM 14,– / $ 3.90

Vol. 6: Group Representations in Mathematics and Physics. Edited by V. Bargmann. V, 340 pages. 1970. DM 24,– / $ 6.60